# Living with Environmental Change

T0347166

Vietnam and the neighbouring countries of Southeast Asia face a diverse set of challenges created by the rapid evolution of their social, economic and environmental systems. The editors of this book argue that any sustainable development strategy must include measures to avert adverse impacts on society and the environment and protect those most at risk.

Using a multidisciplinary approach, *Living with Environmental Change* provides a comprehensive assessment of the factors shaping vulnerability and resilience in Vietnam, and the prospects for sustainable development, as social and economic renovation transforms this enigmatic society. Vietnamese, European and North American authors examine land use, forestry, coastal resources, disaster management, urbanisation and industrialisation, as well as relations between Vietnam and its neighbours over shared resources, in a stimulating mixture of environmental and social science.

The lessons from the past investigated in this important collection will prove to be of pressing relevance to current and future generations. The volume will be of considerable interest to scientists concerned with society's response to environmental change, not just in Southeast Asia but worldwide. Students and policy makers wanting to learn more about interdisciplinary approaches as well as environmental change will also find this book a thought-provoking read.

**W. Neil Adger** is a Lecturer in Environmental Economics in the School of Environmental Sciences and Senior Research Fellow in the Centre for Social and Economic Research on the Global Environment (CSERGE) at the University of East Anglia, Norwich, UK.

**P. Mick Kelly** is a Reader with the Climatic Research Unit in the School of Environmental Sciences, University of East Anglia, and a Senior Research Fellow with CSERGE.

**Nguyen Huu Ninh** is Chairman of the Centre for Environment Research, Education and Development, Hanoi, Vietnam and the Director of the Global Environment Programme of the Hanoi University of Science.

## Routledge Research Global Environmental Change Series

Also available in the Routledge Global Environmental Change Series:

**Timescapes of Modernity**
*Barbara Adam*

**Reframing Deforestation**
*James Fairhead and Melissa Leach*

**British Environmental Policy and Europe**
*Edited by Philip Lowe and Stephen Ward*

**The Politics of Sustainable Development**
*Edited by Susan Baker, Maria Kousis, Dick Richardson and Stephen Young*

**Argument in the Greenhouse**
*Mick Mabey, Stephen Hall, Clare Smith and Sujata Gupta*

**Environmentalism and the Mass Media**
*Graham Chapman, Keval Kumar, Caroline Fraser and Ivor Gaber*

**Environmental Change in Southeast Asia**
*Edited by Michael Parnwell and Raymond Bryant*

**The Politics of Climate Change**
*Edited by Timothy O'Riordan and Jill Jaeger*

**Population and Food**
*Tim Dyson*

**The Environment and International Relations**
*Edited by John Vogler and Mark Imber*

**Global Warming and Energy Demand**
*Edited by Terry Barker, Paul Ekins and Nick Johnstone*

**Social Theory and the Global Environment**
*Edited by Michael Redclift and Ted Benton*

# Living with Environmental Change

## Social vulnerability, adaptation and resilience in Vietnam

Edited by **W. Neil Adger, P. Mick Kelly and Nguyen Huu Ninh**

Routledge
Taylor & Francis Group

LONDON AND NEW YORK

First published 2001
by Routledge

2 Park Square, Milton Park, Abingdon, Oxon OX14 4RN
711 Third Avenue, New York, NY 10017, USA

*Routledge is an imprint of the Taylor & Francis Group an informa business*

First issued in paperback 2016

Transferred to Digital Printing 2005

Typeset in Baskerville by Steven Gardiner Ltd

*British Library Cataloguing in Publication Data*
A catalogue record for this book is available from the British Library

*Library of Congress Cataloging in Publication Data*
Living with environmental change: social vulnerability, adaptation and
resilience in Vietnam / edited by W. Neil Adger, P. Mick Kelly and
Nguyen Huu Ninh.
    p.   cm.
    Includes bibliographical references and index.
1. Vietnam – Economic conditions – 1975–. 2. Vietnam – Environmental
conditions. 3. Sustainable development – Vietnam. I. Adger, W. Neil.
1964–. II. Kelly, P. Mick, 1951–. III. Ninh, Huu Nguyen, 1954–.
HC444.L58 2001
333.7'09597 – dc21    00-047054

ISBN 978-0-415-21722-4 (hbk)
ISBN 978-1-138-99548-2 (pbk)

# Contents

# Figures

# Tables

# Boxes

# Contributors

**W. Neil Adger** is a Lecturer in Environmental Economics in the School of Environmental Sciences at the University of East Anglia and a Senior Research Fellow in the Centre for Social and Economic Research on the Global Environment. His research has focused on climate change mitigation and adaptation, natural resource management and environmental economics. He is the author of *Land Use and the Causes of Global Warming* (with Katrina Brown) and co-editor of *Climate Change Mitigation and European Land Use Policies* and *The Economics of Water and Coastal Resources*.
Dr W. Neil Adger, School of Environmental Sciences, University of East Anglia, Norwich NR4 7TJ, United Kingdom. E-mail: n.adger@uea.ac.uk

**David Drakakis-Smith** was Professor of Economic Geography at the University of Liverpool at the time his contribution to this book was written. He had undertaken extensive field research on developmental issues, particularly in Singapore, Vietnam, South Africa and Zimbabwe. His recent books include *Geographies of Development* (with Robert Potter, Tony Binns and Jennifer Elliott), *Uneven Development in South East Asia* (edited with Chris Dixon), and *Economic and Social Change in Pacific Asia* (edited with Chris Dixon). An extensively revised edition of his widely acclaimed *Third World Cities* was published by Routledge in 2001. Sadly, David died in December 1999 and will be sorely missed by family, friends and colleagues. Enquiries regarding his chapter in this volume and his other work should be sent to his co-author, Andrea Kilgour.

**Duong Van Ni** is a researcher at the Mekong Delta Farming Systems Research and Development Institute at Cantho University, Vietnam, and recieved his PhD from Royal Holloway Institute for Environmental Research, University of London. His main research areas are soil and water processes in tropical wetland areas, especially in relation to land use. These are applied to the design of sustainable farming systems and ecosystem management plans, towards the goals of improving the livelihoods of farmers while maintaining environmental quality.
Dr Duong Van Ni, Farming Systems Research and Development Institute, Cantho University, Cantho, Vietnam and Royal Holloway Institute for

Environmental Research, Royal Holloway University of London, Huntersdale, Callow Hill, Virginia Water GU25 4LN, United Kingdom. E-mail: enrm@ctu.edu.vn

**Hoang Minh Hien** is a meteorologist with the Vietnamese Hydrometeorological Service, specialising in remote sensing, tropical cyclone prediction and research on climate variability and change. He was one of the first Vietnamese scientists to study the El Niño Southern Oscillation phenomenon. He has worked closely with the Center for Environment Research Education and Development since the early 1990s and has been involved in a series of climate studies since that time, most recently in collaboration with the University of East Anglia.

Dr Hoang Minh Hien, Satellite Meteorology Division, Hydrometeorological Service, 4 Dang Thai Than, Hanoi, Vietnam. E-mail: hmh@netnam.org.vn

**Rhys Jenkins** is a Reader in Economics in the School of Development Studies, University of East Anglia, and a former Dean of the School. He has written extensively on issues of globalisation and industrialisation in Latin America and Asia and is currently researching the links between globalisation and pollution.

Dr Rhys O. Jenkins, School of Development Studies, University of East Anglia, Norwich NR4 7TJ, United Kingdom. E-mail: r.o.jenkins @uea.ac.uk

**P. Mick Kelly** is a Reader with the Climatic Research Unit in the School of Environmental Sciences, University of East Anglia, and a Research Fellow with the Centre for Social and Economic Research on the Global Environment in that department. His research areas include causes of natural and anthropogenic climate change, the use of climate information, and climate and development, focusing on vulnerability and adaptation. Author of over one hundred scientific publications, he is Course Director of the MSc in Climate Change and Deputy Director of Graduate Studies for the University of East Anglia.

Dr P. Mick Kelly, Climatic Research Unit, School of Environmental Sciences, University of East Anglia, Norwich NR4 7TJ, United Kingdom. E-mail: m.kelly@uea.ac.uk. Web: http://www.cru.uea.ac.uk/tiempo/

**Andrea Kilgour** is a lecturer in Human Geography in the Department of Geography at the University of Liverpool and a British Academy research associate. Her research has focused on aspects of sustainable urbanisation and economic change in Southeast Asia, and concentrated specifically on urban living conditions and environmental problems at the household level in Vietnam.

Dr Andrea L. Kilgour, Department of Geography, Roxby Building, University of Liverpool, Liverpool L69 3BX, United Kingdom. E-mail: kilgour@liverpool.ac.uk

**Cecilia Luttrell** is a researcher and postgraduate student with the Centre for Social and Economic Research on the Global Environment in the School of

Environmental Sciences at the University of East Anglia working on social issues of local resource management in Vietnam. With a training in geography and forestry, she has worked in natural resource management and community development issues in Ghana, Indonesia, East Africa and the Pacific.
Cecilia Luttrell, CSERGE School of Environmental Sciences, University of East Anglia, Norwich NR4 7TJ, United Kingdom. E-mail: c.luttrell@uea.ac.uk

**Edward Maltby** is Director of Royal Holloway Institute for Environmental Research, and Professor of Environmental and Physical Geography in the Department of Geography, both the Institute and the Department being part of Royal Holloway, University of London. His main research areas relate to wetland ecology and management: soil and ecosystem processes, functions and values, changes due to management and human activities, and impacts of agricultural reclamation and other human activities on soil and peat development. This expertise is applied especially to world conservation strategy, water use, ecosystem and catchment management, and the design of supportive scientific research.
Prof. Edward Maltby, Royal Holloway Institute for Environmental Research, Royal Holloway University of London, Huntersdale, Callow Hill, Virginia Water GU25 4LN, United Kingdom. E-mail: e.maltby@rhbnc.ac.uk

**Andreas Neef** is a Research Fellow with the Institute of Agricultural Economics and Social Sciences in the Tropics and Subtropics in the Department of Development Theory and Policy at Hohenheim University. His research focuses on land tenure issues, sustainable agriculture and participatory approaches. He has undertaken empirical research in Yemen, West Africa and Southeast Asia. He has recently coordinated an interdisciplinary research programme of the University of Hohenheim on 'Sustainable Land Use and Rural Development' in Northern Thailand and Northern Vietnam.
Dr Andreas Neef, Institute of Agricultural Economics and Social Sciences in the Tropics and Subtropics, Department of Development Theory and Policy, Hohenheim University (490 A), 70593 Stuttgart, Germany. E-mail: neef@ uni-hohenheim.de

**Ngo Cam Thanh** is a researcher at the Center for Environment Research, Education and Development, Hanoi, Vietnam, with a background in ecological sciences. She has been involved in several projects on global environmental change in Vietnam and is currently contributing to a Dutch–Vietnamese study of the Red River Delta.
Ngo Cam Thanh, Center for Environment Research Education and Development, A01, K40, Giang Vo, Hanoi, Vietnam. E-mail: cered@hn.vnn.vn. Web: http://www.cru.uea.ac.uk/tiempo/annex/cered/

**Nguyen Hoang Tri** is a Lecturer at the National University of Vietnam in Hanoi and a research scientist with the Center for Natural Resources and Environmental Studies. Author of over fifty scientific publications, his research areas include mangrove–human ecology and integrated modelling of socio-economic and biogeochemical processes in the coastal zone. He has under-taken a series of global change projects in the Red River Delta, including acting as Principal Investigator with the SARCS/WOTRO/LOICZ IGBP Core Project for Vietnam.

Dr Nguyen Hoang Tri, Mangrove Ecosystem Research Division, Center for Natural Resources and Environmental Studies, National University of Vietnam, No. 7 Ngo 115, Nguyen Khuyen, Hanoi, Vietnam. E-mail: nguyenhoangtri@hn.vnn.vn

**Nguyen Huu Ninh** is the Chairman of the Center for Environment Research Education and Development, Hanoi, Vietnam and the Director of the Global Environment Programme at Hanoi University of Science. Recognised inter-nationally as an expert on global change and Vietnam, he has conducted a series of projects on global environmental change and its impact on Vietnam and the wider Southeast Asia region and has contributed to various international programmes. He is coordinator of the Indochina Global Change Network and has published articles and books in various environmental fields.

Dr Nguyen Huu Ninh, Center for Environment Research Education and Development, A01, K40, Giang Vo, Hanoi, Vietnam. E-mail: cered@hn.vnn.vn Web: http://www.cru.uea.ac.uk/tiempo/annex/cered/

**Nguyen Thanh Binh** received his Master of Science degree in Conservation Biology from the University of Minnesota and was then accepted into the Doctoral Program in Economics at Georgetown University. He worked for several years with Oxfam-Ireland in Vietnam on the social and ecological effects of development projects. His research interests include watershed management, the economics of water use and the role of non-governmental organisations working on resource development issues.

Nguyen Thanh Binh, Department of Economics, Georgetown University, Washington, DC 20057–1045, USA. E-mail: nguyenb@gusun.georgetown.edu

**Davide Pettenella** is working at the University of Padova where he is teaching forest products economics and marketing of mountain products and services. He is also a Research Fellow with the Fondazione ENI 'Enrico Mattei'. He has published 130 papers in the field of forest economics as a result of his research activities and fieldwork carried out within programmes financed by the European Commission, Food and Agriculture Organization, European Forestry Institute, World Bank and by Italian national and regional institutions.

Dr Davide Pettenella, Dipartimento Territorio e Sistemi Agro-forestali, Agripolis, I–35020 Legnaro PD, University of Padova, Italy. E-mail: dpettene@agripolis.unipd.it

**Phan Nguyen Hong** is widely recognised as Vietnam's leading expert in mangrove ecology. Actively promoting and participating in mangrove rehabilitation projects, he has worked for many years on the science on the mangrove and coastal biodiversity and has played a major part in a number of national research projects as well as studying the long-term effects of herbicides used on mangrove forests during the second Indochina war. He is author of over sixty scientific publications in English and French, including, with Hoang Thi San, *The Mangroves of Vietnam*, published in 1993 by IUCN.

Prof. Phan Nguyen Hong, Mangrove Ecosystem Research Division, Center for Natural Resources and Environmental Studies, National University of Vietnam, No. 7 Ngo 115, Nguyen Khuyen, Hanoi, Vietnam. E-mail: merd@ netnam.org.vn

**Roger Safford** was involved in this work as a researcher on tropical ecosystems at Royal Holloway Institute for Environmental Research, University of London. His main area of interest is tropical biodiversity and its conservation and management, with particular experience in the forests and wetlands of the Madagascar region and the Mekong Delta. He is now with BirdLife International.

Dr Roger Safford, BirdLife International, Wellbrook Court, Girton Road, Cambridge CB3 0NA, United Kingdom E-mail: roger.safford@birdlife.org.uk

**Chris Sneddon** is an Assistant Professor in the Environmental Studies Program and Geography Department at Dartmouth College. His research interests include the political ecology of river basin development, the politics of cooperation in international river basins and environment and development in South-east Asia. His doctoral dissertation focused on the socio-ecological transformation of a river basin in Northeast Thailand and conflicts over water occurring on multiple spatial scales.

Dr Chris Sneddon, Department of Geography, Dartmouth College, 6017 Fairchild Hall, Hanover, New Hampshire 03755–3571, USA. E-mail: christopher.s.sneddon@ dartmouth.edu

**Tran Viet Lien** is a past Director of the Climate Research Center in the Institute of Meteorology and Hydrology. His research areas include climate application (building, forestry, human health, energy), statistical climate, climate change impacts, environmental change and its impacts, and land–ocean interaction in the coastal zone. Author of over fifty scientific publications he has also participated in postgraduate-level education at Hanoi University and the Institute of Meteorology and Hydrology.

Dr Tran Viet Lien, Climate Research Center, Institute of Meteorology and Hydrology, Lang Trung Street, Dong Da, Hanoi, Vietnam. E-mail: lien@ crc-imh.ac.vn

# Preface

This book is about change, the one constant of our post-modern civilisation. In particular, it is about how the people of Vietnam are responding to change, to the many and varied social, economic and environmental trends affecting their lives as their nation proceeds down its distinctive new road of economic renovation, *doi moi*, against a backdrop of rapid globalisation.

Globalisation, in this context, refers first and foremost to the increasing mobility of capital and information and the ever-tighter linkages between both economies and societies around the world. But environmental problems are also being globalised. National frontiers have never presented a barrier to environmental change. But impacts that were once largely confined in spatial scale, a local manifestation of resource abuse, have become a threat of global proportions as the seas are contaminated, biodiversity is lost and the atmosphere is polluted on an unprecedented scale.

Many of the nations of Southeast Asia have developed their economies significantly over the past half century, with increasing wealth based on financial openness and the rich human resource. In the drive for modernisation, though, degradation of the physical environment, that other great resource of the region, has been overlooked. But such neglect is no longer viable as we enter the twenty-first century.

The authors of this book present an overwhelming case that no country can isolate itself from the impacts of the development process on its environmental resource base. To achieve sustainable development, effective management of the physical environment, integrated with the needs and aspirations of the people, must be a high priority. At the World Summit for Social Development in Copenhagen in March 1995, the Government of Vietnam defined its development strategy as 'for the people and by the people' aiming to 'place human beings at the centre of development and to promote the potential of individuals and communities as well as of the whole nation'.

Close to 80 per cent of Vietnam's population live in rural areas and most of these people are heavily dependent on the country's natural resources. Given this dependence, the vulnerability, resilience and adaptability of Vietnamese society – its ability to respond effectively to stress and to change – must also play a critical role in shaping prospects for the future. Understanding what determines the ability to respond to change, whether that change be environmental, cultural or socio-economic in origin, is, we would argue, another key element in developing a sustainable and equitable development trajectory that will meet the myriad challenges of the present century.

This book has its genesis in a research project funded by the UK Economic and Social Research Council under its Global Environmental Change Programme. The Global Environmental Change Programme, which ran from 1990 to 2000, sought to demonstrate the linkages between the social and natural world in facing the challenges of global environmental change. We are grateful for the support and interest of programme staff, particularly of the Programme Director Professor Jim Skea, in the implementation of the project and the dissemination of our findings.

The research project, conducted by the University of East Anglia in Norwich and the Center for Environment Education and Development in Hanoi, focused on understanding vulnerability to climate trends, making use of the response to tropical cyclone impacts as an analogue for long-term change. The research results highlighted the different patterns of vulnerability within Vietnamese society as well as the rich tapestry of response strategies available to the population.

We acknowledge the invaluable contribution of our collaborators in this research at the Vietnam National University (particularly Professor Duong Duc Tien), the Mangrove Ecosystem Research Division of the Center for Resources and Environmental Studies (Dr Nguyen Hoang Tri and Professor Phan Nguyen Hong), the Vietnamese Hydrometeorological Service (Dr Hoang Minh Hien) and the Climate Research Center (Dr Tran Viet Lien). We also thank our interpreters and research assistants, Ngo Ngoc Thanh, Ha Thanh, Huynh Thu Ba and Luong Quang Huy, Sarah Granich for her diverse and constant assistance, and the local government officials and householders who gave their time and efforts in participating in the research process. Specific mention is due to the People's District Committee of Xuan Thuy for their renowned hospitality.

In order to present the findings of the project to a wider, policy-oriented audience, the conference *Environmental Change and Vulnerability: Lessons from Vietnam and the Indochina Region* (http://www.cru.uea.ac.uk/tiempo/floor0/briefing/hanoi/ecv.htm) was held in Hanoi in April 1998 with funding from the Economic and Social Research Council. While this volume does not represent the proceedings of the Hanoi conference, it was inspired by discussions at that meeting, at which many of the authors were present. The conference brought together leading academics and policymakers from Vietnam and other countries in Southeast Asia and from Australia and Europe.

We would like to take this opportunity to thank Professor Dao Trong Thi, Vice President, Vietnam National University, Hanoi, and His Excellency David Fall, British Ambassador to Vietnam, for supporting the meeting, the British Council and The Netherlands Foundation for the Advancement of Tropical Research (WOTRO) for additional funding for specific participants, and the World Bank Resident Mission in Vietnam, particularly Andrew Steer, for support and a very hospitable reception.

The conference participants called for greater attention to be paid to the challenge of integrating natural and social sciences to enhance understanding of the processes of environmental change and its impact on society, while recognising the role of basic research in both domains. The participants also emphasised that the sustainable management of natural resources is dependent on a diverse range of conditions being met, including considerations of equity and empowerment as well as informed decision-making and institutional responsibility. They stressed that a diversity of solutions based on a range of experiences is required. As this book demonstrates, there can be no single blueprint for sustainable development.

This book also builds on a series of collaborative projects between the University of East Anglia, the Center for Environment Research Education and Development in Hanoi and Hanoi University of Science (National University of Vietnam) initiated by Mick Kelly, Nguyen Huu Ninh and Sarah Granich in 1991. Alongside the Economic and Social Research Council, we thank the Swedish International Development Cooperation Agency, the University of East Anglia Innovation Fund, the British Academy Committee for Southeast Asian Studies, and the John D. and Catherine T. MacArthur Foundation for past and continuing support for this work.

We have, over the years, received much valuable feedback on our research work at seminars and conferences and in informal discussions with colleagues in the United Kingdom and in Vietnam. Too many to be named individually, we would like to take this opportunity to thank them all. As far as this book is concerned, we thank Claire Garth for help in finalising the text and Philip Judge for patiently preparing the graphics.

Finally, we would like to dedicate this book to the people of Vietnam. This volume will, we hope, make a small contribution to the further realisation of the tremendous potential of this rich and vibrant nation.

W. Neil Adger, P. Mick Kelly and Nguyen Huu Ninh
June 2000

# Part 1

# Vietnam and environmental change

The chapters in this section set the scene for the later accounts of sectoral issues concerning the natural resource base and the wider development process. What emerges from these accounts is a perspective on Vietnam as a nation in a state of continual flux over recent centuries, shaped by the need to cope with stress, with change and with conflict. Beginning with an introduction to the position of Vietnam with regard to social and environmental trends at the start of the twenty-first century, the section continues with a theoretical discussion of the link between vulnerability, resilience and adaptation in the face of environmental change, illustrated with case study material. An account of the current state of the Vietnamese environment, including the threat posed by global change, follows and the concluding chapter of this section presents a historical review of key environment and development issues.

# 1 Environment, society and precipitous change

*W. Neil Adger, P. Mick Kelly and*
*Nguyen Huu Ninh*

## Adapting to change

Vietnam is an enigma. It stands at the beginning of the twenty-first century as a nation that has much to celebrate in terms of equitable social and economic development. Many advances towards this goal have been achieved in the most recent decades. But it remains one of the poorest countries in the Southeast Asian region in terms of per capita income. Vietnam is a geographically large and disparate country, comprising lowland coastal plains and upland mountain areas. Yet, despite a history of years of divisive war and then international embargo, the government has, since reunification in 1975, retained a high degree of legitimacy and popular appeal. The social and political organisation of Vietnam is, in fact, its most confounding enigma.

Along with China, Vietnam represents for many economists and political scientists the most striking example of an 'open door' economy. The transition to market economics was effectively initiated in Vietnam in the early 1980s following critical years of crisis in the economy. The government of Vietnam, although disunited as to how to proceed, followed at that time a path of market-led reforms which ultimately became the process of *doi moi* (literally, new change or renovation), in parallel with reform movements in the USSR and China. *Doi moi* concerns the renovation of Vietnam's economy and society, under the leadership of the Communist Party and State, and has three key elements: the move towards a market-oriented economy with State management; the 'democratisation of social life', developing the rule of law in a state of the people, by the people and for the people; and an 'open-door' policy with promotion of cooperation and relations for peace, independence and development with all nations (Vietnamese Communist Party 1991: 147; United Nations 1999: 1).

The relationship between the rural and urban economies of Vietnam in the early part of the transition and the more recent reforms is crucial in explaining the nature of the present-day economic structure and institutions. Vietnam remains a predominantly rural and agrarian nation, with over 76 per cent of its population of over 76 million people absolutely reliant on the sustainability of the natural resource base on which food production and livelihood security depends. The reformers realised early on that the rural populous and the agricultural economy were the key to the gradual rolling back of the state as the fundamental controller of the factors of production. The present level of economic growth experienced in Vietnam, averaging 8 to 10 per cent through the 1990s, has come about in association with the privatisation of state-owned industries and of major product and marketing organisations in Vietnam, with price reform, and with major changes in property rights in the agricultural sector. The apparent macro-economic success of transition in Vietnam, paralleled with that of China,[1] masks a number of crucial microeconomic and social impacts of the process of renovation, not the least of which are observed in the rural sectors of the economy and in the management of natural resources.

At the same time, Vietnam, like the rest of Southeast Asia, faces significant environmental challenges. Threats are apparent at many different spatial- and timescales, ranging from local resource degradation and the adverse consequences of deforestation and other land use changes through urban air and water pollution to the emerging threat posed by climate change and other local manifestations of global environmental change.[2] In broad terms, the environmental stresses facing the people of Vietnam are the product of over-exploitation and mismanagement of resources, of the legacy of colonialism and war, and, increasingly, of global forces (Vo Quy 1997). Limiting the effect of environmental degradation on individual livelihoods and on broader prospects for sustainable economic growth by reducing vulnerability, enhancing resilience and promoting adaptive strategies is urgent and critical if the well-being of the farmers, fishers, forest dwellers and urban populations of Vietnam is to be secured.

This book presents a portrait of Vietnam at a time of precipitous social and environmental change, exploring the interface between environment and development issues in a series of studies based on current research in that country. With a long history of coping with hazards, adaptation has been the watchword for Vietnam in reacting to stress in the past. The nation is, in many respects, well-prepared to confront the escalating challenges of the future. The overarching theme of this book is the response of Vietnamese society to present-day change, focusing on the process of adaptation and the constraints and opportunities that will determine future sustainability and livelihood security.

It is argued that the key to understanding the process of adaptation is to focus on institutions (Adger and Kelly 1999). It is necessary to examine both the institutional architecture, through studying issues such as legitimacy, inertia and dynamic adaptation, and the resilience and flexibility of institutions in managing environmental risk. The term institution, as used here, is not constrained to what is normally thought of in this political economy context, namely those formal institutions of the state. The term institution is used more widely to include informal social and cultural norms as well as formal political structures. Examining the institutional mechanisms of the management of individual and collective environmental risk involves careful analysis of the political economy of resource use and must concern a wide range of aspects of contemporary social science. Adaptation is socially mediated and differentiated and can take place through mechanisms which can be characterised as social learning and policy learning.

The relevant institutional types associated with the issues examined in this book are those of state, market and civil organisations. State, market and civil society can be contrasted by their mechanisms. States enforce by regulation or legal sanction. Markets convey price signals as incentives. Civil organisations are based on bargaining, cooperation and persuasion (de Janvry *et al.* 1993).

In seeking to understand the process of adaptation it is also helpful to consider the related concepts of vulnerability and resilience. Analysis of vulnerability as a social phenomenon also has a long tradition within cultural geography and the critical questions of food security and famine (Watts and Bohle 1993). It is related to the study of criticality (a concept applied spatially at different scales) and to security (Kasperson *et al.* 1996). The term social vulnerability, as used in this book, defines the exposure of groups of people or individuals to stress as a result of the impacts of environmental change. Stress, in the social sense, encompasses disruption to groups or individuals' livelihoods and forced adaptation to the changing physical environment. Social vulnerability, therefore, encompasses disruption to livelihoods and loss of security. For vulnerable groups, such stresses are often pervasive and related to the underlying economic and social situation, both of lack of income and resources, but also to war, civil strife and other factors (see Chambers 1989). In this framework, social vulnerability, and what has been termed the architecture of entitlements, determine the extent to which adaptive action is constrained and the availability of resources upon which adaptive measures can be based (Watts and Bohle 1993; Adger and Kelly 1999; Kelly and Adger 2000).

Resilience increases the capacity to cope with stress and, hence, is an antonym for vulnerability, as used here. In essence, individuals, households, communities and institutions at all levels in society are constrained with respect to prospects for adaptation and adaptive strategies by the resilience of the social and natural systems in which they are embedded. The greater their resilience, the greater the opportunities for absorbing external shocks and adapting successfully to rapid social and environmental change. The less resilient and developed the social system, the greater the vulnerability of social groups or institutions at all levels to externally imposed change.

Resilience can, in fact, be defined in many ways. It is often considered in terms of the buffer capacity of a system, the ability of a system to absorb perturbations, or the magnitude of disturbance that can be absorbed before a system changes its structure by changing the variables and processes that control behaviour. By contrast, other definitions of resilience emphasise the speed of recovery from a disturbance, highlighting the difference between resilience and resistance, where the latter is the extent to which disturbance is actually translated into impact. It is important to note that these definitions are widely used in ecology and refer to ecological processes usually at the ecosystem scale. It is argued by many ecologists that resilience is the key to biodiversity conservation and that diversity itself enhances resilience, stability and ecosystem functioning (Tilman 1997). Ecological economists also argue that resilience is the key to sustainability in the wider sense (e.g. Common 1995).

A key issue in examining social change is that of the resilience of institutions. Institutions can be persistent, sustainable and resilient depending on a range of parameters. Institutional resilience is determined by legitimacy, agenda setting and the selecting of environmental risks which resonate with the institutions' agenda, and the maintenance of social capital (see Adger 2000b). Social capital is taken here to mean the existence of integrating features of social organisation such as trust norms and networks. Resilience facilitates adaptive behaviour and the co-evolution of institutions with their environment in a manner which enhances legitimacy and sustainability. The cultural context of institutional adaptation, and indeed the differing conceptions of human environment interactions within different knowledge systems, are central to the resilience of institutions. These contexts tend to be overlooked in considering equity and economic efficiency in the sustainable use of natural resources (Gadgil *et al.* 1993; Brown 1997).

An example of the resilience of institutions can be found in the ability of institutions of common property management to cope with stress. Social capital, ecological resilience and social resilience are all tested when upheaval and stress are placed on institutions. Commonly managed coastal resources are being degraded throughout the world through the breakdown of property rights or inappropriate privatisation (see Berkes and Folke 1998). Nowhere is this more clear than in the case of coastal resources such as fisheries and mangroves, in urbanisation and pollution issues, and in the forces of economic globalisation, all of these are critical issues for Vietnam.

The issues of vulnerability, resilience and adaptation resonate with social change and major restructuring of the former centrally planned economies. The single most important factor distinguishing Asian from Eastern Bloc reform in the move from planning to new hybrid economies is the emergence of networks and institutions as the driving forces of change. These networks link entrepreneurship, collective action and the remnants of government in novel ways (see Giddens 1996; Grabher and Stark 1998; Smart 1998). Institutional and governance issues emerge as critical in many chapters of this book; the institutions which manage collective risk in both the environmental and economic spheres are, in fact, a hybrid of post-socialist planning and of traditional Vietnamese collective action and culture. This experience perhaps adds weight to the perspective that there is developing a uniquely Asian vision of the development process, if not to the rejection of the 'end of history' argument and the inevitability of the rise of neo-liberalism.

The various chapters in this book use novel and evolving approaches to the society–environment nexus that highlight the interface between institutional, economic and cultural perspectives on different scales ranging from individual decision-making in the face of poverty and scarce resources to the role of the Vietnamese state in both domestic and international policy processes. Little, if any, distinction is made between local and global environmental change, since the objective is to examine how individuals and institutions manage environmental stress and risk, however and wherever it is perceived and experienced. Moreover, the present day is the major concern on the grounds that, if current problems cannot be resolved, it matters little what the future holds in store.

This introductory chapter provides an outline of the key constraints and opportunities facing Vietnam at the turn of the millennium, and highlights the precarious position of many members of Vietnamese society at this time. It also introduces the principal themes of the book, related to the rapid environmental and political changes facing Vietnam and its position in Southeast Asia in the early twenty-first century. The context is the management of rapid economic growth, in line with other Southeast Asian 'tiger' economies, alongside the, often painful, adjustment towards a market oriented system that typically faces the former centrally planned economies. A context, it is argued, that must also encompass the uncertain, and sometimes unforeseen, consequences of environmental change, driven by both local resource use and by external forces. It is the effective management of the interaction between environment change and development prospects that necessitates, as is argued throughout this book, a focus on the process of adaptation; a process which can be observed throughout the historical evolution of society and environment in this enigmatic country.

## Asian development and economic crisis

Vietnam is often perceived as a tiger cub, a younger cousin of the East Asian tiger economies. Aspirations to the status of a full tiger economy has come with rapid economic growth sustained throughout the 1990s through a mixture of labour productivity gains, openness to foreign direct investment and reform of various sectors of the domestic economy (see Riedel 1993). For some countries in the region, the phenomenon of economic growth over the past three decades is regarded as nothing short of miraculous. The East Asian 'miracle', coined by the World Bank (1993a), refers to the perception that the economies of South Korea, Taiwan, Singapore and Hong Kong, as the first-tier newly industrialising economies (NICs), and Malaysia, Thailand and Indonesia, as second-tier NICs, have achieved remarkable development rates, with sustained economic growth and large investments in human capital accompanied by an equitable distribution of income and welfare across these societies.

Despite the shocks to these economies since the regional financial crisis that began in 1997, the prospects for such growth remain high. The causes of that financial crisis were, it can be argued, primarily associated with deregulated financial systems and the vast inflows of international financial assets. Optimism over the future does not mean, of course, that the national financial systems of the Southeast Asian region does not need further regulation. Rather, we would argue, in line with some commentators, that the crisis may mark a distinct move away from the doctrines of neo-liberalism towards a recognition of the need for strong, but accountable, governments to ensure effective economic management in the region (see, for example, Wade 1996, 1998). Accountable government has implications for environmental management and all areas of public policy, as well as more narrowly for the financial sector.

A dilemma for Vietnam, and the other Asian transition economies such as China and Laos, is whether economic growth can be sustained without a widening of the representativeness of social institutions in decision-making and democratic structures. A related issue is whether economic liberalisation in Vietnam would necessarily lead to social and political change. Analysis of the other East Asian tiger economies suggests that there is no simple answer to either question. It can be argued that an authoritarian regime in South Korea, for example, enabled private investment and high educational attainment, while ruthlessly defending the interests of private property in that country over the past thirty years (Pyo 1995). Yet economic revolution in South Korea largely excluded public participation in decision-making. Perhaps it is only when macro-economics goes badly wrong that the limitations of authoritarian development are demonstrated.

## *Opening the economy and the challenge of environmental change*

The role of public participation and institutional accountability is an important issue, not only because of its effect on the overall political economy, but also because of the many environmental threats facing Vietnam and other countries in a similar situation. Of all the dilemmas facing Vietnam, in fact, perhaps one of the greatest is managing the environmental and consequent social implications of its development path. The experience of its neighbours in coping with rapid economic growth in terms of pollution, escalating urbanisation and the strain on renewable and non-renewable resources points to some salutary lessons for Vietnam. In particular, the resulting changes in property rights, the evolving locus of decision-making and the shifting role and, often, structure of the institutions that manage environmental risk, particularly in the rural sector, are posing immense challenges.

The present political economy of Vietnam can be seen as a uniquely Vietnamese compromise, drawing deep on the cultural and social context in which Vietnamese society claims security, equity and autarky as its principal demands. The bureaucratic interests in the status quo have also led to what Kolko (1997: 30) describes as 'eclectic opportunism'. Whatever the interpretation of the radical change brought about by the current government of Vietnam, there is no doubt that it has spawned a complex and intriguing mix of pragmatic politics and economics.

It can be argued that the crux of the present economic and political transition as far as the effective management of environmental change is concerned is the altering distribution of, and access to, human, economic and natural resources. East Asian countries, in general, initiated rapid industrialisation programmes with relatively evenly distributed wealth and land assets at the outset. Some would argue, in fact, that the 'miracle' may be somewhat of a mirage, since it is this initial wealth allocation that largely explains the even distribution of income in the present day (You 1998). Regardless, as will be demonstrated throughout this book, equitable wealth distribution in Vietnam, along with the strong traditional and historical basis for the institutions that manage water, irrigation, land, access to forests, and broader interactions with the environment, are the key parameters enabling adaptation to environmental change.

A further irony for Vietnam in this context is that, despite decades of political and economic isolation following reunification, it can no more be isolated from global environmental change than it can from the vagaries of a globalised market economy. Just as Vietnam has felt the draught of the Asian economic crisis, so it will also feel the ill winds of global resource scarcity, loss of biodiversity and even the literal ill winds of changes in the global climate regime. There are, therefore, direct linkages between the 'grand' environment debates and the 'grand' development debates in Vietnam, as elsewhere. The conceptual model outlined in Chapter 2, along with examples from many other chapters in this book, highlight these linkages and point to lessons for Vietnam, the newly and not-so-newly industrialising economies of Asia, and for the newly emerging societies of the former centrally planned world.

## Cultural and historical contexts of environmental management

The political history of Vietnam is punctuated by upheaval and revolution, tempered, over the past ten centuries, by a predominantly unified culture and language. There are distinct periods in the history of Vietnam that have shaped present-day institutions and society. These are the period prior to European colonisation after the coming into existence of Vietnam as a nation; the French colonial period from around 1859 to 1954; the communist period which include the anti-colonial wars of independence and the American war; and the most recent period of economic liberalisation, characterised as the *doi moi* period from around the mid 1980s. A chronology of the major political events which define these periods is presented in Table 1.1, with more detailed information on the most recent reform process outlined in Table 1.2.

The identity of Vietnam is in large measure due to the common language of the lowland sedentary agricultural people of the Mekong and Red River Deltas and the intervening coastal lowlands. Modern Vietnam also comprises a small proportion of ethnic groups from the mountainous regions on the borders of modern Laos and Cambodia. Within the culture of the dominant lowland Vietnamese, social structure is relatively homogenous and village level organis- ation and systems of land tenure were similar across the lowland regions in the period prior to European interference. The village level organisation was a complex mixture of patriarchy and kinship with Confucian elements, that dominated the allocation of communal lands, and a strong system of government at regional and even national level (see Wiegersma 1988). Thus, rather than being strictly feudal, the system of political control under both Chinese colonial rule and under independence could be characterised as having some signs of 'private' ownership, with the basic unit of production being the independent village commune, and a centralised monarchy (Nguyen The Anh 1995). As indicated in Table 1.1, the latter part of this period was characterised by national rule by a

*Table 1.1* A chronology of major political changes in Vietnam till reunification in 1975

| | |
|---|---|
| *Pre-French Colonial Period (Chinese colonialism and independence)* | |
| 111 BC–905 AD | Chinese colonial rule with numerous revolts |
| 905–1427 | Alternate independence and Chinese colonial rule up to the Ming dynasty |
| 1427–1786 | Independence under Le dynasty including division of north and south Vietnam under two viceroys from 1627 |
| 1774 | Revolt in northern provinces led by Tay Son brothers over high taxes |
| 1786–1859 | Unification under Tay Son and Nguyen dynasties |
| *French colonial period* | |
| 1859 | First French invasion of Cochinchina |
| 1867 | Concession of all of Cochinchina to France by Hue Court |
| 1867–83 | Expansion of territorial interests and trade monopolisation by French until all of Vietnam is under French rule, under Annam, Tonkin and Cochinchina |
| 1884–1927 | Anti-colonial resistance in various regions of Vietnam with periodic revolts |
| 1927 | Formation of Vietnamese Nationalist Party under Ho Chi Minh |
| 1930–31 | Nghe–Tinh soviet movement in central Vietnam |
| 1940 | Japanese invasion of Indochina with agreement from French administration |
| *'Communist' period* | |
| 1945 | Surrender of Japan and Declaration of Independence by Ho Chi Minh (2 September 1945) |
| 1946–54 | Franco-Vietnamese War concluded with victory at Den Bien Phu |
| 1954 | Division into North and South Vietnam under Geneva Agreement |
| 1960 | Formation of National Liberation Front of South Vietnam |
| 1965–68 | US ground troops in Vietnam and bombing of northern Vietnam |
| 1975 | Reunification of Vietnam |

Source: Adapted from Hy Van Luong (1992), Popkin (1979) and others.

mandarinate, sometimes as part of Chinese colonial rule and at other times partially or exclusively independent. During this period, the Emperor would be regarded as the protector of the nation and would perform symbolic duties. He would, for example, plough the first furrow each year for the new rice crop, thereby confirming the state's position as responsible for national welfare whatever the natural conditions (Wiegersma 1988).

In the period prior to European colonisation, the central government of Vietnam, particularly at the regional level, had strong control over land tenure and infrastructure issues such as water regulation in the lowlands for lengthy stretches of time. The first three government departments established in Vietnam covered flood fighting, agriculture and civil defence (Benson 1997; see also Chapter 10). Despite numerous revolts and political upheavals in the course of these centuries, Vietnamese society retained its underlying patriarchal and partly Confucian social order.

*Table 1.2* A chronology of major political changes in Vietnam since reunification in 1975

| | |
|---|---|
| *Unified period* | |
| 1975–79 | Expansion of collectivised agriculture in southern Vietnam |
| 1979 | Chinese invasion of northern Vietnam |
| 1979 | Limited introduction of household contract system in agriculture |
| 1981 | National application of household contract system under Directive 100; cooperatives and work point system still operating |
| 1986 | *Doi moi* introduced following Party Congress |
| *Renovation period* | |
| 1988 | Agricultural reforms under Decree 10. Collectivisation of farms in Mekong Delta. Land allocation on lease system. Cooperatives to act as service operations to household agriculture |
| 1988–93 | Land ownership still with the state. A rental market for land emerges even though illegal |
| 1993 | Land Law instigates 20 year and longer leases of agricultural land |

Source: Adapted from Fforde and de Vylder (1996) and others.

The French expansion of its colonies into Vietnam from the 1850s onwards heralded a shift in the operation and governance of Vietnam, and in the institutions of government and of land management in particular. The French colony was expanded piecemeal through a series of invasions and treaties and the French divided the Vietnamese area formally into the three states of Tonkin and Annam in the north and central, and Cochinchina in the south, areas which had only recent precedent in Vietnamese history (Wiegersma 1988). The imposition of French colonial rule, it has been argued, did not dismantle local hierarchies in terms of the populations at the village level, but they did throw into fundamental conflict the role of these hierarchies, in terms of the institution of private property within the village system. Essentially, as argued from most historical perspectives on the evolution of peasant revolution (e.g. Marxist, liberal, neo-classical, see reviews by Scott 1976; Popkin 1979; Wiegersma 1988; Nguyen The Anh 1995), colonialism led to the formal institutionalisation of capitalism for the first time in Vietnam. This became manifest in the creation of private landlord classes who controlled land and rented it to landless peasants, thereby replacing collective labour with waged labour. The origins of the present system of government stem from the overthrow of the French colonial power and the resonance of the multi-layered Vietnamese society with some forms of communal action which became mobilised into a peculiarly Vietnamese communism, as discussed in Chapter 4.

The post-colonial period can be characterised as a period when the government system was reinforced by the apparent harmony with underlying cultural influences, given nationalism and the external forces which threatened it at this time. The system of government adopted under the Communist collectivisation was particularly effective in dealing with the threat of external forces associated with war, political isolation and economic stagnation (see Lacoste 1973; Thrift and Forbes 1986).

## Socialism, the market economy and the determinants of policy change

### *The context of reform*

Following the reunification of Vietnam in 1975 after more than three decades of war, Vietnam entered a period of evolution of its socialist practice, providing a present-day legacy of economic and political structure that fundamentally affects how the environment is managed. Here, we focus on three particular aspects. First, the economic growth of the most recent *doi moi* period, following years of stagnation, is outlined, paying particular attention to the agrarian sector in this predominantly agricultural nation. Second, the legacy of human resources is discussed. Vietnam has invested heavily in health and education during the period of central planning and, hence, demonstrates favourable indicators of human development. This positive legacy may be threatened by the current process of transition. The liberalisation of labour markets is also important in the management of environmental resources, since regional migration and resource flows are now common. The growth of Hanoi and Ho Chi Minh City threatens the fabric and is stretching the services of these cities. Finally, the evolution of the distribution of wealth and income during this period and the consequences of the recent opening up of the Vietnamese economy are described.

### *The economy and the agrarian sector*

At the turn of the century, the Government of Vietnam is committed to the modernisation and industrialisation of the nation and to 'people-centred' development (a development process that is implemented by the people and for the people); rural development and poverty reduction are also major priorities (United Nations 1999; United Nations Development Programme 1999). The basis of the government's development strategy is to create an enabling environment of sustainable high economic growth, political, social and economic stability, and equity (a decent minimum standard of living and equal opportunities for all), thereby helping the people to help themselves (United Nations Development Programme 1999: 6).

Supported by the process of *doi moi* and the lifting of the trade embargo, growth rates of close to 10 per cent were achieved during the 1990s. The relative performance of the Vietnamese economy over the early 1990s compared to other countries facing similar challenges is shown in Figure 1.1. The major crisis of the Vietnamese economy in the mid-1980s had largely been turned around due to the contribution of the agricultural sector to increasing national output. Nevertheless, with increasing investment in the industrial and service sectors, the sectoral composition of GDP is shifting away from agriculture. During the 1990s, the share of agriculture fell from 40.6 to 23.8 per cent while that of industry rose from 22.4 to 34.3 per cent (United Nations Development Programme 1999). But the basis of the economy is still agriculture. Close to 70 per cent of the workforce are

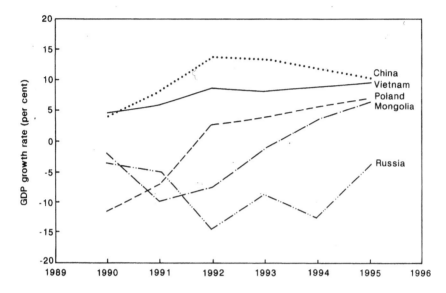

*Figure 1.1* Gross Domestic Product growth rate, 1990–1995, for selected 'transition' economies
Source: World Bank (1996b: 173).

employed in agriculture and forestry and agriculture remains one of the major sources of export earnings (United Nations Development Programme 1999). The agrarian sector is, however, undergoing a rapid evolution.

In the period up till 1981, agriculture was exclusively based, at least in the lowlands of northern Vietnam, on the collectivised system where village level cooperatives organised the labour allocation and distribution of external inputs and outputs. This period was short-lived, primarily due to the crisis of falling output in the late 1970s (Figure 1.2), which heralded the introduction of a system of household contracts in 1981, relinquishing control of at least part of the factors of production to the household level. The subsequent period had incremental but significant changes in the legal and institutional framework in agriculture (Table 1.2). The Sixth Party Congress in 1986 introduced reforms which were to become known as *doi moi*, encouraging an agriculture-based economy. State-owned enterprises in the manufacturing and other sectors have not been scrapped, but have been opened to market forces. The fast-evolving situation of land tenure and control has been influenced by a resolution of the National Assembly in 1988 (Resolution No. 10) which formalised a full contract system of household responsibilities in agriculture, and by a Land Law in 1993 which, in effect, allows long-term lease of land from the state and allows these leases to be tradeable. At present, all land is owned by the state and collectives retain control over the distribution of inputs and provision of irrigation.

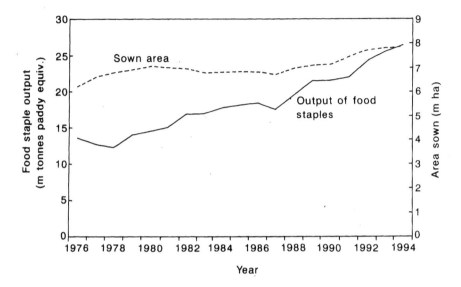

*Figure 1.2* Area under crops and food staple output for Vietnam, 1976–1994
Sources: Vietnam General Statistical Office (1995) and Fforde and de Vylder (1996).

The present-day situation can be characterised in terms of rapid increases in marketed agricultural output, facilitated by incremental changes in the distribution system and higher real prices to farmers as well as incentives to invest in private agricultural land. In this way, avoiding 'big bang' economic liberalisation in the agricultural sector, in the sense generally applied to transition economies (de Melo *et al.* 1996), has brought about some immediate economic benefits in agricultural production. But this process has also had complex social and political consequences within all sections of society, particularly the rural agrarian economy.

The period of decollectivisation is involving radical changes in the institutional framework of resource use and the increased exposure of local level resource users to external forces in the policy environment. The major institutional changes which have occurred involve the atomisation of land tenure in the progression from the household contract system of the early 1980s to effect privatisation through a long lease control of agricultural land by individual households. In the agriculture sector, there have been dramatic increases in agricultural production and marketing since 1990, as evidenced both at the national and at district levels, partly due to increased use of external inputs, and capital, and partly to management changes. Outside the land use sector, the state owned enterprise sector has been emasculated and the country opened to foreign investment. Fforde and de Vylder (1996: 238–9) argue that rapid economic growth in Vietnam has induced a role for the state unlike that in most other newly industrialised countries. The central role of the state in Vietnam, they observe, is moving from being

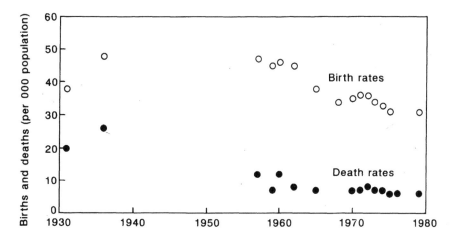

*Figure 1.3* Estimates of birth and death rates (births and deaths per 000 population) for North Vietnam, 1930–1979
Source: Bryant (1998: 237).

redistributional to a strategy which they call accumulatory. They argue that this has been tempered, however, by the gradual move towards liberalisation, unlike that experienced in other centrally planned economies.

### Population and human resources

Vietnam is rich in human capital. It lies 122nd in rank in the Human Development Index of 1998 (United Nations Development Programme 1998), despite an income level of US $1200 per capita in purchasing power parity terms and ranking 148th in terms of this indicator. This situation reflects high levels of investment in education, social development and health, within the constraints imposed by lack of income and isolation from the international community over recent decades. This richness in human capital enhances prospects for sustained economic growth. The political and economic institutions of government, at least in the north of the country, shaped the demographic structure of Vietnam over the past half century such that it has mortality and fertility rates near the world median. This is an impressive achievement for a country with one of the lowest income per capita rates in the world (Bryant 1998). Even market liberalisation, which has had some significant detrimental impacts on health service provision (Ensor and San 1996), does not seem to have stemmed the continued mortality and fertility declines (Bryant 1998).

The demographic record for Vietnam is shown in Figure 1.3, with some estimates from the 1930s, based on the classic work of Gourou (1955) on human ecology in the Tonkin Delta, and more recently on analysis of census information. This profile suggests that a population transition akin to those of many developing countries is already under way (see Adger and O'Riordan 2000, for a review). Vietnam's success must be credited in large part to the delivery of an effective primary health care programme from the 1960s onwards. This effectiveness was a direct result of low labour mobility and the hierarchical political structures in charge of all factors of production and social planning (Bryant 1998).

The process continues. In July 1998, the Parliamentary Committee for Social Affairs and the United Nations Population Fund, for example, highlighted the need for 'accounting for spontaneous migration when designing a comprehensive socio-economic development plan' (United Nations Family Planning Association Press Release, 8 July 1998, Hanoi, Vietnam), and the related need to 'actively support access to ownership or use of land and access to water resources especially for family units, to make and encourage provincial investment to enhance rural productivity, to improve rural infrastructure and social services such as education, health, reproductive health in rural areas'.

It has been argued that stemming rural–urban migration accompanied by economic growth based on increased value-added in agriculture is the best, and perhaps only, sustainable development trajectory open to Vietnam (e.g. Timmer 1993). This is particularly relevant in the light of the current Asian economic crisis. Indeed, economic growth based on a model of enhancing urban productivity as the principal mechanism for tackling poverty has the potential to jeopardise the sustainability of that urbanisation process (see Drakakis-Smith 1996; and also Chapter 12). Despite the government's migration policies, developed provinces are still attracting more immigration and less developed provinces more out-migration (Anh Dang *et al.* 1997).

### Equity and economic development

Equity and development are important issues when considering prospects for the future because changes in the distribution of wealth and assets in Vietnam, as elsewhere, have related implications for environmental management, for the ability to cope with external change, and for the rights of access to resources in times of stress. Kolko (1997) argues that the explicit goal of social justice of the Government of Vietnam since reunification has been eroded by the transition process. This analysis, however, may focus too much on income within the formal sector of the economy, missing the potential for cultural and institutional changes and feedbacks within modern Vietnamese society.

Increasingly skewed distribution of income may be offset by strengthening and re-emergence of kinship ties within Vietnamese society (as argued, for example, by Hy Van Luong 1993). In addition to formal changes in Land Law referred to above, there is evidence of a re-emergence of local level institutions associated with local collective action and with economic activities. The liberalisation process is fuelled by the interactions between land use, resources and social change and has allowed the development of nascent civil society which has far-reaching consequences (Malarney 1997).

The former centrally planned economies in general exhibit lower income

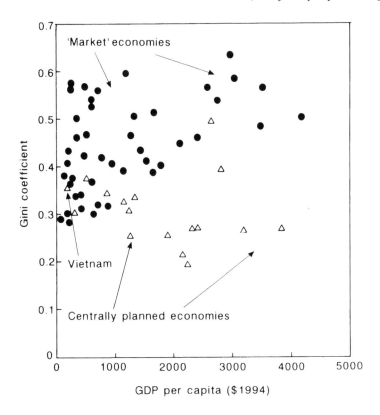

*Figure 1.4* Relationship between income inequality (as measured by the Gini coefficient) and GDP per capita for selected 'market' and former centrally planned economies

Source: Data from World Bank (1996b).

inequality than countries with similar levels of income per capita, with inequality measures rising in the period of liberalisation towards market oriented economies (summarised, for example, in World Bank 1996b). National level data on income inequality the former centrally planned economies exhibit lower income demonstrates that inequality in general than countries with similar levels of income per capita. This is demonstrated in Figure 1.4, showing the relationship between income per capita and the national level Gini coefficient summary measure of inequality for income levels calculated by the World Bank (1996b) for around sixty countries.

From the low base in Figure 1.4, inequality in the former Soviet Republics has risen in the past decade to levels equivalent to the world average of middle income countries. The countries of central and eastern Europe now have inequality levels similar to those of western Europe but lower than the former Soviet Republics. These changes are explained by the World Bank (1996b) as the consequences of changes in particular sectors of the economy such as energy and banking where real incomes have risen sharply and education and health where real incomes have fallen dramatically. By contrast, China and Vietnam are both said to be experiencing rising inequality because of rises in inter-regional inequality. In other words, there are particular regions and growth poles in both China and Vietnam (the large coastal cities of both countries) where economic growth is concentrated. Rising inter-regional inequality is exacerbated through regulations on internal migration and restrictions on the mobility of labour.

## Towards the future

Vietnam and the other countries of Southeast Asia face an uncertain future. The recent turmoil in the financial systems of the region bears testimony to this uncertainty in the economic realm. Then there is environmental change, another threat that cannot be predicted with confidence. Take global warming, for example. The possibility of climate change and sea level rise as a result of increasing concentrations of atmospheric greenhouse gases is established beyond serious doubt, but precise forecasts of effects on particular areas are not yet available (Houghton *et al.* 1996). And even when the impact of environmental degradation can be well-defined, it is often not possible to identify effective solutions with confidence.

Over years to come, Vietnam must respond to the external pressures of globalisation and the vagaries of the international economic order, as well as steering the internal process of renovation and averting the adverse effects of environmental change. Effective management of this process will call on the tested capacity of the Vietnamese government and people to find inventive and pragmatic solutions as fresh challenges emerge. Effective management of the environmental, social and economic future will depend on how the inherent adaptability of Vietnamese society is channelled in the first decades of the new century. Given the contested and offsetting outcomes, enigmas and ironies that characterise the present situation of Vietnam, there is clearly a critical need to better understand the processes by which adaptation has occurred in the past, is taking place at present and can be facilitated in the future.

## Notes

1  Muldavin (1996; 1997) has argued that in China the rapid economic growth also experienced in its rural economy is brought about through the 'mining of communal capital' (1997: 579) and that this phenomenon is not a transitional phase but a characteristic of the new hybrid economy.

2  Brookfield and Byron (1993) and Parnwell and Bryant (1996) set these global environmental challenges in a Southeast Asian regional context. King (1998) and contributors demonstrate that the historical legacy of colonialism is a powerful influence on present-day institutions and resource use in the region.

# 2  Social vulnerability and resilience

*W. Neil Adger and P. Mick Kelly*

## Introduction

To understand how Vietnamese society responds to the challenge of environmental change, it is important to develop a conceptual framework for examining the process of impact, reaction and adaptation. Implicitly or explicitly, this framework is made use of throughout the chapters of this book. The basis of the framework is that the adaptive options open to any social grouping are constrained by the resilience of the human and natural systems that comprise or define that grouping. A high level of resilience implies greater opportunities for absorbing external shocks and successful adaptation to both social and environmental change. Low resilience means the vulnerability to externally imposed change is greater, i.e. adaptive options are limited. But resilience and vulnerability are not pre-determined. Nor are individuals or institutions passive in these processes. Social vulnerability and resilience are determined by a host of complex social processes and economic factors, from access to resources through to informal and formal social security, insurance and social capital. In essence, these determinants are related to the concept of entitlements and access of individuals or groups to resources. Resilience and vulnerability can be observed at different scales, but they are essentially relative concepts.

The impetus for this framework grew from a consideration of the external challenges facing Vietnam associated with the threat of global climate change (Adger 1999b; Adger and Kelly 1999; Kelly and Adger 2000). Focusing on the social and economic determinants of vulnerability to climate change, we have argued that information on vulnerability to current climate variability and hazards allows the identification of adaptive strategies aimed at enhancing resilience that represent 'win–win' options, with benefits in the present day as well as the long-term future. In effect, the response to long-term environmental change is facilitated and constrained by the same architecture of entitlements as adaptation to other, more immediate social and environmental stresses. In a series of case studies and theoretical discussions, we have demonstrated that this approach reduces the conflict between the demands placed by immediate development needs and those that would ensure long-term environmental protection.

The framework outlined in this chapter builds on those insights by arguing that resilience and vulnerability are relevant concepts in examining how societies and individuals evolve in the face of wider social and environmental change. The framework developed here resonates with other theoretical insights on development and environment interactions (Watts and Bohle 1993; Norgaard 1994; Berkes and Folke 1998). We emphasise both social and economic determinants of resource use and resource degradation, as well as highlighting policy-relevant interventions and solutions to environmental challenges (Blaikie and Brookfield 1987; Pearce and Warford 1993). This approach provides a robust, policy-relevant rationale for impact assessment. It recognises that society is continually responding to environmental stress, and constantly evolving such responses.

## Vulnerability and resilience

### *The architecture of entitlements*

The central insight brought by social scientists to the study of vulnerability is that the state of vulnerability is socially differentiated. Vulnerability to environmental change is not the same for different populations living under different environmental conditions or faced with complex interactions of social norms, political institutions and resource endowments, technologies and inequalities. Moreover, it has been demonstrated that the social causes of vulnerability often evolve on much more rapid temporal scales than the environmental changes that interact with these processes.

We base our examination of social vulnerability to environmental change on an understanding of the human use of resources. Following Sen (1990) and others, we consider that the extent to which individuals, groups or communities are 'entitled' to make use of resources determines the ability of that particular population to cope with and adapt to stress. The concept of entitlements, which extends beyond income or other material measures of well-being, has been developed in analyses of vulnerability to food insecurity and famine (Sen 1990). Extending this, examining social vulnerability to environmental change through an entitlements approach means consideration of the availability and distribution of entitlements, the means by which entitlements are defined, contested and, therefore, change over time. This is, and the wider political economy of the distribution and formation of entitlements, a complex of factors which together form the construction we term the 'architecture of entitlements'.

We are concerned with providing a guide to policies likely to reduce levels of vulnerability, enhance resilience and facilitate adaptation. We are striving, therefore, to define the social construction of vulnerability and resilience, how different characteristics, phenomena or processes reduce or heighten levels of vulnerability and resilience over time, and how they differentiate these states within a population. Adaptation is, of its very nature, in continual evolution, not least because the technological and institutional factors which shape vulnerability and resilience are themselves in a state of constant flux. It is this dynamic aspect of the process of adaptation and the underlying factors that determine adaptive behaviour that it is most important to capture, rather than any measure of vulnerability, resilience or adaptive potential taken at a particular point in time. From a policy point of view, the critical challenge is to detect, then reinforce, modify or offset, trends in the factors that determine vulnerability, resilience and adaptive behaviour as they emerge.

### Defining vulnerability

Social vulnerability is a negative state endured by groups or individuals. In the broadest sense, vulnerability occurs because livelihoods and social systems are exposed to stress and are unable to cope effectively with that stress. Though most evident in times of crisis, vulnerability is a chronic and pervasive state related to the underlying economic and social situation, not only in terms of lack of income and resources but also with respect to a range of factors determined by government policies, societal trends, and so on.

More precise definitions of vulnerability, suitable for operationalisation, can be found in the food insecurity, or famine, and natural hazards literature (e.g. Sen 1981, 1990; Watts and Bohle 1993; Bohle *et al.* 1994; Blaikie *et al.* 1994; Cutter 1996; Ribot *et al.* 1996; and Hewitt 1997; see also review by Adger 1996). Watts and Bohle (1993), for example, argue that vulnerability to food insecurity can be defined in terms of: first, the exposure to stress and crises; second, the capacity to cope with stress; and, third, the severe consequences of stress and the related risk of slow recovery. From a natural hazards perspective, Blaikie *et al.* (1994) define risk as consisting of two components. The first component, equivalent to Watts and Bohle's exposure, is a measure of the natural hazard. The second, vulnerability itself, is equivalent to capacity and is, they argue, largely determined by socio-economic structure and property relations. Blaikie *et al.* (1994) clearly separate what we may term the biophysical and the social dimensions, defining vulnerability in terms of the human dimension alone as 'the capacity to anticipate, cope with, resist, and recover from the impact of a natural hazard' (Blaikie *et al.* 1994). The biophysical component, the exposure or measure of the hazard, is formally outside their definition of the concept of vulnerability.

If we trace the linguistic roots of the term vulnerability then we find support for this latter perspective. The Collins English Dictionary (Second Edition, 1986) defines vulnerability as, *inter alia*, the 'capacity to be physically or emotionally wounded or hurt'. The origin of this word lies in the Latin *vulnus*, meaning 'a wound', and *vulnerare*, 'to wound'. Specifically, the word vulnerable derives from the Late Latin *vulnerabilis*, the term used by the Romans to describe the state of a soldier lying wounded on the battlefield, i.e. already injured therefore at risk from further attack. The relevance to the present discussion is that vulnerability, in this classic sense, is defined primarily by the prior damage (the existing wound) and not by the poorly defined, future stress (any further attack). By analogy, then, the vulnerability of any individual or social grouping to climate stress can be defined by their existing state, their capacity to react to future stress, in line with the definition employed by Blaikie *et al.* (1994). In studying vulnerability from this

*Table 2.1* Characteristics of vulnerability and their measurement

| Vulnerability indicator | Proxy for: | Mechanism for translation into vulnerability |
| --- | --- | --- |
| Poverty | Marginalisation | Narrowing of coping strategies; less diversified and restricted entitlements; lack of empowerment |
| Inequality | Degree of collective responsibility, informal and formal insurance and underlying social welfare function | *Direct*: concentration of available resources in smaller population affecting collective entitlements *Indirect*: inequality to poverty links as a cause of entitlement concentration |
| Institutional adaptation | Architecture of entitlements determines exposure; institutions as conduits for collective perceptions of vulnerability; endogenous political institutions constrain or enable adaptation | Responsiveness, evolution and adaptability of all institutional structures |

Source: Adger and Kelly (1999).

perspective, the emphasis must clearly be on identifying existing damage which might limit the capacity to respond, rather than an assessment of impacts, sensitivity and adaptive options in some speculative future.

It is important to recognise that the state of social vulnerability does not equate directly to, say, the level of poverty or any other single characteristic of an individual or group as there are many, diverse factors involved in determining the capacity to respond. Indicators of vulnerability include poverty, marginalisation and access to resources; resource dependency and diversity; inequality and marginalisation; and the appropriateness of institutional structures for enhancing resilience (Table 2.1). Poverty is related directly to access to resources and the process of marginalisation. Though income does have limitations as an index of poverty (see Baulch 1996), it is readily quantifiable and is an important indicator in our work. It should be noted that lack of poverty, wealth, is not in itself a guarantor of security as resources are mediated through property rights and so on.

It is helpful to split social vulnerability into the two distinct aspects of individual and collective vulnerability. Individual vulnerability is determined by access to resources and the diversity of income sources, as well as by the social status of individuals or households within a community. The collective vulnerability of a social grouping is determined by institutional and market structures, such as the prevalence of informal and formal social security and insurance, and by infrastructure and income. While these individual and collective aspects can, for convenience, be disaggregated, they are, of course, intrinsically linked through the political economy of markets and institutions.

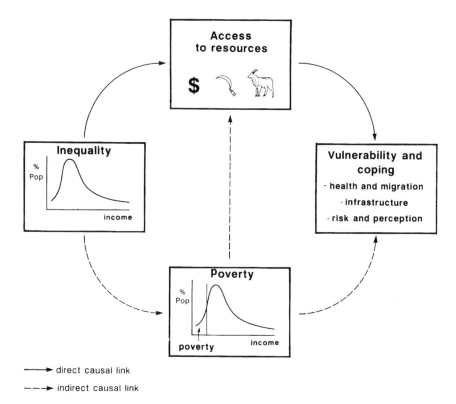

*Figure 2.1* How inequality and vulnerability are linked. Direct links are through skewed
access to resources. Indirect linkages between inequality and vulnerability are
through impacts of inequality on poverty

The collective aspects of vulnerability involve interaction at various scales,
from a single community to a nation. The level of infrastructure, institutional
preparedness, and other factors important in the implicit collective vulnerability
of a nation, region or community may not be accurately reflected in measures of
economic activity. Increasing inequality over time within a population, or between
different parts of the population, increases collective vulnerability to climate
change. Such changes in inequality are linked to the reduction of communal
allocation of resources and the pooling of risk, and other social phenomena
associated with the 'moral economy'. In addition, inequality and vulnerability
linkages are associated with relationships between inequality, diversification of
income sources and poverty. In other words, inequality affects vulnerability *directly*
through constraining the options of households and individuals when faced with
external shock; and *indirectly* through its links to poverty and other factors.

The direct links between inequality and vulnerability, as demonstrated in the
upper part of Figure 2.1, concern the concentration of resources in fewer hands,

constraining coping strategies based on private resources for households faced with external stresses. As has been shown under conditions of drought in agricultural societies, both income (immediately accessible resources) and wealth (disposable capital assets) are important in coping strategies (Watts 1991; Davies 1996). Hence, distribution directly affects the ability of households as part of the community, to cope with the impacts of extreme events. The impact of skewed access to resources can be ameliorated in all social situations by the effectiveness of institutions.

The indirect link between inequality and vulnerability, in the lower part of Figure 2.1, occurs because skewed accumulation is associated with increased levels of poverty, and hence insecurity. Once again, this is a complex issue whereby rising inequality does not necessarily *cause* poverty, but both poverty and inequality are jointly associated with *constraints* on coping strategies. An example of such constraints is where wealth concentration confines access to credit to certain sectors of the population, thereby *reinforcing* income poverty and enhancing vulnerability.

The importance of the indirect link between inequality and vulnerability is illustrated by the common argument that reducing inequality in the long term reduces poverty. There is, however, no underlying theory that predicts the relationship between inequality and poverty (Reardon and Taylor 1996). In terms of the ability of households within a population to cope with external stresses and the causes of vulnerability, the diversity of income sources is also an important factor. At the micro-level, there is no fundamental pre-determined relationship between inequality, poverty and income diversification. Inequality in the economic sense is usually conceptualised in terms of the distribution of wealth or in terms of the distribution of income.

Poverty, the use of resources, and the distribution of wealth and income within a population are all institutionally determined. It is formal political institutions that devise and implement the legal enforcement of property rights, for example, and all economic structures can be conceptualised as dependent on the institutional structure. Analysis of the political economy is, therefore, an essential component of any assessment of vulnerability, i.e. on constraints on the ability to respond to stress. Assessment of vulnerability, then, must rely on examination of the structures of institutions and of constraints on their evolution, and on the constraints they exert on individuals. The cultural context and the social differentiation of entitlements are not constrained in our analysis to those institutions of the state but extend more widely to include both formal political structures and the more diffuse 'rules of the game' and social and cultural norms.

### Defining resilience

In the case of resilience, too, there are numerous, competing applications of what, in the most general sense, is a measure of the ability of a system to respond to, or recover from, a perturbation imposed by external force or stress. The term has, perhaps, been most commonly used in ecological studies though even here definitions vary. In the ecological context, the resilience of an ecological system relates to the functioning of the system, rather than the stability of its component populations, or even the ability of the system to maintain a steady ecological state (Common 1995; Holling *et al.* 1995; Gunderson *et al.* 1997). Resilience in ecological systems is not easily observed, and there seems at present to be no agreed relationship, for example, between the diversity of ecosystems and their resilience. Thus, many tropical terrestrial ecosystems have stable and diverse populations but are relatively low in resilience, while similar ecosystems in temperate regions with apparently low diversity can exhibit greater resilience. Coastal and estuarine ecosystems, for example, are typically of low species diversity since they experience periodic physical changes and have a high degree of organism mobility. Yet such ecosystems can be highly resilient because of their high levels of functional diversity.

Social and ecological systems are themselves linked, in ways which Norgaard (1994) and others have likened to synergistic and co-evolutionary relationships. Thus, the resilience of a social system is itself related in some way, still undefined, to the resilience of the ecological systems on which it depends. This is most clearly exhibited within social systems dependent on a single ecosystem or a single resource. Taking the concept of resilience from the ecological sciences and applying it directly to social systems is, however, contested in its implicit underlying assumption that there are no essential differences between socialised institutions and ecosystem behaviour (Adger 2000b). The term has, however, a valuable use in describing the ability of social systems to respond to change and it is frequently used in, for example, institutional analysis.

Institutions have been described as persistent, sustainable and resilient depending on a range of parameters, including legitimacy (O'Riordan *et al.* 1998), agenda setting and the selecting of environmental risks which resonate with the institutions' agenda (Cantor and Rayner 1994), and the maintenance of social capital. Social capital is taken here to mean the existence of integrating features of social organisation such as trust, norms and networks (e.g. Pelling 1998). In considering institutional resilience, it is important to bear in mind that this characteristic is not simply a matter of the economic relations in which institutions engage. Institutional resilience is a function of the cultural context of institutional change and adaptation and, indeed, the differing conceptions of human–environment interactions within different knowledge systems.

There is a relationship between resilience and stability in the societal context, though stability here must incorporate the possibility of adaptation. In other words, a stable system is not necessarily fixed in time, content, role or structure but can manage change without undue upheaval. The stability of a social system can, in itself, be a factor contributing to induced innovation and technological development, both of which are components of the adaptive process. Further, it has been argued that sustained economic growth is dependent on capturing positive externalities from investment in human capital (Hayami and Ruttan 1985; Stern 1995). Both of these sources of economic growth are encouraged by stable social and economic circumstances. There is also increasing evidence that sustained economic growth is promoted by equitable distribution of assets within populations, for various reasons. The linkages between growth and equitable income distribution are, for example, based on Keynsian arguments of enhancing aggregate demand (Kim 1997), and on the arguments that the economic productivity of the workforce is being jeopardised by the consequences of large-scale inequality (Persson and Tabellini 1994). In the societal context, then, there is a link between equity and resilience, as there is between inequity and vulnerability.

It must be recognised that resilience, like vulnerability, cannot be considered in absolute terms. For example, migration can be an important factor in determining levels of resilience, but is a phenomenon the presence of which can be cited both as evidence for instability and as a component of enhanced stability and resilience, depending on the type of migration examined (Adger 2000b). Migration and population movement are important aspects of social stability and, in many circumstances, contribute to livelihood security and resilience at the household level through remittances as these provide opportunities for diversification and reduction in resource dependency. However, when a 'resilience threshold' is crossed, large-scale migration of whole households as part of emergency coping strategies is demonstrably an element of social instability. In the literature on food security, this kind of forced adaptation represents an extreme manifestation of vulnerability (Watts and Bohle 1993).

## Analysing social vulnerability and resilience in Vietnam

### *The architecture of entitlements in practice*

To illustrate the use of this framework, and in particular the concepts of vulnerability and resilience, the remainder of the chapter describes the findings of studies undertaken in Vietnam at two different scales. First, vulnerability to environmental change at the scale of Xuan Thuy District,[1] a coastal district in Nam Dinh Province, is outlined, based on field research and household surveys in that district. Residents of Xuan Thuy District are vulnerable to present-day typhoons and other natural hazards, and a summary of the research reported by Adger (1999b) highlights the usefulness of the concepts of vulnerability and resilience in explaining how people in this district live with and adapt to these circumstances. This case study illustrates the social differentiation in many of the development processes highlighted throughout this book. Second, these local manifestations of vulnerability are put in the national context by highlighting regional discrepancies in socio-economic indicators. When policy change, for example, in land tenure, agricultural pricing or health provision occurs, some sections of the population are more resilient and less vulnerable than others. The following section shows how liberalising the export market for rice has resulted in differential impacts and vulnerabilities not only across the various regions of Vietnam but also between rural and urban areas.

## *Vulnerability to hazards: Xuan Thuy District, Nam Dinh Province*

Vulnerability in Xuan Thuy was researched through household-level surveys and interviews with key informants, allied to analysis of present and potential climate regimes to define exposure.[2] The focus was vulnerability to tropical storm impacts. Xuan Thuy is an agricultural district in Nam Dinh Province on the fringe of the Red River Delta in northern Vietnam. The district is physically vulnerable to present-day climatic hazards due to its topography, proximity to the coast, and the present condition of its physical infrastructure. The Red River Delta, along with the Mekong Delta in the south of Vietnam, is a fertile agricultural region and, up until the 1990s, the production system in the northern Delta was predominantly organised in collectivised farms, with their attendant emphasis on collective security as well as production. The agrarian economy of Vietnam operates through a formal and sophisticated system of social security facilitated, even in the post-collectivisation era, through local-level government institutions.

Analysis of the household-level data from Xuan Thuy study reveals the complex mesh of factors that shape the vulnerability of a community as social and economic trends reinforce, transform or weaken existing patterns of risk. If the households and sectors of society in Xuan Thuy are classified by their major livelihood source, differentiated vulnerability to climate variability is readily apparent (Table 2.2). At the two extremes are those households who are presently engaging in aquaculture and those reliant on salt production.

Aquaculture has been a major beneficiary of the economic renovation that has occurred in Vietnam over the past decade, but it is an activity inherently susceptible to the impact of coastal storms. The economic returns to aquaculture are high for those with the capital and legal entitlement to engage in this activity, and the sector is the major cause of rising inequality in the district. At the other end of the spectrum, the salt-making communities have low incomes, a less diversified resource base and contribute less to inequality than to share of income. They are marginalised in the processes of land allocation and exhibit lower social resilience as manifest through permanent out-migration and other phenomena. The salt-making community would appear, then, to be more vulnerable to climate variability and change, and time series information for the locality suggests that the relative positions of these groups have been exacerbated by recent economic trends.

Detailed analysis of the household-level data reveals a number of key characteristics and trends. Overall, Xuan Thuy is, in rural Vietnamese terms, a relatively wealthy and productive district with a low incidence of absolute poverty and might be considered, from this isolated perspective, less vulnerable in the context of rural Vietnam as a whole. Poorer households are particularly dependent on a narrower range of resources and income sources and are thus more vulnerable, in the context of the local population, as they have reduced access to resources for coping with extreme events, such as credit sources, and are more reliant on activities such as salt-making which could potentially experience a significant

*Table 2.2* Internal social differentiation observed and the causes of vulnerability explained at the local level, Xuan Thuy, Vietnam

| Vulnerability characteristic | Observations | Mechanism for translation into vulnerability |
|---|---|---|
| Poverty | Poverty incidence in Xuan Thuy: one fifth of population below poverty line on the basis of basic needs | Perception of marginalisation in land allocation and access to communal resources by salt-making communities |
| | Poverty and income in two social groups: Salt-making communities – low incomes with low variance Aquaculture and rice based communities – higher income with high variance | Landlessness – higher rate of permanent migration among poor |
| Inequality | Estimated income inequality in Xuan Thuy higher than for rural Vietnam | *Direct vulnerability link:* concentration of productive capital and entitlements in fewer hands |
| | Contribution of income sources to present inequality in Xuan Thuy: aquaculture contributes most, followed by wages and remittances, agriculture and salt-making | *Indirect vulnerability link:* increased inequality and economic change not enhancing absolute poverty so not indirect link to vulnerability |
| Institutional adaptation | *Formal:* powerful communes (including those in aquaculture) diverting District resources for individual benefit | Increased impact of climate extremes exacerbated through neglect of collective action |
| | *Informal:* re-emergence of post-collective credit and rural institutions | Offset by adaptation and policy learning and by informal civil society mechanisms |

Sources: Adger (1999b) and Adger and Kelly (1999).

impact in the face of coastal flooding and other natural hazards. The distribution of resources within the district is relatively even compared to many agrarian societies, but is less even than in other parts of rural Vietnam. Moreover, underlying inequality is increasing due to the emergence of capital-intensive commercial activities, principally aquaculture, in the period since market liberalisation. Finally, the data reveal that the increasing dependence on aquaculture is having complex effects on levels of vulnerability. On the one hand, it should increase the overall wealth of the district with trickle-down effects benefiting the population as a whole but, on the other hand, it is heightening levels of inequality and is tying up capital in an inherently risky venture which, setting aside storm impacts, has been shown to be unsustainable in many parts of the world (Folke and Kautsky 1992).

The parallel analysis of institutional issues reveals how access to decision-making is a critical factor in this instance (Adger 2000a). The government institutions of the district have taken advantage of increased autonomy as a result of *doi moi*, but have become less directly influential in resource use since agricultural land has been allocated to private individuals and private enterprise encouraged. There has been a reduction in the resources available for sea dike maintenance as the conversion of the previous labour-based system to a tax-based system has permitted the diversion of finances away from dike maintenance and into, for example, road building. The inland communes are not aware of this shift in investment in collective security; they are persuaded by the coastal communes that the maintenance programme is being maintained at former levels and gives sufficient protection. In this way, formal institutions are seeking to maintain their resources, powers, and their authority in a time of rapid change at the expense of collective security. The research also shows that informal institutions have offset some of the negative consequences of market liberalisation and the reduction of the role of government by evolving collective security from below, for example, through risk spreading in credit unions, particularly in fishing communities.

The Xuan Thuy study demonstrates the difficulties in generalising about levels of vulnerability even in a relatively small community. We do not consider it wise to attempt to determine the overall impact on levels of vulnerability of the *ensemble* of socio-political trends that are currently underway. The framework that has been developed does, however, permit the identification of causal linkages and, the key policy goal, of measures that might compensate for adverse effects and reinforce beneficial consequences. Drawing on a series of case studies undertaken in northern Vietnam, we have identified a number of strands of wider applicability concerning the promotion of measures that might improve the situation of those increasingly at risk as a result of recent societal trends (Kelly and Adger 2000):

- poverty reduction clearly must be a priority, though that alone may not be sufficient to ensure the wider access to resources necessary to reduce vulnerability;
- risk-spreading through income diversification can be promoted in a number of ways and, again, will assist most the poorer members of the community;
- the loss of common property management rights represents a serious erosion of the ability to resist stress and, where it cannot be avoided, compensatory measures should be implemented; and,
- the reduced efficiency, or loss, of forms of collective action or investment affects the community as a whole and this process warrants careful monitoring with efforts to promote the development or resuscitation of other, perhaps traditional, forms of community security.

At a deeper level, the underlying causes of vulnerability must be tackled, addressing directly the reasons for the maldistribution of resources that determines patterns of vulnerability.

## Vulnerability to policy change: rice export liberalisation

Access to resources and the incidence of poverty are key determinants of levels of vulnerability and resilience. The causes and manifestations of poverty in Vietnam are heterogeneous, as illustrated in Table 2.3. This shows a set of socio-economic indicators for the major regions of the country. The indicators are made up of social development parameters such as service provision and childhood nutritional status as well as economic indicators and measures of the productivity and diversity of the agricultural sector and rural economy (covering the majority of the population in these regions). The indicators are based on analysis of the household- and community-level surveys of the Vietnam Living Standards Survey, a nationwide sample survey carried out in 1992 and 1993 (summarised in Vietnam State Planning Committee and General Statistical Office 1994). Lessons regarding the vulnerability and resilience of these regions can be deduced by examining various poverty indicators and considering the underlying diversity of the resource base.

Wiens (1998) argues that the observed differentiation in the incidence of poverty in Vietnam is explained by both geography and by social and historical factors. Several major constraints in resource endowments, such as soils, climate and water availability, lead to differential development patterns between the regions – these are the 'geographical' factors. The major deltas of the Mekong and Red Rivers have highly fertile alluvial soils and widespread irrigation systems to enhance agricultural productivity. With high population densities, these areas encourage low-cost provision of infrastructure such as health and education thereby continually outstripping the highland regions in agricultural performance as well as social indicators. Regions such as the South East and the Red River and Mekong Deltas also include large urban areas, Ho Chi Minh City and Hanoi, which, as demonstrated in Chapter 12 for Hanoi, encourages diversity of income recognised as 'off-farm' income in Table 2.3. By contrast, the topography of the North Central Coast region ensures that agricultural production is often water deficient as a result of being in the rain-shadow of the mountains on the border with Laos and that irrigation is not technically feasible.

But not all regional differentiation is explained by geography. Ethnic minorities in Vietnam are, in general, marginalised from decision-making and have less access to other resources (see Chapter 7 and Rambo *et al.* 1995). This situation is reflected in high poverty incidence in the Central Highlands region and the Northern Uplands. The Central Highlands also has the highest share of population born outside the region (45 per cent) (Wiens 1998). The Central Highlands is a frontier area, absorbing population from other parts of the country. Frontier farming populations have been shown in many parts of the world to experience particular deprivations and challenges regarding lack of social networks and institutions for social resilience as well as low incomes (Brown and Muchagata 1999). These factors often exacerbate vulnerability to external changes in land tenure or the impact of flood or drought.

Given this significant differentiation between regions in Vietnam, social and

*Table 2.3* Regional and geographic variations in socio-economic indicators of rural economy, poverty and development in Vietnam

| | Northern Uplands | Red River Delta | North Central | Central Coast | Central Highlands | South East | Mekong Delta |
|---|---|---|---|---|---|---|---|
| *Rural economy (all VND 000)* | | | | | | | |
| Income per household in rural areas | 5 520 | 5 351 | 8 212 | 3 883 | 7 217 | 4 752 | 8 208 |
| Per capita food consumption | 620 | 624 | 841 | 571 | 715 | 687 | 740 |
| Agricultural output per ha | 47 392 | 64 366 | 65 347 | 44 776 | 53 696 | 39 225 | 283 205 |
| Crop sales per household | 525 | 550 | 2108 | 418 | 3466 | 401 | 2687 |
| Off-farm wage income per household | 283 | 334 | 1 699 | 211 | 766 | 777 | 1 355 |
| *Non-income development indicators (%)* | | | | | | | |
| Adult literacy | 87.6 | 91.6 | 90.9 | 85.2 | 68.3 | 92.5 | 84.8 |
| Electricity supply | 35.0 | 76.4 | 35.6 | 58.3 | 23.6 | 65.6 | 24.1 |
| Potable water | 0.6 | 19.0 | 0.5 | 19.2 | 0.8 | 38.1 | 17.0 |
| Sanitary toilet | 71.3 | 70.4 | 49.8 | 51.0 | 57.6 | 68.8 | 12.3 |
| Childhood nutrition status (children with stunted growth 0–5 years) | 58.5 | 53.7 | 58.4 | 46.1 | 52.4 | 29.8 | 44.5 |

Sources: Dollar and Glewwe (1998) and Wiens (1998) based on Vietnam Living Standards Survey.

Notes
Potable water is defined as water obtained from private water taps, public standpipes and tubewells.
Sanitary toilets are defined as flush toilets and latrines.
Stunted growth for children defined as height for age Z score of less than –2 (see Ponce *et al.* 1998).

political changes implemented at the national level can have significant regionally differentiated impacts on vulnerability, resilience and well-being. For example, liberalisation of rice markets may have significant impacts on rural and urban households. All over the world, national governments engage in policies and intervene in the markets for staple foodstuffs. The reasons for intervention may ultimately be to maintain low prices for the majority of the population, particularly articulate and politically important urban populations, but at the expense of lower producer prices for farmers. This phenomenon, termed urban bias (Lipton 1977), results in differential impacts of food policy on nutrition and economic well-being (Mellor and Ahmed 1988; Sahn 1988; Timmer 1993; for example). Changes in such policies can, however, also have differential impacts.

Mellor (1978) found that, in India, higher food prices resulting from liberalisation resulted in economic gains to net sellers of food but in losses to net buyers such as urban consumers and, importantly, landless rural households.

The government of Vietnam presently limits rice export through a quota system to ensure domestic supply. The quota system is assumed only to impact on a minority of farming households since only one third of farms are net suppliers of rice and 95 per cent of all rice exports originate from the Mekong Delta region (Minot and Goletti, 1998). The export quota, in fact, reduces domestic rice prices to consumers as well as to producers. But agricultural production has risen dramatically in the 1990s as a result of liberalisation of labour and land markets and the adoption of high yielding varieties (Pingali and Vo Tong Xuan 1992; Pingali *et al.* 1997). The liberalisation of rice exports has, therefore, the potential to realise significant economic benefits. But at what cost to the vulnerable?

Minot and Goletti (1998) simulate the impact of liberalising the export of rice from Vietnam on urban and rural dwellers and on the regions. They find that the base situation backs the government policy: urban dwellers currently benefit from the lower prices and, indeed, less than half the farmers, even in the Mekong and Red River Delta regions, are net sellers of rice (Table 2.4). The impacts of liberalisation would be to raise rice prices (bottom section of Table 2.4) by around 12 to 19 per cent across the regions. This would raise production as farmers would have an incentive to plant and sell more, and reduce domestic consumption, with the net impact on overall surpluses shown in the last row of Table 2.4.

But when the economic impacts of these changes are examined, the picture of vulnerability and impact of such policy changes appear to be more complex. Table 2.5 shows that the real income of Vietnamese households rises by 3 per cent. Poorer households (defined in Minot and Golletti's (1998) study as the poorest 25 per cent of households) are better off still at 3.5 per cent on average. But the real incomes of urban households fall (1.7 per cent) because they are net purchasers of rice and, indeed, poor urban households are the worst affected because they spend a higher proportion of their income on food. Minot and Goletti's (1998) assessment shows that the rice growing areas would gain the most. But they also examine the impact on vulnerable households who are both poor and are net buyers of rice. A greater proportion of the vulnerable populations to this policy change are, in fact, the rural poor as in rural Vietnam these households tend to be the lowest income sections of society and, hence, spend the greatest proportion of their income on food. This example of the impact of a national policy change demonstrates that poverty, livelihood diversity and, hence, resilience to change are highly heterogeneous across regions and social groups in Vietnam.

## Opportunities for intervention?

The process of *doi moi*, economic renovation, provides a valuable opportunity to observe how rapid socio-political change can influence levels of vulnerability and adaptive possibilities. What emerges is an impression of a rich and continually-evolving process of response and adaptation which affects a particular community

*Table 2.4* Rice production, consumption and sales by region and the impact of liberalising rice exports

|  | Northern Uplands | Red River Delta | North Central | Central Coast | Central Highlands | South East | Mekong Delta |
|---|---|---|---|---|---|---|---|
| *Present situation* | | | | | | | |
| Rice production as % of income | 28.8 | 42.5 | 32.6 | 19.3 | 19.3 | 14.3 | 37.8 |
| Rice consumption as % of income | 34.8 | 34.8 | 33.9 | 25.6 | 32.6 | 19.0 | 24.3 |
| Net sales of rice as % of income | −6.0 | 7.7 | −1.3 | −6.3 | −13.4 | −4.8 | 13.6 |
| Net sellers of rice | 24.1 | 45.9 | 31.7 | 18.4 | 10.0 | 16.6 | 43.1 |
| Zero net position in rice | 14.1 | 8.7 | 9.5 | 5.7 | 7.5 | 2.3 | 1.1 |
| Net buyers of rice | 61.8 | 45.4 | 58.8 | 75.9 | 82.5 | 81.0 | 55.2 |
| *Simulated impact of eliminating rice export quota (% change from with-quota situation)* | | | | | | | |
| Rice price | 12.4 | 13.4 | 14.9 | 16.7 | 17.0 | 19.0 | 19.4 |
| Production | 6.3 | 6.9 | 7.1 | 4.8 | 5.1 | 8.7 | 8.7 |
| Consumption | −6.6 | −7.3 | −7.7 | −12.3 | −12.8 | −18.4 | −16.1 |
| Rice surplus | −27.8 | 192.9 | −70.1 | −122.1 | −30.7 | −39.3 | 23.2 |

Sources: Minot and Goletti (1998). Present situation based on Vietnam Living Standards Survey.

*Table 2.5* Distributional effects between urban and rural areas of eliminating Vietnam's rice export quota

|  | Vietnam (all households) | Urban | Rural |
|---|---|---|---|
| Population (% of total) | 100 | 20 | 80 |
| Average change in real income (%) | | | |
| • Average | 3.0 | −1.7 | 4.1 |
| • Poor | 3.5 | −2.9 | 3.9 |
| • Non-poor | 2.8 | −1.6 | 4.2 |

Source: Minot and Goletti (1998).

in diverse ways. In Xuan Thuy District, for example, the economic changes accompanying *doi moi* have increased the opportunities for income generation (as in the other coastal communities that we have studied), while increasing differentiation within those communities (see also Hirsch and Nguyen Viet Thinh 1996). In related developments, the rights to common resources such as the mangrove forest have been lost and institutional changes have reduced the efficacy of various forms of collective action and investment. These trends have their major impact on the poorer members of the community, who rely on the mangrove ecosystem to supplement their livelihood and on collective action to provide protection against stress and support in times of hardship. Even a single policy shift, such as the hypothetical rice export policy change discussed above, may generate a complex set of consequences across a population.

We draw four general conclusions. First, we believe that any attempt to ensure a sustainable response to environmental change by facilitating adaptation must be capable of encompassing the diversity of responses to change apparent in this chapter and throughout this book. The need for sensitivity to cultural context is, of course, self-evident.

Second, it should be recognised that adaptive strategies, as well as the physical and societal contexts within which they exist, are constantly changing and evolving so any policy intervention must itself be flexible and capable of swift modification. This requirement implies a need for gradual or step-by-step implementation, supported by continual monitoring and evaluation.

Third, to the extent that generalisation is possible, it is the poorer members of any community that are most vulnerable to change, whether social or environmental, and these people should be the focus of any policy intervention. There is strong evidence that, by reducing inequity across a community, the community as a whole should benefit.

Finally, we note that a precautionary response to future environmental threats must build on the existing capacity of communities to adapt to change by reducing vulnerability and enhancing resilience to present-day stress. Vietnam, a nation which has successfully confronted environmental and political stress throughout its history, responding with speed and imagination, is well-placed to provide positive examples of effective adaptive strategies.

## Acknowledgements

We thank Katrina Brown, Andrew Jordan, Catherine Locke, Tim O'Riordan and Kerry Turner for discussions over the past years in developing the concepts discussed here.

## Notes

1  The bulk of Xuan Thuy District has recently been renamed Giao Thuy District.
2  Observing the proxy indicators for vulnerability involved fieldwork in 1995 and 1996, investigating indicators in eleven communes (village level administrative units) through household survey, based on a stratified area sample, and semi-structured interviewing of key informants. Data for analysis of institutional adaptation and of institutional inertia in the treatment of present climate extremes in Xuan Thuy District were collected through semi-structured interviews of commune level officials and from households within these communes, as well as discussions at the district level (see Adger 1999b).

# 3   Managing environmental change in Vietnam

*P. Mick Kelly, Tran Viet Lien,
Hoang Minh Hien, Nguyen Huu Ninh
and W. Neil Adger*

## Introduction

There can be no doubt that Vietnam's commitment to modernisation and industrialisation brings with it a serious obligation to environmental protection if sustainable development is to be achieved. In striving to maintain ecological health, Vietnam has to manage not only the many internal processes that may lead to environmental damage but also pressures from outside, regional pollution and resource conflicts, global environmental change and any adverse effects of the process of economic globalisation. Nevertheless, as Tran Thi Thanh Phuong, writing as Deputy Director for Planning and International Relations of the National Environment Agency, Ministry of Science, Technology and Environment, has observed, 'the ongoing reform process [*doi moi*], together with the fact that Vietnam has a unique opportunity to learn from the experiences of its neighbours, creates a good opportunity for the development and introduction of effective environmental management policies and instruments' (Tran Thi Thanh Phuong 1996).

The purpose of this chapter is to provide an overview of the present state of Vietnam's environment and to identify the main trends that will shape the ecological future of this nation. We begin by defining Vietnam's natural resources and the major aspects of environmental decline that have affected this country over the second half of the twentieth century, highlighting causal factors. We then document Vietnam's construction of a legal and institutional infrastructure to support environmental protection and ecological conservation. Finally, we identify critical issues that must be dealt with if a sustainable future is to be realised.

## The state of the environment

### Natural resources

Vietnam is a land of geographical contrasts. Lying on the East Sea (the South China Sea) and bordered by China, Laos and Cambodia, the country spans 16 parallels of latitude resulting in diverse geomorphological, hydrological and climatic conditions. The climate varies from warm, moist tropical in the south to moist subtropical in the north which is subject to cold outbreaks from temperate latitudes during the winter months (Figure 3.1). The seasonal cycle in temperature is weak in the south but much stronger in the north. Rainfall averages 1500 to 3000 mm a year over the country with the highest levels experienced in central Vietnam. The rainfall is highly seasonal, with the wet season extending from May to October. The coastal discharge of the country's rivers is about 900 billion m³ a year, with 75 to 80 per cent of the flow occurring during the flood season. Both the local wind systems and ocean currents along the coast undergo a seasonal reversal as the monsoon flow shifts.

Vietnam has a land area of 331 690 km² with a coastline over 3200 km in length. The northwest of the country is mountainous, an extension of the Yunnan Plateau in China. The neighbouring lowland area, the Red River Delta, is one of the two major population centres. Central Vietnam consists of the Annamite Chain and a coastal lowland fringe. At its narrowest, this part of Vietnam is only tens of kilometres wide, giving the country its distinctive 'S' shape. The second major population centre is the Mekong Delta in the south. About three quarters of the territory is covered by mountains or hills, with almost a third of these extending above 500 m. The hydropower potential of this wet, mountainous country is considerable. There are around 11 million ha of arable land (about one quarter of the land area), lying for the most part in the two delta regions and the Western Plateaux in central Vietnam, and 16 million ha of forested land. The nation is rich in mineral resources such as coal, oil, gas, iron, copper, tin, bauxite, gold, silver and precious stones and has been classed as the sixteenth most biologically diverse country in the world (World Conservation Monitoring Centre 1994). There are over 80 protected areas covering about 1.1 million ha of land (3 per cent of the territory). Extensive marine fisheries are a major source of export earnings.

Vietnam may be a country rich in natural resources, but it also experiences a wide range of natural hazards. Tropical cyclones, which frequently visit the coastline and have the potential to wreak considerable damage, are the prime threat (Chapter 10). Heavy rainfall, saltwater and freshwater flooding, drought, heatwaves and sub-freezing conditions are also frequent hazards. Heat and humidity create favourable conditions for the propagation of contamination and disease. The switch between El Niño and La Niña, the warming and cooling of the eastern Pacific Ocean, modulates the frequency and severity of a number of weather- and climate-related hazards (Nguyen Duc Ngu 2000; see also Chapter 10). Earthquakes are experienced in mountainous areas, though away from heavily populated areas. Environmental damage is increasing the frequency or scale of certain hazards such as rain-induced flooding and landslides, as forest cover is reduced and erosion occurs, and saltwater flooding in coastal areas, as the protection afforded by mangrove forests is lost (Chapter 9).

### Problem areas

Analysts agree in identifying the main environmental difficulties that Vietnam has experienced during the second half of the twentieth century (see, for example,

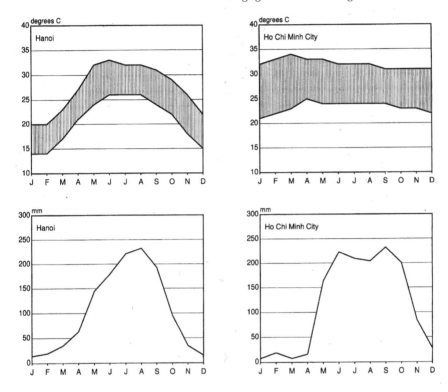

*Figure 3.1* Seasonal cycle in monthly mean temperature (minimum and maximum) and total rainfall for Hanoi and Ho Chi Minh City

Vietnam National Resources and Environment Research Programme 1986; Vietnam State Committee for Sciences *et al.* 1991; Le Thac Can and Vo Quy 1994; World Conservation Monitoring Centre 1994; Cao Van Sung 1995; World Bank 1995a; Panayotou and Naqvi 1996; Vo Quy 1997; United Nations 1999). Though issues clearly overlap, we divide the major problem areas into four categories for convenience: deforestation; resource degradation (as a result of scarcity, irrational use and overexploitation); loss of biodiversity and genetic resources; and environmental pollution. Where possible, we identify the root causes of these problems, recognising that relative contributions may not be defined with certainty.

### Deforestation

Vietnam's forest cover stood at around 58 per cent of the land area during the nineteenth century (Le Thac Can and Vo Quy 1994). As the low-lying deltas and coastal plains were cleared, deforestation spread into the hills and mountains. By the early 1940s, the remaining forest area was about 44 per cent. It dropped only slightly during the subsequent decade, to 40 per cent in the early 1960s. The rate of deforestation then increased with the forest cover dropping to 29 per cent by the mid-1970s and 24 per cent in 1983.[1] According to official statistics, around 26 per cent of Vietnam's land area, 9 million ha, was covered by natural forest in 1997 (United Nations 1999: 78). The loss of forest vegetation leads to soil degradation, nutrient loss and erosion where soils liable to degradation are cultivated. As a result, most deforested areas eventually become barren – 40 per cent of the country, 13 million ha, is now classified as unproductive wasteland (Vo Quy 1997).

Vietnam's forests play a vital role in supporting livelihoods and protecting the environment (Panayotou and Naqvi 1996: 273). They provide a home for around 20 million people, all dependent to some extent on forest resources. Fuelwood collection has been estimated at 22 million tons a year and food and medicines are taken from the forest. Alongside direct economic use such as timber extraction, the forests maintain environmental and ecological health by limiting erosion, protecting watersheds, reducing flooding and providing a range of habitats for Vietnam's diverse wildlife. Factors cited as contributing to deforestation include the effects of war, logging practices, forest fires, clearance for agriculture, encroachment, shifting cultivation,[2] fuelwood collection and overgrazing, with population growth, poverty, overexploitation and mismanagement as underlying causes (Vietnam National Resources and Environment Research Programme 1986; Vietnam State Committee for Sciences *et al.* 1991; World Conservation Monitoring Centre 1994; Vo Quy 1997; Dubois and Morrison 1998; United Nations 1999: 78).

Since the 1980s, the Government of Vietnam has been undertaking an ambitious programme of reforestation in order to reverse the trends of the past half-century. In the early 1990s, priorities were large-scale reforestation, encouraging farmers to plant tree crops and fruit crops and to develop agroforestry schemes, and popular tree-planting movements, with the practice of shifting cultivation strongly discouraged under a sedentarisation programme. Following reviews in the early 1990s (Nguyen Quang Ha 1993), reforestation goals were set for the year 2000, including a target of 40 per cent forest cover, protection forests and nature reserves were to be established, production forests were to replace exploitation of natural forests, and forestry activities were to become more integrated with agriculture and fisheries with more involvement of local people (World Conservation Monitoring Centre 1994). Since the late 1990s, direct approaches to reforestation have been supplemented by policies aimed at promoting better management of forest land, flowing from the process of *doi moi*. For example, local people are being given more responsibility for forest management (though see Rambo *et al.* 1995 and Chapters 6 and 7 for a critical discussion of recent trends).

*Resource degradation*

Resource degradation is a serious problem in a number of areas: land, soil, water, fisheries and other ecosystems, minerals, and so on. Here, we focus on land, soil, water and fisheries – key components of social and economic health.

Productive land is now a scarce resource. In 1945, with the population at 25 million, the total land resource per capita was around 1.3 ha, with 0.2 ha exploited for agriculture. By the early 1990s, the land resource per capita had dropped to 0.47 ha, with only 0.11 ha for agriculture (Le Thac Can *et al.* 1994). In parts of the Red River Delta, the land available for agriculture per head of population is as low as 0.06 ha (World Bank 1993c). Only 7 per cent of the potential arable land in the Red River Delta has not been intensively exploited; 8 per cent in the Mekong Delta. Though land remains to be brought into cultivation in upland regions, achieving this means that difficult environmental and ecological problems have to be dealt with and financial and labour constraints overcome (see Chapters 6 and 7).

With a population of just over 76 million people at the end of the twentieth century, Vietnam is the twelfth most populous country in the world (United Nations 1999: 37). The average population density is 1180 people/km$^2$ in the two major deltas. But rapid demographic change is underway. The population growth rate has dropped from 2.1 per cent during the 1980s to 1.7 per cent over the 1990s and is projected to drop to just under 0.8 per cent by the period 2019–24, resulting in a population of 100 million in the year 2024 (United Nations 1999: 37). Fertility and mortality rates have decreased significantly to levels comparable with more developed countries in the region, breaking the link between levels of economic development and demographic characteristics (Johannson *et al.* 1996; see also Bryant 1998). The demographic transition cannot, however, be isolated from wider social and policy changes. There is evidence, despite a stated 'access for all' policy and a commitment to intensify family planning programmes during the 1990s (United Nations 1999: 37), of reduced provision of health services, particularly for poor rural households in particular regions (Ensor and San 1996), that may, in the long run, undermine the sustainability of the transition.

Should population growth, in itself, be perceived as a problem, as leading inexorably to environmental degradation? As Le Trong Cuc and Rambo point out:

> The Vietnamese are doing a remarkably successful job of managing the very limited resources of the [Red River] delta to meet human needs . . . Environmental degradation, particularly water pollution and reduction of wild species, is severe, but existing agricultural practices display a high level of sustainability. Yet only 50 years ago, the Red River Delta, with a population density then only one-half of present levels, was considered by its French rulers to be a hopeless case for rural development, an area condemned to perpetual poverty by its mass of tradition-bound peasants struggling to survive on minuscule plots of land.
>
> (Le Trong Cuc and Rambo 1993: ix)

Rather, it is the combination of population pressure with a host of other factors, internal to and external to the locality and to the country, that generates environmental degradation (Blaikie 1993; Dasgupta 1995; Hartmann 1998; see also

Hansen 1994 and Lindahl-Kiessling and Landberg 1994). Easing of restrictions on movement under *doi moi* has created new opportunities for employment and may reduce pressure on land and other resources in highly-populated areas (Locke *et al.* 2000), though perhaps at the expense of increased problems in urban centres (Chapter 12).

Soil is being lost through erosion and it is being degraded. Fertile soil is continually swept out to sea along the many rivers that cross Vietnam's agricultural plains. Nguyen Quang My (1992) estimates an annual loss of alluvium soil nutrients equivalent to 500 000 tons of nitrogen, 300 000 tons of phosphorous and 300 000 tons of potassium. Alongside deforestation, unsustainable farming practices, including poor irrigation management, are a prime cause of deterioration in this resource. As one example, the National Plan for Environment and Sustainable Development cites the migration of lowland Vietnamese into the uplands without adequate knowledge of the new agro-ecosystems upon which they must base their livelihoods (Vietnam State Committee for Sciences *et al.* 1991: 41). Rates of soil loss are particularly high on cultivated slopes (slopes steeper than 15 degrees comprise over half the cultivated area), exacerbated by high rainfall and crops such as groundnut and cassava that do not reduce surface water flows (Panayotou and Naqvi 1996: 270).

Water resources are not limited, with rainfall plentiful in most years. Rivers and canals cover 653 566 ha, lakes 394 000 ha, ponds 56 000 ha and marshlands 85 000 ha (Vo Quy 1997). Irrigated agriculture is the keystone of the nation's well-being. Nevertheless, this resource too is being degraded (Panayotou and Naqvi 1996; Vo Quy 1997; United Nations 1999). Access to safe drinking water is limited compared to neighbouring countries such as Thailand (Panayotou and Naqvi 1996: 274). Deforestation is increasing sedimentation rates, reducing the efficiency of canals and reservoirs. The loss of forest cover is also affecting river flow, exacerbating the severity of flooding and increasing variability in flow (United Nations 1999: 79). Pollution by agrochemicals, industrial outflows and domestic sewage, particularly in the larger cities, is reducing water quality (Vo Quy 1997). In the Mekong Delta, saline intrusion and sulphate acid leaching, partially the result of human activity, are causing increasing problems with groundwater supplies. Design weaknesses in water supply and irrigation systems and ingrained attitudes towards what has been a 'free' resource compound these problems. Competition between sectoral interests (hydropower, flood control and irrigation) has resulted in sub-optimal investments; and failure to coordinate across administrative boundaries has created downstream problems (United Nations 1999: 79).

Finally, we consider fisheries in the coastal zone. As saltwater mixes with nutrient-rich freshwater in the estuaries of the deltas and along the coast, spawning and breeding grounds are created for a diverse range of aquatic resources: shrimp, crab, lobster, oyster and many species of fish. Marine fisheries provide about 30 per cent of animal protein consumed in Vietnam (Vo Quy 1997). Over one hundred species of economic value have been identified and fisheries products are a major source of export earnings. In the mid-1990s, around one million tons of marine fish were caught annually, about 50 000 tons for export.

*Table 3.1* Comparison between the number of species in Vietnam and in the world

| Taxa | Number of species in Vietnam (SV) | Number of species in the world (SW) | SV/SW (%) |
|---|---|---|---|
| Mammals | 265 | 4 000 | 6.8 |
| Birds | 800 | 9 040 | 8.8 |
| Reptiles | 180 | 6 300 | 2.9 |
| Amphibians | 80 | 4 184 | 2.0 |
| Fishes | 2 470 | 19 000 | 13.0 |
| Plants | 7 000* | 220 000 | 3.2 |
| Mean percentage of global biodiversity | | | 6.2 |

Source: World Conservation Monitoring Centre (1994).

Note
*May be as high as 12 000.

Nevertheless, riverine and marine pollution and over-fishing have placed a number of species at risk (Le Thac Can and Vo Quy 1994). Despite increased fishing intensity, catches are not rising at a corresponding rate. Part of the problem is lack of investment and technology to support deep-sea fishing, so activity is concentrated in estuaries and the near-shore zone; the source of 80 to 90 per cent of total marine production is shallow coastal waters. Fishing equipment can be overly destructive with mesh size too small and explosives used (Le Thac Can *et al.* 1994). On the coastline, mangrove is being cleared for commercial aquaculture with the consequent loss of a vital habitat (Chapter 9). The area used for coastal aquaculture rose from 67 000 to 288 600 hectares between 1985 and 1994 (United Nations 1999: 80). Increased pollution associated with the development of tourism is a growing concern in major resort areas.

*Loss of biodiversity and genetic resources*

The variety of habitats throughout Vietnam supports a rich diversity of species and ecosystems (Table 3.1). Vietnam's forests contain the highest bird and primate diversity in mainland Southeast Asia (Panayotou and Naqvi 1996: 275). Over one thousand medicinal plants have been recognised (World Conservation Monitoring Centre 1994). Vietnam shows the highest level of endemism in the Indochina region; the proportion of endemic forms has been estimated at 40 per cent over the whole country (Thai Van Trung 1970). Vietnam contains significant populations of a number of Asia's rarest animals including the Javan Rhinoceros, Asian Elephant, Tiger, Eld's Deer, Kouprey, Crested Argus and Green Peacock (World Conservation Monitoring Centre 1994). Large mammalian species, such as the Giant Muntjac and the striped rabbit from the genus *Nesoagus*, are still being discovered in Vietnam's forests and the border regions with Laos (for example, Surridge *et al.* 1999).

An inevitable consequence of deforestation and the degradation of natural resources in general is that biodiversity and the pool of genetic resources come under threat. In the early 1980s, the Government of Vietnam reacted to this state of affairs with the inventorying of the country's flora and fauna and the listing of endangered species warranting official protection. Of the total land area, 3 per cent was then designated as protected areas in the mid-1980s and a national strategy for conservation was developed which, as described in the following section, formed the basis for a broader framework for environmental protection developed during the early 1990s. A series of government decisions, instructions, regulations and laws has attempted to control unsustainable logging (including a ban on the export of unfinished timber), and the hunting and trade of various species, though with mixed success (World Conservation Monitoring Centre 1994). The World Conservation Monitoring Centre reports that management of protected areas in the early 1990s was 'almost universally poor', largely because management teams were relatively new, lacked training and experience and were not well-paid. Moreover, law enforcement standards were weak and 'even when cases are taken to court, the police and magistrates do not feel the offences of forest cutting or wood stealing are serious' (World Conservation Monitoring Centre 1994). Reserve management has moved on since that time, though the conflict between nature protection and the interests of local people remains to be resolved (Vo Quy 1997).

*Environmental pollution*

Vietnam is rapidly acquiring a number of the less attractive trappings of industrialisation as levels of pollution rise and the urban environment comes under increasing pressure. While comprehensive environmental standards are now in place, along with requirements for environmental impact assessment, laxity in monitoring and enforcement mean that companies continue to operate environmentally-damaging technology (United Nations 1999: 80). According to a study by the Vietnamese National Environment Agency and Asia Development Bank (1998), 275 000 tonnes of hazardous industrial waste was, in the mid-1990s, generated and, for the most part, dumped without treatment. Of this total, around 110 000 tonnes should have been treated physically or chemically, 45 000 tonnes required incineration and 120 000 tonnes a secure landfill.

Underlying the process of modernisation and industrialisation is the growth in energy demand and generation. England and Kammen (1993) review the state of energy resources in Vietnam in the early 1990s, considering environmental risks associated with the development of the energy sector. Biomass has traditionally been the main source of energy at the domestic level. Nguyen Anh Tuan and Lefevre (1996) track the evolution of household energy sourcing as fuelwood supplies become limited and 'modern' fuels such as coal provide a more easily available alternative, with electricity attractive to the more affluent. This is a trend that is likely to be more and more common in years to come. Hydropower is already a major source of commercial energy in Vietnam and the untapped potential is considerable (England and Kammen 1993). Hirsch and Nguyen Viet Thinh (1996) examine the social and environmental effects of the Hoa Binh

Dam, the largest in Southeast Asia, which, when completed in the mid-1990s, supplied nearly half of Vietnam's electrical energy. There is considerable concern regarding the impact of large dams, as elsewhere in the region (cf. Goldsmith and Hildyard 1984; McCully 1996). The critical issue is one of scale. Emissions resulting from energy generation having been growing since the 1980s. Nguyen Anh Tuan (1997) documents the effect of trends in commercial and traditional energy on emissions of carbon dioxide ($CO_2$), sulphur dioxide ($SO_2$) and the oxides of nitrogen ($NO_x$), showing a doubling of $SO_2$ releases resulting from commercial energy production between 1980 and 1992 and a 66 per cent increase in $NO_x$ emissions over that period. The study highlights the substantial contribution of residential biomass consumption to the nation's carbon emissions (see following section).

A key issue for Vietnam, as any other country at this stage in the process of modernisation and industrialisation, is whether environmental control will become more effective as economic development proceeds. The question is recognised at the highest level:

> I have read many reports of authoritative scientists, professors and experts in the field of environment, including those of our country. They have provided undeniable facts and figures and convincing scientific proofs, sounding an alarm of the dangers of environmental destruction. Economically, these problems may lead us up a blind alley. While the environment may be destroyed with no development, it may be more destroyed with development. This is what concerns the strategic planners and policy makers in our country.
>
> Ha Nghiep, former Scientific Advisor to the General Secretary
> of the Communist Party of Vietnam (Ha Nghiep 1993)

This issue, central to the challenge of sustainable development, is discussed further in Chapter 11 and, in one guise or another, in many other chapters of this book.

Mirroring the development of major cities throughout the world, increasing population and affluence is stretching the ability of the urban infrastructure to cope with the ever-accelerating demands placed upon it. The first phase of the modern development of Vietnam's cities occurred during a period of warfare and shortages. Consequently, much of the infrastructure for sanitation, drainage, water supply and sewage could not be replaced and remains weak or ineffective (Vo Quy 1997). Moreover, many factories run by state enterprises were set up without much concern for zoning, next to heavily-populated areas. By the late 1990s, a number of cities were slowly relocating polluting industries out of residential areas and new industries, particularly those with foreign investment, were being set up in industrial zones on the fringes of major urban centres (United Nations 1999: 80). Throughout the 1990s, as the bicycle was replaced by motorbikes and traffic of all kinds increased, pollution levels around major roads increased leading to a series of studies of ways in which the problems experienced by other major cities in the region such as Bangkok could be avoided. The continued use of old, poorly maintained vehicles, particularly for freight transport, is a particular problem. The urbanisation process in Vietnam is discussed in further detail in Chapter 12.

## Environmental protection and biodiversity conservation

According to the Center for Environment Research, Education and Development (1993), there have been three phases in Vietnam's response to environmental problems since reunification in 1975. From 1975 to 1980, the focus concerned the rehabilitation of environmental damage caused by war; from 1981 to 1990, resource and management issues related to socio-economic development and population growth; and, from 1991 onwards, resource and management issues related to sustainable development. The account that follows is largely drawn from the Center for Environment Research, Education and Development (1993) and Tran Thi Thanh Phuong (1996), supplemented by more recent information from the United Nations (1999).

### *The period of rehabilitation*

During the years from 1943 to 1975, continuous warfare wrought serious damage to Vietnam's environment. Large areas of forest, agricultural land and settlements were cleared by bulldozers and bombing displaced three billion cubic metres of earth. From 1965 to 1975, an estimated 72 million litres of herbicide was sprayed on forest and agricultural land, affecting 1.7 million ha of land in total (Le Thac Can *et al.* 1994; Vo Quy 1997). Herbicides destroyed 36 per cent of the mangrove area in southern Vietnam (Phan Nguyen Hong and Hoang Thi San 1993: 97). Loss of wildlife, erosion and soil degradation often followed removal of the forest cover. Mutagenic effects, obstetric problems and birth defects occurred in the human population as a result of dioxin exposure and, in the long-term, there is a likelihood of increased incidence of cancers (Schecter 1991). As recently as the mid-1990s, a study[3] found dioxin contamination related to Agent Orange in fish from fishponds created close to the former Aluoi Valley airbase in southern Vietnam (Hatfield Consultants 1998). It was concluded that the 'levels found would trigger a consumption advisory process (i.e. recommendations on maximum human consumption levels) and possibly prohibitions against consumption if they were from a location in Canada or other Western juris-dictions'. The investigators found a consistent pattern of contamination within the food chain leading to significant human exposure and uptake.

After reunification, priority was given to the restoration of settlement and agricultural areas (Center for Environment Research, Education and Development 1993). Initial activities included dealing with unexploded bombs, repairs to constructions of all kinds (houses, schools, hospitals, water supply and other forms of infrastructure), improvements in land management and land use and tree planting. In contaminated areas, houses, farmland and water supplies had to be disinfected. Later, restoration of forests took place though conservation and the creation of conditions favouring regeneration, artificial reforestation and reintroduction of wildlife. Some former forested areas were converted into agricultural land.

### The post-war development phase

During the 1980s, Vietnam entered a new period of economic growth with the development of agriculture, forestry and fisheries and the start of serious industrialisation. Agricultural production intensified in the deltas and coastal plains of northern and central Vietnam. Rice farming expanded in the Mekong River Delta. Rubber, coffee, tea and pepper plantations were developed in the central plateaux and southern midlands. With the increasing demand for energy resulting in the construction of large power plants and urbanisation and industrialisation rates accelerating, new environmental problems were created. The loss of habitats, species and ecosystems was becoming increasingly evident. In 1981, aware of these trends, scientists from universities and research institutes proposed an interdisciplinary research project on the rational use of natural resources and environmental protection. Supported by the Ministry for Higher Education and the State Committee for Sciences, the Natural Resources and Environment Research Program (NRERP) began in mid-1981. Two hundred scientists participated in the research.

In 1985, the Council of Ministers issued a Decision concerning 'Activities on Basic Investigation, Rational Utilization of Natural Resources and Environmental Protection'. This set down two basic principles: first, development should be based on the two fundamental strengths of the nation, natural resources and labour; and, second, development should proceed side-by-side with environmental protection. Relevant ministries, such as agriculture, forestry, fishery, mineral resources and water supplies, were directed to carry out surveys of resource state and environmental conditions, and provincial administrations were also instructed to compile inventories and to evaluate their local resources. Research institutions undertook studies of effective measures for rational resource use and limiting waste and pollution.

In 1985, the NRERP, with assistance from IUCN (now the World Conservation Union), proposed a National Conservation Strategy for Vietnam (Vietnam National Resources and Environment Research Program 1986). The Strategy's recommendations covered: the maintenance of ecological processes and life-support systems; the preservation of genetic diversity by the development of protected areas, identification of protected species, hunting regulations, trade controls and *ex-situ* conservation; the sustainable use of renewable resources; the maintenance of environmental quality for human life; and the international dimension of conservation in Vietnam. The Strategy also proposed cross-sectoral activities and the distribution of responsibilities and cooperation between central government departments as well as a role for non-governmental organisations.

---

*Box 3.1* Action programmes established by the National Plan for Environment and Sustainable Development

1    Urban Development and Population Control
2    Integrated Watershed Management
3    Integrated Coastal and Estuarine Zones Management
4    Protection of Wetlands
5    Maintenance of Genetic Diversity
6    National Parks, Protected Areas and Wildlife Reserves
7    Pollution Control and Waste Management

---

Vietnam's first protected area, Cuc Phuong National Park, was established in 1962. As a most tangible sign of the efforts of the early years of the decade, in 1986, over 80 more were set up, selected to represent the country's major ecosystems as well as a number of sites of historical or scenic significance (World Conservation Monitoring Centre 1994). By the end of the decade, the combination of policy development, legislation and scientific advice had 'paved the way for practical activities . . . at various levels, carried out by central government ministries, provincial and district administrations as well as by individual farmers and cooperatives' (Center for Environment Research, Education and Development 1993).

### Sustainable development

In the late 1980s, recognising that the pace of development could only accelerate and aware of the 'new' concept of sustainable development, the Government of Vietnam called on Vietnamese scientists, with assistance from foreign experts and international organisations,[4] to develop a national environment action plan. The result, the *National Plan for Environment and Sustainable Development 1991–2000: A Framework for Action* (Vietnam State Committee for Sciences *et al.* 1991) was approved for submission to the Council of Ministers at the 1990 Conference on Environment and Sustainable Development held in Hanoi. The National Plan built on the earlier National Conservation Strategy and provided the basis for a comprehensive framework for environmental planning and management that emerged over the following years as well as identifying areas in which priority action was required to deal with immediate problems (Tran Thi Thanh Phuong 1996).

The initial framework established by the National Plan includes institutional, legislative and policy components, responsible for the orientation of environmental authority, developing policy and law, environmental monitoring and information management, setting strategies for sustainable development, environmental impact assessment, and disaster management. Seven action programmes

*Table 3.2* Major international environmental conventions and agreements ratified by the Government of Vietnam

| Year | Environmental convention or agreement |
| --- | --- |
| 1948 | Agreement for the Establishment of the Indo-Pacific Fisheries Commission |
| 1971 | Convention on Wetlands of International Importance especially as Waterfowl Habitat (Ramsar Convention) and 1982 Protocol |
| 1972 | Convention Concerning the Protection of the World's Cultural and Natural Heritage (World Heritage Convention) |
| 1973 | Convention on International Trade in Endangered Species of Wild Fauna and Flora (CITES) |
| 1973/78 | MARPOL Convention for the Prevention of Pollution from Ships |
| 1977 | United Nations Environmental Modification Convention (ENMOD) |
| 1982 | United Nations Convention on the Law of the Sea |
| 1985 | Food and Agriculture Organization International Code of Conduct on the Distribution and Use of Pesticides |
| 1985 | Vienna Convention for the Protection of the Ozone Layer, 1987 Montreal Protocol and 1990 and 1992 Montreal Protocol Amendments |
| 1988 | Agreement on the Network of Aquaculture Centres in Asia and the Pacific |
| 1989 | Basel Convention on the Control of Transboundary Movements of Hazardous Wastes and their Disposal |
| 1992 | Agenda 21, the Action Plan from the UN Conference on Environment and Development (Earth Summit) |
| 1992 | Convention on Biological Diversity |
| 1992 | United Nations Framework Convention on Climate Change |
| 1994 | United Nations Convention to Combat Desertification |

Source: Tran Thi Thanh Phuong (1996), United Nations (1999).

Note
Vietnam is also party to nuclear accident and bacteriological and toxic weapons agreements that have environmental implications.

were devised (Box 3.1) and two support programmes concerned with education, training and awareness and international and regional cooperation. The Plan also recommended research and data collection priorities as well as pilot areas in which the feasibility of the Plan could be tested.

In 1991, the NRERP was relaunched as the National Research Program for Environmental Protection and Sustainable Development, including four sectoral research activities on environmental monitoring, environmental engineering, ecosystems management and socio-economic issues. Having joined the Ramsar Convention only in 1989, a wetland site in Xuan Thuy District in the Red River Delta (cf. Chapter 9) was approved as an international Ramsar site. Forestry policy was the subject of two major reviews during the first years of the 1990s, the Tropical Forestry Action Plan and the Forestry Sector Review (cf. Vietnam Ministry of Forestry 1991), and Decision 327 of the Council of Ministers regarding forest protection, restoration of barren lands and the sedentarisation of shifting cultivators also resulted in policy adjustment (World Conservation Monitoring Centre 1994). In 1992, the Vietnamese Association for Nature Conservation and Environmental Protection took part in the Earth Summit in

Rio de Janeiro. In 1993, Vietnam ratified the Convention on Biodiversity. A Biodiversity Action Plan was subsequently developed and approved in 1995.

During the early 1990s, the Law on Environmental Protection (LEP) was developed, with support from the IUCN Law Centre in Bonn, to advance implementation of the National Plan. It was submitted by the National Environment Agency, which had been active through the 1980s, to the National Assembly in December 1993 and enacted in January 1994 by a Decision of the President (Tran Thi Thanh Phuong 1996). The Law on Environmental Protection sets principles and describes measures to limit environmental pollution and degradation. It defines the responsibilities of the Ministry of Science, Technology and the Environment (MOSTE), established in 1992 as the successor to the State Committee for Sciences, at the national level and of the People's Committees at the provincial and city level. The National Environment Agency was designated as the implementation wing of MOSTE in 1993, with responsibility for studying and formulating policy, inspecting laws and regulations, taking protection measures, environmental impact assessment, pollution control, environmental monitoring and organising and guiding public activities. At the provincial and city level, the Departments (now Offices or Directorates) of Science, Technology and the Environment were to carry through MOSTE's responsibilities. Each relevant ministry also has an environmental protection infrastructure of some form.

During the mid-1990s, the government initiated the development of a National Environment Action Plan for Vietnam. This new Plan had three main programme areas: protect, rehabilitate and manage productive resources; control urban and industrial pollution; and strengthen institutional arrangements for environmental management. In 1998, the Politburo issued a directive on 'Strengthening Environmental Protection in the Period of Industrialisation and Modernisation,' intended to support implementation of the Law on Environmental Protection. Enforcement of the Law had been weak and, with few incentives to change and a lack of knowledge, companies had not respected the standards that it set. The Directive encouraged cost-effective waste minimisation and cleaner production and provided assistance to the provincial Departments of Science, Technology and Environment and environmental institutions to improve effectiveness (United Nations 1999: 80).

The Law on Environmental Protection and the National Plans had set a broad framework for environmental protection. During the late 1990s, these initiatives were strengthened and depth was added by sectoral measures in a number of areas (United Nations 1999). In 1998, for example, the National Assembly approved the Water Resources Law, establishing a framework for comprehensive and integrated water resources management and the Master Plan for Fisheries to the Year 2010, prepared by the Ministry of Fisheries, defined measures to enhance the capacity to protect the nation's fisheries and natural resource systems affected by the fisheries industry. The Five Million Hectare Reforestation Programme, also approved in 1998, represents a major governmental commitment to sustainable forest management. As the century turned, the National Plan for Environment and Sustainable Development was being updated.

# New challenges for the early twenty-first century

## The present-day

While developments since the mid-1980s have resulted in the rapid creation of a national framework for environmental protection and associated infrastructure, implementation continues to be hampered by a series of difficulties. In the mid-1990s, Tran Thi Thanh Phuong (1996) cited weak administrative and institutional capacity, poor regulatory enforcement, lack of knowledge and financial resources and inconsistent application of the law, compounded by rapid rates of industrialisation and capital accumulation and population growth. He noted that Vietnam's political system means that, on the one hand, regulations can be applied effectively in the absence of private property rights as regards natural resources (cf. the discussion in Chapter 5) and the State can intervene directly to limit pollution. On the other hand, the system also contains 'incentives to pollute' in that most raw materials are treated as free or open access goods, lack of private ownership can lead to a lack of protection, pollution control appears to be non-productive and not a priority of administrators more concerned with increasing economic growth whatever the environmental cost. The main conclusions emerging from the latest United Nations System review of environmental protection capacity that was prepared in close collaboration with the Government of Vietnam (United Nations 1999) are presented in Box 3.2. Areas warranting attention include the need for improvements in integrated planning and prioritisation of the environment, data availability, institutional capacity, public awareness, protection strategies, disaster preparedness and financing. They echo Tran Thi Thanh Phuong's earlier assessment.

We can distinguish between three categories of weakness in, or constraints on, Vietnam's system for environmental protection: (i) resource limitations (lack of financial, technical or human resources); (ii) weaknesses that result from the fact that the system is relatively new (lack of experience, undeveloped institutional capacity, organisational difficulties); and (iii) what might be termed systemic weaknesses such as cultural attitudes. Of the three, it is the last category that presents the most cause for concern when considering the longer-term future. Despite the principles that have been established, economic growth or the exploitation of limited resources too often becomes the overriding priority to the exclusion of environmental considerations. Unless the cultural attitudes underlying this weakness alter, the effectiveness of the environmental protection system will continue to be undermined.

## The future

How well is the framework and infrastructure developed to protect Vietnam's environment equipped to deal with the new challenges of the early twenty-first century? To date, the development of the system has been oriented towards national environmental problems and difficulties, driven by concern about the

*Box 3.2* Key issues that must be addressed if environmental protection is to be strengthened in Vietnam, according to the 1999 United Nations Common Country Assessment

### Integrated planning and prioritising the environment
The need for integration of economic, social and environmental factors in the development planning process is still not well understood by the Government. The recent directive on strengthening environmental protection is a step in the right direction, but environmental protection agencies remain understaffed and poorly funded. For example, the National Environment Agency has only 70 staff and an annual budget of just US$1.5 million. The main priority of the Government continues to be economic development, modernisation and industrialisation. This emphasis has brought many short-term benefits but, in many instances, long-term costs to the environment.

### Data for environmental planning
A reason for the low profile of environmental conservation is that policy-makers are not sufficiently aware of these long-term costs because high-quality baseline data are lacking. Thus, even if environmental protection were a high priority, it would be difficult to set priorities since environmental trends are not well understood. There is no regular environmental monitoring programme that produces reliable and comprehensive state-of-the-environment reports, or even an agreed set of indicators. Thus, decision-makers are not fully aware of negative environmental trends and the hidden costs of development.

### Institutional capacity for environmental planning and management
Even if policy-makers were aware of the urgent need to prioritise and integrate environmental issues in their planning, institutional capacities for environmental planning and management remain weak. This is partly a result of the above mentioned limited human and financial resources, but it also results from disincentives and conflicts of interests that are built into the environmental management system. For example, many of the largest natural resource ministries (Ministry of Agriculture and Rural Development, MARD, Ministry of Forestry and the Ministry of Industry) retain obligations both to extract and protect natural resources such as timber, fish and coal. Capacity building is needed both for strategy and policy development – for example, in updating the NPESD – and for implementation.

### Public awareness

Public awareness of environmental issues is still low, and tolerance for environmental degradation is high. These attitudes result partly from the long legacy of central planning, under which State ownership of all natural resources discouraged a sense of responsibility on the part of individuals. The land allocation process is slowly changing those attitudes, but much remains to be done before individuals, civic groups, mass organisations, the mass media and businesses have a strong environmental voice in Vietnam.

### Protection strategies

It is difficult to promote regulatory approaches in Vietnam, due to conflicting and often overlapping areas of responsibility. Thus, support for non-regulatory approaches including voluntary compliance should be a key element of any environmental protection strategy. Improving the awareness of important groups – such as the business community, environmental journalists and women – would help promote voluntary compliance with environmental regulations.

### Disaster preparedness

Vietnam is one of the more disaster-prone countries in the world and suffers from regular typhoons, floods, tropical storms and droughts. In response to such threats, MARD and the Central Committee for Flood and Storm Control are working together to establish strategies for disaster preparedness, prevention and mitigation. There are still a number of areas in which improvements could be made, such as better dissemination of disaster information and improved management capacity for droughts, forest fires and sea-water intrusions.

### Financing

Financing remains a problem. Even if none of the above problems existed, policy-makers would still be faced with the problem of paying for environmental protection measures. In the near term, this is mostly a problem of using existing financial and physical resources more efficiently. In the long run, Vietnam will need to develop more creative ways to ensure that the costs of environmental degradation are internalised and accounted for by those causing the degradation.

Source: United Nations (1999: 82).

effects of population pressure, urbanisation and industrialisation. Consistent with the early stages of the evolution of most national environment strategies during the second half of the twentieth century, the process has been inward-looking. The turn of the century, though, brings with it threats to Vietnam's environment that have their origin outside the nation's border. Some of these threats are regional in scale, for example, transboundary pollution and degradation of shared resources. Others are rooted in systems that are worldwide in extent, global environmental change and the adverse effects of economic globalisation. Issues related to shared resources, in this case the Mekong River, are considered in Chapter 13 of this book and the potential environmental effects of globalisation are discussed in Chapter 11. Here, we review the challenge of global environmental change.

Global environmental change is a generic term covering the problems of climate change induced by human activity, ozone depletion, biodiversity loss and marine pollution (Galloway and Melillo 1998). These problems, because of the spatial extent of the sources and impacts, necessitate an international response and, indeed, are the subject of a series of international conventions or agreements to which Vietnam is party (Table 3.2). To illustrate the challenges presented by global environmental change, we select the threat of climate change and sea-level rise induced by anthropogenic emissions of greenhouse gases, that is, global warming (Houghton *et al.* 1996). Parry *et al.* (1991), Granich *et al.* (1993), Asian Development Bank (1994), Fu *et al.* (1998), McLean *et al.* (1998), Tran Viet Lien and Nguyen Huu Ninh (1999) and Kelly and Adger (2000) present different perspectives on the potential impact on Vietnam and the neighbouring region. We consider two aspects of the problem that have implications for environmental management in Vietnam: national emissions, and vulnerability to impacts.

Vietnam is not, nor has it been, a major contributor to global greenhouse gas emissions. It is one of the lowest consumers of energy in the region; per capita consumption is a third that of Indonesia, a sixth that of China and a thirtieth that of South Korea (United Nations 1999: 81). Per capita carbon dioxide emissions associated with fossil fuel consumption and production and cement manufacture stood at 0.4 metric tons per person in the year 1995; by comparison, the figure for China was 2.7 metric tons per person and for the United States 20.5 metric tons per person (see Chapter 11). (All emissions estimates are from the World Resources Institute inventory at http://www.wri.org/.) High deforestation rates in the late 1980s and early 1990s meant that carbon dioxide emissions associated with changing land use, a total of 40 000 million metric tons in 1991, exceeded fossil fuel and cement-related emissions at that time; the 1992 total was 21 522 thousand metric tons. But this does not alter the nation's low position in the world league of carbon emitters.

Energy demand and generating capacity is, however, growing rapidly (Figure 3.2). It is estimated that electricity generating capacity, 4.6 gigawatts in 1997, will have to rise at 17.5 per cent a year to keep pace with demand (United States Energy Information Administration, Vietnam Country Report, 1998).[5] Electricity generation is based on hydropower (67 per cent in 1997), thermal power (18 per cent) and natural gas (15 per cent). As noted earlier, hydropower

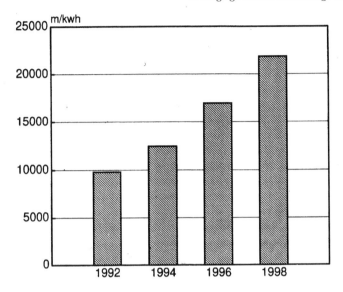

*Figure 3.2* Growth in electricity output during the 1990s
Source: General Statistical Office, as presented in United Nations (1999: Annex).

presents the best opportunity to meet increasing demand for electricity. Nevertheless, increased use of fossil fuel sources is inevitable throughout the energy economy, considerable reserves of coal, gas and oil are available (England and Kammen 1993), and with this must come a rise in emissions of carbon dioxide and other greenhouse gases. In fact, between 1992 and 1995, energy-related per capita carbon emissions rose by a third, according to the World Resources Institute inventory, despite the trend in population. A projection undertaken for the Economic and Social Commission for Asia and the Pacific in 1991 suggests that, by 2010, increased use of fossil fuels could result in carbon emissions three times the 1986 baseline (Economic and Social Commission for Asia and the Pacific 1991); in reality, with economic growth rates higher than anticipated at the time of the study, emissions had already doubled by 1995, suggesting that this projection may well prove too conservative.

Three means of limiting the rise in Vietnam's energy-related contribution to global greenhouse gas levels have been identified (Granich *et al.* 1993). First, make more effective use of basic energy sources, especially coal, by improving technology as well as equipment efficiency and replace the use of coal with other energy sources, such as oil and gas, which release lower amounts of carbon. Second, step up the exploitation of the hydroelectric potential on the national scale, taking account of impacts on the environment. Third, make more use of renewable energy sources such as biomass, used sustainably, solar and wind energy.[6] This will require research on the application of advanced technologies at lower costs.

As far as emissions from deforestation and changing land use are concerned, there is some reason to believe that improvements in forest management might limit trends in this sector. According to a recent assessment, changes in policy, improvements in food security and the allocation of forest land to smallholders may have been responsible for a decrease in the rate of deforestation during the 1990s (United Nations 1999: 78). It is considered that the Five Million Hectare Reforestation Programme, running to 2010, could produce major changes in the forestry sector as local people take more responsibility for forest management and large-scale plantation development and State forestry enterprises play a reduced role. Nevertheless, 'the continuing decline of natural forests is of great concern to the Government' (United Nations 1999: 78). In the agricultural sector, it is difficult to see how growth in emissions, of methane for example, can be avoided given the inevitable intensification of activities.[7]

At present, Vietnam, as a developing country, does not take on any formal target with regard to emissions control as a party to the United Nations Framework Convention on Climate Change. It is, however, required to produce regular emissions inventories and to consider what efforts might be made to limit future emissions growth; this work is being undertaken. It can also take advantage of foreign support for emissions control and sink enhancement[8] measures, either through the Global Environment Facility or credit transfer devices such as the Clean Development Mechanism (Forsyth 1999; Grubb *et al.* 1999). If Vietnam is to develop its economy, a rise in greenhouse gas emissions cannot be avoided but, clearly, a sustainable development path should limit that growth. It is imperative that the implications of any development project for greenhouse gas emissions are considered explicitly. This requirement will necessitate, amongst other things, capacity strengthening and reinforcement of existing institutions.

When considering the possible impact of climate change on Vietnam, a critical issue that must be considered is that of scientific uncertainty. Estimates of the likely climate future vary not only in scale but also in sense. For example, the change in rainfall over Vietnam will be critically dependent on the manner in which the seasonal monsoons respond to the larger-scale change in the climate system. Yet climate models disagree as to whether the Asian monsoonal circulations will weaken or strengthen (McLean *et al.* 1998: 391). The overall scale of the impact on climate is critically dependent on a parameter known as the 'climate sensitivity', a measure of how much the global climate system will respond to a given change in atmospheric greenhouse gas concentrations. Current estimates of this parameter vary by a factor of three (Houghton *et al.* 1996). Taking other relevant factors into account, such as the role of the oceans and related uncertainties, the net result is a factor of two to three uncertainty in any time-dependent projection of global climate change. There is even greater uncertainty when it comes to forecasting regional trends. Granich *et al.* (1993) presented a range of temperature scenarios for Hanoi, based on forecasts of global-mean temperature change following a business-as-usual emissions path and regional patterns derived from a set of general circulation models.[9] At the low end of the range of uncertainty, annual temperatures may rise by a mere 0.5 degree Celsius by 2050. At the other end of the range, temperatures may reach 2.0 degrees Celsius above the 1990 baseline by that time. For rainfall, the projections suggested a moderate increase, by anything up to 15 per cent by 2050, though the uncertainty was considerable and a decline could not be ruled out. Current projections of regional trends are in general agreement with these early estimates (McLean *et al.* 1998).

The uncertainties in the projections of the climate future create two related dilemmas: first, it is difficult to judge what level of priority to attach to responding to the threat of climate change whether by limiting emissions or facilitating adaptation; and, second, as it is difficult to determine just what the long-term impacts are likely to be, planning for adaptation might be considered problematic. It is because of these difficulties that the scientific community has recommended that 'precautionary' measures be taken at this time: measures which can be undertaken at little or no cost or which have benefits in other areas that justify investment. If the case for action on climate proves less strong as time passes and understanding improves, little will have been lost. If it emerges that the threat is, in actuality, as serious as feared at this time, then a step in the right direction will already have been taken. In the case of averting impacts, it has been argued that uncertainty in the climate forecasts can be side-stepped by adopting a 'win–win' strategy, identifying measures which meet present-day needs as well as the longer-term objective of climate protection (Kelly 2000; see also Chapters 9 and 10). The reason why action is justified despite the remaining scientific uncertainties is, of course, that the impact of climate change and sea-level rise on Vietnam could be extremely serious.

Agriculture must be considered the major impact area because of the dependence on it of the majority of the Vietnamese population (Parry *et al.* 1991). Any effect on production in the Red River Delta and the Mekong Delta would have national implications. Some consequences for agricultural production may be beneficial but many would be adverse. Not only might yields and quality be affected but crops may have to be grown in different locations as climate conditions alter. Agricultural production is likely to be affected by sea-level rise, changes in the intensity of, and the potential for, extreme weather events (such as typhoons, droughts, floods and cold outbreaks in northern regions), local climate change (temperature rise, rainfall variation and the potential for desertification in southern areas), pest development, erosion and changes in soil fertility, and levels of increased atmospheric carbon dioxide (Granich *et al.* 1993).

In the coastal zone, sea-level rise, climate change and any increase in storm risk would have a series of impacts. According to one study, an annual investment of US$172 million would be required over the next 20 years to protect the coastline adequately against sea-level rise (Asian Development Bank 1994). Tran Viet Lien and Nguyen Huu Ninh (1999) describe Vietnam's exposure to a one metre sea level rise over the present century as 'critical', citing adverse impacts on agriculture, coastal ecosystems such as the mangrove forest, fisheries and aquaculture, human settlements, transport and infrastructure, and energy systems. They note that cities, ports and other coastal facilities were designed with existing sea-level trends in mind, not any accelerated rate. The river- and sea-dike system is intended to cope with storms rated up to 9 or 10 on the Beaufort Scale; any tendency towards the more frequent occurrence of stronger storms could have catastrophic consequences.

The physical exposure of Vietnam to climate change and sea-level rise is clear. But the degree to which the coping strategies that are already available to the population might minimise impacts has yet to be assessed in any depth and any full assessment of the implications of global warming must also take prospects for adaptation into account. Kelly and Adger (2000) note, for example, that Vietnam's well-developed system of storm protection compensates for the physical vulnerability of coastal zones (see also Chapter 10). There is a clear need to examine the systems and procedures for protecting the Vietnamese population and physical infrastructure against weather and climate hazards as a first step in limiting the impact of any future trends in these events. Moreover, it is important to 'climate-proof' development plans by considering not only potential effects on greenhouse gas emissions levels but also the degree to which they affect levels of vulnerability and resilience. Again, capacity strengthening and reinforcement of existing institutions will be needed to meet these requirements, alongside a more pro-active approach to managing environmental risk.

The issue of climate change highlights a further shift in humanity's relationship with the environment. Even with the most elaborate and effective system of environmental protection, society cannot completely avoid environmental damage, stress and change. It is intrinsic to the development process, however environment-friendly it might be, that there must be some compromise between the needs of humanity and the needs of nature. And it is, therefore, essential that any truly sustainable development process takes full account of the inevitability of adverse impacts on the environment and ensures adequate protection for those vulnerable to environmental risks and hazards. This is not an admission of defeat, but a humble recognition that we have more power to harm the environment than we do to remedy its ills. And we must respect the environment's capacity to turn that damage back upon us – 'If we shoot a pistol at the environment, it will fire a cannon back at us' observed La Ngoc Khue, Vice Minister of Transport (quoted by England and Kammen 1993).

## Conclusions

In spite of the many important legislative and regulatory measures that have been taken to protect Vietnam's environment, environmental degradation continues. Current patterns of natural resource exploitation are not sustainable and as such threaten the long-term economic as well as environmental health of the country.

(United Nations 1999: 82)

It is too easy reviewing the state of a nation's environment to become overly negative about future prospects. As a matter of balance, we should note that, as far as Vietnam is concerned, there are many factors that could offset adverse trends: not least the fact that the country has 'extraordinary human resources: the population is highly educated, hard working, and endowed with ingenuity' (England and Kammen 1993).

Ha Nghiep (1993) identifies four advantages for Vietnam given its late entry into the process of industrialisation: first, access to the experience of predecessors on what to do and avoid, particularly lessons on the role of government in a market economy; second, awareness of the threat so that laws on protection of nature and environment have already been enacted and implemented; third, effective coordination of actions between governments, international organisations and experts; and, fourth, scientific and technological advances that can help promote the process of development at the same time as minimising environmental damage and, as necessary, restoring ecological balance.

With a solid framework for environmental protection and biodiversity conservation in place as a result of the efforts of the past two decades, there is a pressing challenge. Vietnam must, as a matter of urgency, ensure that what is enacted in abstract principle is implemented in reality. This is an absolutely critical issue that cannot be neglected. Too often in the past, careful planning has been let down by ineffectual action.

Finally, it would be inappropriate to end this account of the state of the Vietnamese environment without observing that perceptions of what is or is not acceptable environmental and ecological damage vary. It is up to the Government of Vietnam and the Vietnamese people to make conscious decisions, step-by-step, regarding the trade-off between development goals and environmental concerns, taking account of international standards but evolving a model of sustainable development consistent with national history, circumstances and aspirations.

## Notes

1 These historic figures, from Le Thac Can *et al.* (1994), are not directly comparable with recent official statistics which include industrial plantations and secondary vegetation cover.

2 According to the World Conservation Monitoring Centre (1994), traditional shifting cultivation is no longer widely practised in Vietnam and much of what is referred to as shifting cultivation is actually forest encroachment and clearance.

3 The study was conducted by Hatfield Consultants, based in Vancouver, Canada, with the cooperation of the Vietnamese Government 10–80 Committee, a team of doctors mandated to study the consequences of the chemical used during the war, and the Vietnamese resource management agencies.

4 The United Nations Development Program, the Swedish International Development Authority, the United Nations Environment Program and IUCN assisted in the development of the National Environment Action Plan.

5 Available at http://www.eia.doe.gov/emeu/cabs/vietnam.html

6 The sustainable use of biomass as an energy source is particularly important. According to Nguyen Anh Tuan (1997), biomass constitutes a major component of total carbon dioxide emissions and releases from this source were growing at a rate of 5 per cent a year over the period 1988 to 1992. Biomass, such as fuelwood, bamboo and straw and stubble, provides an important source of energy in daily life and in small industries. Solar

energy is used in the drying of agricultural products and in heating water. Exploitation of the wind resource is difficult as wind speeds are low in inland areas, restricting the use of this resource for pumping and electricity generation. On the coast and islands, the occurrence of tropical cyclones presents serious technical problems. The geothermal and tidal potential is great but little has been done to harness these resources. See Forsyth (1999: 124) for a detailed discussion of renewable energy in Vietnam from the perspective of the climate issue.

7  Rice production is the major source of methane emissions in Vietnam (Granich *et al.* 1993). The degree to which methane is produced depends on a variety of factors. The nature of the production system is important. Wet paddy rice produces methane whilst dry upland rice does not. How the crop is fertilized (whether with organic matter or chemicals) affects the amount of methane produced, as does the manner in which crop residues are disposed of. The rice variety also influences methane production. The most effective approach to reducing emissions would appear to be a strategy based on improving the efficiency and sustainability of rice production, an immediate need. The same is true in the case of the nitrous oxide release resulting from the application of nitrogenous fertilizers. Here, more efficient application could not only reduce nitrous oxide releases but would have very direct benefits in terms of production economics, soil protection and ensuring sustainable yields.

8  For example, taking carbon out of the atmosphere into forests.

9  See Hulme and Brown (1998) for a discussion of the methodology.

# 4 Historical perspectives on environment and development

*Cecilia Luttrell*

## Introduction

A consideration of the ecological, cultural and historical heterogeneity of Vietnam provides an opportunity to examine the differentiated effect of a variety of administrative policies, ethnic groups, cultural traditions and physical features on natural resources and environmental management. The very definition of 'Vietnam' itself is controversial with huge differences between the ethnic groups and their relationships between culture, environment and methods of land use. The conflict in Vietnamese society between the need for a centrally administered economy and the indisputable existence of strong village institutions creates social tensions between the geographical core and the periphery.

This chapter provides an outline of the historical events relevant to resource use and social and environmental change in Vietnam and discusses some of the theories that have been proposed to explain significant historical and modern developments. The discussion focuses on the issue of land reform, traced through the dynastic pre-colonial regimes and the subsequent upheaval of French colonial expansionist policy to the present day, and related developments concerning land concentration, tenancy and the destruction of traditional institutions as a result of the introduction of cash economies. Various theoretical perspectives are put forward and discussed in an examination of the relative importance of historic tradition versus modern changes brought about by colonialism and various economic transitions.

Throughout the discussion, the importance of the control of land is emphasised. Land, and the struggle for its control, is a central tenet of the emergence of a resistance and, later on, of official cooperative systems. The process of resistance is discussed in the context of both historical upheavals associated with capitalism and the divergence of capitalism away from local tradition. Traditional village land management provided important social buffers through the existence of communal land. The development and failure of the cooperative system in post-reunification Vietnam (from 1975) is also examined and the effect of the recent withdrawal of the state on socio-cultural, economic and environmental aspects is focused upon. Events in the forestry sector provide a useful example for the discussion of many of these issues due to conflict which has occurred between upland and lowland interests and the problems faced during collectivisation of the sector and subsequent allocation of land.

Vietnam's population is concentrated in the two core productive areas of the country: the Red River Delta and the Mekong Delta. These two areas are characterised by sharply contrasting ecological and historical differences. The Red River Delta is one of the most heavily populated areas, and is characterised by a history of hunger and social unrest. The Mekong Delta environment was more predictable and favourable with abundant land. In addition, there are serious problems of climatic instability in the north and central regions whereas the south has relatively stable weather.

In his book on the cultural evolution of Vietnam, Jamieson (1993) argues that ecological differences have been important in the different cultural and social organisation of the two regions. Vietnam was divided into the South and North in the seventeenth and eighteenth centuries but culturally and linguistically there are clear north, central and southern divisions. Historically, the Mekong Delta is a frontier region and the Vietnamese only began to settle the upper areas in the seventeenth century and the lower delta areas in the nineteenth century (Jamieson 1993: 5). The south (Cochinchina) was more intensively colonised and administered directly by the French whereas Annam (Central) and Tonkin (North), which were Protectorates were administered indirectly by Vietnamese administrators. These features however only partly explain the differences in regional features and social change which have taken place.

## The origins of Vietnam

### Ethnicity and religion

The Vietnamese have been cultivating rice in the Red River Valley since 200 BC. Hence, the Red River Delta is the ancient heartland of Vietnamese culture, as it is currently understood (Marr 1995). In 208 BC, the Nam Viet Kingdom emerged in coastal China, and was merged with the Red River area soon afterwards and governed as a Chinese province (Jamieson 1993: 8). Although the local culture remained intact, Chinese cultural influences were obvious in such aspects as the Confucian model of government. This resulted in two separate concepts of political legitimacy: the indigenous tradition and the Chinese system of politics and administration (Jamieson 1993: 9). In the ninth century, there was a revival in local cultural traditions, and uprisings became more numerous until 939 when the Chinese were driven out and the Vietnamese gained independence. Conflict has been an important feature of Vietnam's history which is interspersed with Chinese attacks, expansion to the south and peasant uprisings against oppressive regimes (Hodgkin 1981; Thrift and Forbes 1986: 46), as well as environmental stress in the form of heatwaves and droughts, storms and floods. The people of the Red River Delta became accomplished at dealing with a harsh natural environment and quick to mobilise themselves (Marr 1995).

The Ly Dynasty ran from 1009 to 1225 followed by the Tran Dynasty between

1225 and 1400 under which Confucianism began to thrive. The fourteenth century saw economic crisis and peasant revolt and in 1407 Chinese rule returned under the Ming Dynasty. Regulation by the mandarins was a feature which was to continue until the nineteenth century, despite Emperor Le Thai-to's expulsion of the Chinese in 1428 (Jamieson 1993: 9). From the seventeenth century onwards, however, the French Catholic missionaries began to gain influence (Thrift and Forbes 1986: 46). Between 1620 and 1774, the Trinh Dynasty ruled in the North while the Nguyen Dynasty ruled in the South. As a result of the Tay Son Rebellion of 1786, rebel leaders gained control of the whole of Vietnam and this destroyed the balance of power among the noble families (Marr 1971: 16).

The ethnic Vietnamese, *Kinh*, make up 90 per cent of Vietnam but there are also highland groups many of whom are upland rice cultivators (Wiegersma 1988: 2). There are large cultural differences between the Vietnamese and the uplanders (Le Trong Cuc *et al.* 1990). The lowland Vietnamese are historically a unified group throughout the country with a common language. The basic social structure is characterised by a patriarchal extended family, the neighbourhood and the village or commune.

The *Kinh* have tended to separate themselves mentally and physically from the upland people (Marr 1971: 8). The 15 metre contour around the Red River Delta marked the edge of *Kinh* territory due to their fear of malaria and to their lack of knowledge of dryland farming techniques for sloping terrain and shifting cultivation as practised by some ethnic groups. Traditional upland society is 'dispersed, small-scale, mobile, with a yearly rhythm and with social and ritual institutions all aspects which were foreign to the *Kinh* society' (Tran Duc Vien and Fahrney 1996: 79). The term 'Montagnard' (mountaineers or highlanders) has replaced the earlier term of 'mois' or 'savages' used by the French to refer to the upland ethnic minorities. According to Salemink (1991), the French approach to the Montagnards was based on the assumption that the Montagnards been compelled to withdraw when faced with the superior *Kinh*.

The defining feature of the history of upland land use is that of settlement and migration. Under the French rule, much of the Montagnard land was seized and cleared for coffee and rubber plantations, displacing Montagnards or forcing them to work on the plantations. In addition, the missionaries, who faced difficulties in making conversions in these areas in the 1920s, encouraged the migration of the *Kinh*, whom they were more successful in converting, to the highlands (Lang 1996: 40). Until the late 1950s, the highlands were sparsely populated (Tran Duc Vien and Fahrney 1996: 6) but following the French defeat in 1954 the South Vietnamese government announced all of South Vietnam to be sovereign territory and thus the Montagnard lost all claims to land. Encroachment was more rapid in the central Provinces than in the north because of the heavier effect of the war and because of the Southern government's expansionist land settlement policies between 1955 and 1975 (Beresford and Fraser 1992: 6). In 1955, President Diem commanded the settlement of thousands of Catholic refugees from the North in the Central Highlands, allowing them to claim Montagnard land. Later, during the second Indochina war, many areas in the Central Highlands were declared free-fire zones forcing 85 per cent of Montagnards to move from their villages.

In 1975, the resettlement of millions of *Kinh* in the highlands took place under the 'New Economic Zone Strategy' in the hope of encouraging highland development, to ease pressure on lowland agricultural land, to assist with the settlement of shifting cultivators and to try to integrate ethnic minorities into mainstream Vietnamese society (Liljestrom *et al.* 1998: 184; Economic Intelligence Unit 1996) in what has been termed the process of Vietnamisation. Two million people moved into the Central Highlands, Eastern Nam Bo and the Mekong Delta alone (Beresford and Fraser 1992: 6) and the *Kinh* are now the majority in many areas (Tran Duc Vien and Fahrney: 1996: 6). The area of fallow land available for ethnic minorities has reduced (Liljestrom *et al.*1998: 184) there has been an increase in logging (Lang 1996: 41) and in high levels of soil loss and destruction of natural habitat due to the imposition of unsuitable lowland farming techniques (Tran Duc Vien and Fahrney 1996: 6). Much of the blame for such problems has been put onto the ethnic minorities. In addition, there have been movements of uplanders, especially of those at higher altitudes, to valleys and to the Central Highlands under the government scheme of 'Fixed Cultivation and Settlement' which aimed to settle nomadic people.

Present-day Vietnamese religion is a medley of Taoism, Buddhism and Confucianism (Ngyuen Ngoc Binh *et al.* 1975: 24; Rambo 1982: 407; Ho Tai Hue Tam 1985: 25–6). Jamieson (1993: 11) sees the traditional Vietnam worldview as an all-encompassing approach based on the natural order of reason '*ly*' and the 'nature of things' which rationalises the hierarchical order of the cosmos. During Chinese rule in the fifteenth century, Neo-Confucianism was introduced but it was not until the nineteenth century that Confucianism became a predominant force, becoming the basis of national culture during the time of the Nguyen Dynasty. Many scholars have connected traditional Vietnamese society to Confucianism and have shown how the traditional village operated in a mandarin framework where Confucianist values about proper social relationships and the nature of reality are emphasised (Jamieson 1993: 11). Nguyen Khac Vien (1972: 47) saw the later transition to socialism as rooted in Confucianism since it concentrated on political and social problems with its emphasis on collective discipline. The reallocations of land which have occurred throughout Vietnamese history have helped to control the power of individual patriarchs.

Wiegersma (1988: 17) has shown that the emphasis in Confucianism on the patriarchal extended family structure as a labour force is at odds with the importance in traditional Vietnamese society of reciprocal assistance through neighbourhood groups. She, therefore, concludes that property relations were based on a value system which is only partly Confucian. Jamieson (1993: 15) identifies conflicting views in Vietnamese studies between the 'yang' Neo-Confucian traditional cultural system and the 'yin' approach of Buddhism, Taoism, animism and egalitarian village institutions. The 'yang' approach was dominant and accepted by the ruling classes and the 'yin' was emphasised by early French scholars who studied Vietnam after a focus on Chinese studies. Jamieson

(1993: 21), however, sees 'yin' and 'yang' as parallel aspects of one cultural system which is constructed from the tension between active participation in society as advocated by Confucianism and the passivism of Buddhism and Taoism which suggest a removal from activity in society.

### Communal traditions: north and south

The village, in traditional lowland Vietnamese society, was an important landowner. The extent of communal land varied. In many areas, village land was reallocated every few years and some land was rented to villagers to supply the village with an income (Jamieson 1993: 21). Only people from the village could possess or work on village land. Le Thanh Khoi (1955: 225) refers to the institutionalised hierarchy of the traditional commune and points out that communal land was allotted by the mandarins, officials and old men among the commune, not on an equal basis but according to social position, seniority and title. Under the Nguyen Dynasty in the nineteenth century, the king could bestow ownership for communal land, but had the right to claim it back at his will (Ngo Vinh Long 1973: 5).

Traditional collective regimes were broken down to a large extent by capitalist markets and the emergence of private property. In 1860, all landlords were commanded to give one third of their property to communal land, and the sale and ownership of communal land was prohibited (Bray 1986: 171). Many peasants, however, were not aware of these reforms and central government had little power over the land-owners. The French also tried to expropriate communal lands by permitting village officials to sell it (Morel 1912) so that the amount of communal land dropped dramatically. Before 1945, in Hai Van commune in the Red River Delta, for example, the land was theoretically controlled communally and the village leader made annual allocations to families. In reality, though, only that land which had not been appropriated by landlords was shared (Houtart and Lemercinier 1984: 21).

The amount of communal land differed across the regions. Early in the twentieth century, 20 per cent of the cultivated land in the Red River Delta was communally owned (Gourou 1955: 385). In Quang Tri province, almost the whole cultivated area was communal (Nguyen Van Vinh 1961) whereas, in the south, communal land was scarce (Rambo 1973). Other estimates show that in pre-colonial times, the area of communal lands in the north, central and southern regions covered 20, 26 and 2.5 per cent of the total land area respectively (Gourou 1940). By 1931, these areas were 20, 25 and 3 per cent, respectively (Nguyen Van Vinh 1961).

The historically low level of communal land in the South can be partly attributed to its seizure by rich landlords (Gourou 1940). In addition to this, life has been more secure in South Vietnam, less rigid, more accepting of individualism and difference and has had less fixed population features typical of a 'frontier environment' (Rambo 1973). In north and central Vietnam, villages are close-knit and communal with more independence (Jamieson 1993) and land holdings were cushioned by communal land. Woodside (1976) and Duiker (1976) argue that the communists based themselves firmly on tradition. The Vietnamese have strong historical conventions of the collective and family work at the village level which the socialists were able to build on. Ho Chi Minh used the words *Xa Hoi Hoa* to symbolise the word socialism: *Xa* meaning 'traditional village' or 'commune'; *Hoi* – 'society'; and *Hoa* – ' the act of putting society under a new system' (McAlister and Mus 1970). The existence of communal land eased collectivisation in 1975 and the collective aspect remained strong alongside individual and family interests (Fforde 1988), and an acceptance of the legitimate role of the nation state: 'Thus one can talk of Vietnamese culture being simultaneously highly collective and highly individualistic' (Fforde 1990: 112).

## Land reform

Most of the Red River Delta has been under intensive rice cultivation for centuries. By the thirteenth century, ancient dikes which prevented the Red River from flooding, were reinforced, reclamation of salt marshes was taking place by the fourteenth century and canals were being built in the fifteenth century (Bray 1986: 94). The Vietnamese have effectively been administered by central government since the fifteenth century and it is suggested that centralised control in Vietnam is essential because of the canal and dike systems (and the underlying environment) on which the lowland area of the country is dependent, and to resist attack against the Chinese, the hill-tribes and the Cham and Khmer to the south (Wiegersma 1988: 2–3). The Emperor was seen as the protector of agriculture and, therefore, the consequences of drought, flood or bad harvest could be blamed on him and the peasants had the right to revolt (Le Thanh Khoi 1955). Despite this strong central authority, society has always been rooted in the rural village economy. Hence, the development of modern Vietnamese institutions can only be understood when studied in both of these contexts. Land-tenure systems have undergone a multitude of dramatic upheavals during the history of Vietnam and the following overview highlights the importance of land tenure and land reform as the arena in which social and environmental issues are played out.

### Pre-colonial land tenure

During the Ming Chinese invasion and subsequent period of control (1407–27), there was much social upheaval. Land was concentrated in the hands of supporters of the Chinese regime (Marr 1971: 14) and a large number of peasants were rounded up and sent to work opening mines and cutting timber in the mountains (Ngo Vinh Long 1993: 5). After the expulsion of the Ming, Emperor Le Thai-to (1428–33) introduced the *quan dien* (equal field) land system to redistribute land more equally (Phan Huy Chun, 1821; Tran Trong Kim 1964). Individual land was redistributed every four years according to the number of labourers in the family, and land was redistributed when they died. Ngo Vinh Long

(1993: 5) has, however, pointed out that only people of equal status and rank received an equal area of land in this period.

By the eighteenth and nineteenth centuries, land was concentrated in the hands of the elite, although these inequalities were somewhat reversed during the Trinh Dynasty in the seventeenth century (Bray 1986: 171). During the Nguyen Dynasty of the eighteenth century, all land belonged to the king and people paid taxes to use it (Luong Duc Thiep 1944; Dao Duy Anh 1951: 54). The Vietnamese mandarin Nguyen Cong Tru executed two large-scale reclamation projects in Thai Binh and Ninh Binh in 1828 and reclaimed land was distributed to groups of 10 households in lots of 100 mau (Bray 1986: 41) (1 mau covers 0.36 hectares). The land reforms of 1839 converted half the private land in Binh Dinh to communal land and that helped to address problems of corruption among local officials (Ngo Vinh Long 1993: 9).

*Colonial land tenure*

Peasant rebellions had been occurring since the beginning of the nineteenth century in response to repression, floods, droughts and crop failures (Ngo Vinh Long 1973: 4). The establishment of French control over the Vietnamese countryside in the middle of the nineteenth century caused further turmoil and had a significant impact on the ownership of land in Vietnam. Before the French period, the largest landowners were the communes and most peasants cultivated their own plots as private land or as part of communal land, rural tenancy was rare (Fforde 1983: 24).

During French colonial rule, there was a dramatic shift from a 'rural social structure, based primarily upon small peasant cultivators of their "own" land, to one based upon a mixture of heavily indebted and rack-rented tenants and a poverty-stricken landless rural proletariat' (Fforde 1983: 41). French policy ensured the predominance of the cash economy which concentrated land in the hands of fewer landlords and tenancy (Popkin 1979). The poor were adversely affected by the introduction of monetary taxation and, by the 1930s, a landless labouring class was developing and the communal organisations were disappearing (Fforde 1983; Gourou 1955: 383).

The French introduced different policies in the northern (Tonkin), central (Annan) and southern (Cochinchina) regions. After 1888, both French and Vietnamese were permitted to own land in Tonkin and the port areas of Annan and size limits on land ownership were abolished (Long 1973: 13). The Mekong Delta was settled under the landlord tenancy system and this became the means by which the Delta was managed (Sansom 1970: 18). The Mekong Delta had very low population levels when the French took control because disease, fighting and poor agricultural land had kept numbers low (Bray 1986: 94). The French brought medical and engineering skills which enabled the widespread opening up of new land, with massive drainage programmes and settlement between 1868 and 1930 (Sansom 1970: 21). In 1867, only 200 000 hectares of alluvial land along the Mekong and Bassa river banks were cultivated (Gouvernement General de l'Indochine 1930) but, by 1930, 2.21 million hectares had been cultivated (Ngo Vinh Long 1973: 1). The French used skilled peasants from Tonkin but the methods used were less intensive and productive than in Tonkin (Robequain 1939: 96). Many settlers were eventually forced into tenancy as a result of debt from having to borrow so much money (Gourou 1940: 285–6).

The French acquired complete command over the Mekong Delta in 1862 and all land which was not settled was confiscated. The 1863 statutes granted rights of ownership and compelled all peasants to return to abandoned land to claim it (Ngo Vinh Long 1973: 11). Supporters of the regime, mostly Catholics, were, however, granted much of the land which had been abandoned when the French invaded. Large areas were also granted to the French settlers who then rented it out.

During the 1930s, economic conditions of the Mekong Delta declined (Sansom 1970: 45), population increased and new land was no longer available. Small tenants were pushed by landlords into a poorer rural proletariat class to serve the export market and the numbers of landless peasants grew as land became increasingly concentrated in the hands of a few, and peasants were forced to become tenants, share-croppers and agricultural wage labourers (Ngo Vinh Long 1973: 44). Sharecropping was an unknown phenomenon before French times. The emergence of the Viet Minh (1946–54) and the Viet Cong (1960–7) must be seen against this context of changing tenancy conditions and poverty. It is said that French rule managed to 'destroy the traditional equilibrium of village life, undermine the scholar-gentry class and block the growth of an indigenous bourgeoisie' (Economic Intelligence Unit 1996) and thus assisted communism to become an effective challenge.

*Collectivisation*

By the final years of French rule in the 1940s, land ownership was hugely unequal. In North Vietnam, one third of agricultural land was owned by 5 per cent of rural households, the Catholic church and French citizens (Kerkvliet and Seldon 1998: 38). In South Vietnam, 3 per cent of landowners owned 45 per cent of the cultivated area and 75 per cent of the rural population were landless (Kerkvliet and Seldon 1998: 38; Quang Truong 1987). After the downfall of French rule, between 1945 and the end of the 1970s, redistributive land reforms were carried out followed by collectivisation and the agricultural cooperative programme which was introduced by Ho Chi Minh's government in 1954 (Hy Van Luong 1992: 7). This led to the abolition of tenancy and hired labour.

Collectivisation, involving a shift of authority over land and labour from the household to a centralised authority in the form of the agriculture cooperative (*hop tac xa bac cao*), restricted household production and controlled markets. The collectives had to buy and sell from the state at official prices, and local officials organised and allocated labour in production groups (*doi san xuat*) which specialised in different types of farming. Official statistics from 1960 state that 86 per cent of North Vietnam's rural population were members of a cooperative, and most of these retained some rights to the land (Kerkvliet and Seldon 1998: 40). In 1969, this percentage had increased to 92 per cent of households (Quang Truong 1987). Until the 1970s, cooperatives tended to be made up of 50 families but, by the late 1970s, the amalgamation of cooperatives into larger units, resulting in one cooperative of about 5000 families per commune, was common (Liljestrom *et al.* 1998: 8).

In the South, the process of land redistribution was more complicated due to the war, and it occurred later. The Viet Minh, realising the significance of control over land, launched a campaign against landlords, many of whom withdrew to the cities; large areas of the Mekong Delta were seized between 1946 and 1953 (Sansom 1970: 55). By the late 1940s, much of the land in the area which had been controlled by the Viet Minh since 1945 was redistributed or rents were not being paid (Fforde 1984: 10; Bray 1986: 194; Kerkvliet and Seldon, 1998: 39). In many of these areas, traditional communal institutions resumed much of their previous importance as the Viet Minh were able to take control even before the Land Reform Law was officially passed in 1953 (Moise 1983). In most areas, land was seized and disbursed to landless peasants. The Land Reform Ordinance 57 (1956), however, only allocated 30 per cent of the paddy in South Vietnam, thus leaving 88 per cent per cent of the population without any benefit (Sansom 1970: 57). This lack of effective equitable land allocation was one of the factors leading to the Viet Cong campaigns in the Mekong Delta between 1960 and 1966 and to the redistribution of land, resulting in two hectares per household. After reunification, the new government built on this feeling in the south by encouraging the move towards land reform (Nguyen Thu Sa 1990; Sansom 1970; Callison 1983). Collectivisation, however, was not favourably received in the south as it was in the rest of the country.

Examining the revolutionary processes of modern Vietnam within the context of capitalism and in relation to the local indigenous setting provides an insight into how the present land-use structure has evolved. The previous account demonstrates that the colonial period resulted in huge social and economic shifts due to the development of tenancy property and capitalist markets that came into conflict with traditional communal action. This commercialisation of agriculture and new systems of tenancy and taxation threatened the stability of the peasant and caused fierce opposition. These changes meant that much of the wealth and access to resources was now outside the control of the village, and, for example, the loss of village lands undermined the protective service of the community. 'The commune, no longer owning communal land, no longer assures, as it had traditionally, assistance to inhabitants who find themselves in need' (Dumarest 1935: 206).

The progressive loss of communal control over local common resources such as forest, wasteland and pasture was detrimental to the village economy and placed a heavy burden on the peasants. This loss occurred due to a combination of reasons ranging from over-population pressure, to colonial legislation which authorised local officials to claim communal lands and to colonial taxes which meant there were no longer free rights to products. Murray (1980) claims that the maintenance of the protective functions of the village was essential for the wage labourers from rural areas who were compelled to migrate to find work in the urban centres of production while the village acted as a social insurance system and was, therefore, an important part of the capitalist system. He suggests that it was a deliberate policy of the French to maintain some features of the village in order to support the migrant labour system. The patterns of land tenure, class relations and village organisation which developed after the French arrived emerged in reaction to colonial rule and the influences of the capitalist world.

Based on events in late colonial Burma and Vietnam, Scott's (1976) seminal analysis of peasant-based resistance to colonialism sees the peasant unrest as involving a notion of social justice and some moral agreement among both rich and poor that all had the right to survive. Scott examined the Nghe-An and Ha Tinh Soviets of 1930–1 and concluded that peasants perceived social justice to include secured subsistence from the landlord classes as a right. This right included sufficient income in return for their labour and certain standards concerning public management of property. Peasant action can, therefore, be explained in terms of a moral imperative. Peasants supported the communists because of their subsistence welfare concerns. This subsistence ethic is at the centre of Scott's analysis of peasant politics. He argues that in circumstances where subsistence options are reduced, the peasants are forced to enter a market economy and hire out their labour. Capitalist conditions in colonial Vietnam created a situation in which the poor felt that basic needs for survival were endangered, conditions ripe for rebellion.

The vast divergence of capitalist imperialism from local tradition in central Vietnam in the 1930s fuelled the ideological support and provided the organisational resources for the revolutionary movement: 'The structures of both the capitalist imperialist systems and the indigenous social formation have shaped historical events in Vietnam. These structures are in turn shaped by historical events' (Hy Van Luong 1992: 19). Hy Van Luong (1992) also suggests that one must not only look at class but at the indigenous socio-cultural setting and the 'rich texture of the native systems of rules and meanings' to understand rural-based nationalism and rebellion in Vietnam during colonialism.

## Forests and forestry

### *History of forest use and policy*

History has seen increasing demands on the forests of Vietnam (Eeuwes 1995: 71). The disappearance of forests has been a long-term process, often accompanying the migration of the *Kinh* whose rice culture has caused the clearance of much lowland forest. The French and lowland colonists encouraged the creation of plantations and the clearing of land for agriculture. Even until the mid 1980s, the assets of land clearance were still being emphasised by the government (Beresford and Fraser 1992: 6). War has been a major cause of deforestation, especially in the South before 1975 and more recently in the Chinese border region. During the

American intervention, 5 per cent of the forest was destroyed and 50 per cent damaged, mostly by defoliation, the negative impacts of which are still being suffered (see Chapter 3). From 1976 to the 1980s, there was a huge impetus for highland and wetland reforestation.

After 1959 in the north and 1976 in the south, the cooperative movement was at its strongest. Between the 1960s and 1980s, forest policy aimed to help build up the socialist state, and forestry production was organised in forestry enterprises or under the local district administration. Between 1977 and 1981, this process was characterised by centralised investment and administration over large-scale exploitation or monocultural plantations. Investment for afforestation came from the State and the forest enterprise. A high rate of success did not profit the Forestry Enterprise, and its management did not provide any sort of motivation to its workers (To Dinh Mai 1987: 132 and 1991).

Deforestation has been identified by the Vietnamese as the main environmental problem facing the country together with the associated issues of soil erosion and the purification of water supplies (Beresford and Fraser 1992: 4; see Chapter 3). Of the rural population, 80 per cent depend on agriculture, fishing, forestry and resource-based activities (Irvin 1996: 192). Figures of forest cover and deforestation rates vary between sources, but typically they reveal vast destruction, particularly since the 1940s. Figures on the change in forest cover range from 43 per cent (Vo Quy 1997) to 65 per cent (Irvin 1996: 92) in the mid 1940s to a cover of 27 per cent (Irvin 1996: 192) to 28 per cent (Do Dinh Sam 1994) in the mid 1990s with an estimated current rate of 200 000 hectares a year (Economist Intelligence Unit 1996: 40). The main cause of destruction since 1975 has been slash and burn, agricultural clearing, small-scale logging and state logging activities (considerable despite Vietnam's low degree of integration into the world market).

In areas where forestry enterprises took over the land, households often remained as the unofficial managers. Official user rights were given back to households with the official reallocation of forest land to the communes therefore sanctioning customary rights to forest land (Liljestrom *et al.* 1998: 27). After the creation of the socialist state came the announcement that all land, forest and water belonged to the people, that is to the State, which in effect meant that traditional users were no longer free to use forest resources. However, the proclamation that resources belong to the people was interpreted by farmers as meaning that the resources belonged 'to the pagoda' and, therefore, were available to all (Liljestrom *et al.* 1998: 18).

Traditional practices of shifting cultivation have been heavily threatened by government policy. Since 1968, there has been a strong government endeavour to stop slash and burn and to focus on the preservation of long-term fertility. In 1986, the Ministry of Forestry forbade shifting cultivation in 'protected' and 'special use forest' (Vietnam Ministry of Forestry 1987, 15–20) and the National Plan for the Environment (see Chapter 3) showed concern over shifting cultivation and tried to constrain such movements in the affected areas (Irvin 1996: 195). The 1991 Forest Law restricts, in effect, shifting cultivation because it only assigns forest land for stable long-term use (Eeuwes 1995: 67). The introduction of market economy has seen the clearance and selling of many traditional shifting cultivation areas (Do Dinh Sam 1994).

In the case of the four forestry enterprises examined by Liljestrom *et al.* (1998), all had received areas of land far greater than they could manage. There was much destruction and many unresolved conflicts of interest between the local people and the State forestry. During the mid 1980s and early 1990s, workers in the agricultural cooperatives and the forestry enterprises encountered hardship and worry about an insecure food supply (Liljestrom *et al.* 1998: 244). When the State stopped supplying subsidies, the forestry enterprises had to produce their own income which led to over-exploitation and, ultimately, the need to stop all production. Forest Protection legislation in 1991 set a prohibition on log exports and in 1993 there was a blanket ban on the export of timber products, which was later dropped.

In the 1980s, with the economic situation worsening, land degradation became a significant issue, with the state particularly anxious about forest protection because of its function in watershed protection. Important transformations took place in the late 1980s in forestry planning as the objective shifted from that of production to 'rehabilitating and developing the forest resource based on socio-economic points of view and environmental protection' (Nguyen Quang Ha 1991: 2). This represented a move from 'traditional' to 'people's' forestry, shown clearly in a statement by a Ministry of Forestry official that 'deforestation cannot be stopped without the participation of the local population' (To Dinh Mai 1987: 135). National administrators began to deal with powerful local autonomy differently and local revisions of forestry guidelines were no longer seen as a difficulty (Fingleton 1990: 15). In 1990, reformulation of policy at the district level was said to be 'most suitable' (Vietnam Ministry of Forestry 1990: 33).

In 1988, a plan to allocate two hectares of forest lands for five years for management by households was developed, giving the households a 15–20 per cent share of the harvest value. Contracts which were signed with the cooperatives were now transferred to the villages. Two categories of forest lands are now disbursed to households by the People's Committee: cooperative-held forest and forestry enterprise forest. The cooperative forest tended to be less productive than the Forestry Enterprise areas. In 1993, the government permitted longer-term (20–50 years) transfers of cooperative forest land to households. Allocation was usually based on *de facto* existing land use but agreements seemed to be variable depending on the household's capacity, the size and condition of the forest, forest management, the village policy towards land-holding and products accountable to the village. The state maintained authority over land-use and harvesting procedures.

There have been problems associated with these changes. After allocation was introduced, much of the forest land remained under management of the cooperatives as few households wanted to take on the burden of managing distant natural forest areas. Afforestation requires big investment and households require help with capital in order to survive economically. Land for house plots and home gardens was frequently assigned to those who owned it pre-collectivisation, which in some cases were extensive areas, and therefore the land allocation process intensified the traditional elite groups. Extensive areas of hill land have, therefore, been converted from *de facto* open access to private management. This may be good for the management, but those who do not own land no longer have access to firewood and fodder which was available when the land was under the cooperative system.

Other researchers have found more positive impacts in upland forest areas: 'Everyone we interviewed, well-off and poor alike, said without hesitation that their lives are better off now than they were under the cooperative period' (Liljestrom *et al.* 1998: 325). A study by Terry Rambo and Le Trong Cuc (1996b) concludes that, since 1989, there has been an encouraging improvement both in the condition of the environment and in the rural economy in upland areas. Biological and landscape diversity has increased dramatically especially on hill-slope lands and the forestland has shifted from monocultures of eucalyptus to enrichment with other species. Forest covers the once bare hills, the rice yield has improved and there are now surpluses to sell. The shift from collective to household ownership seems to have resulted in biological and economic productivity in all agro-ecosystem constituents.

## Explaining decollectivisation

### How decollectivisation took place

Even by the early 1970s, there was evidence of flaws in the cooperative system. Ngo Vinh Long (1973) showed that some rural households earned 40 times more than poorer households and that a new class of cadres, 'new local despotism', had developed. Between 1976 and 1980, the economy grew at less than 2 per cent, rice production fell and, in 1978, 1.4 million tonnes of grain had to be imported (Watts 1998: 459). It had become clear that the way in which the cooperatives had applied equal distribution of land had deprived farmers of any incentive because their income no longer depended on the quality or quantity of their work (Liljestrom *et al.* 1998: 8).

Cooperatives had become an ownerless regime and many households, particularly in the south, were unofficially eluding the legal land structure and had short-term land contracts as well as belonging to cooperatives. In addition, the cooperatives began to designate farm work to farmers on a contractual basis (Liljestrom *et al.* 1998: 8). The management committee of the cooperative also had extensive rights to decide the access and use of resources and so more than the allowable 5 per cent of cooperative land was often leased out to individuals (Kerkvliet 1995b: 116; Fforde, 1990: 116). Transformation came from the grass roots and, by opposing the process, villagers were slowly able to bring the system to a halt (Scott 1986): 'Just as millions of anthozoan polyps create, willy-nilly, a coral-reef, so do thousands upon thousands of individual acts of insubordination and evasion create a political or economic barrier reef of their own' (Scott 1986: 8).

The 1978 recession was followed by widespread food shortages, particularly in 1980. In the north, there was a slump in agricultural production and the south suffered a crisis in agriculture after a rapid collectivisation programme in the late 1970s which was not successful (Kolko 1997: 89). Between 1979 and 1980, efforts to increase agricultural production resulted in significant changes for agricultural collectives. The sixth Plenum of the Communist Party introduced the concept of the 'contract system' which would devolve production decisions and some land rights to households. The emphasis of the Plenum was on the industrialisation and modernisation of agriculture. Other issues raised were price reform, the right to retain surpluses, and self-financing systems in state owned enterprises. This development was followed in 1981 by Directive 100 which was the first formal government initiative in favour of the household economy and sanctioned the 'output contract' (*khoan san pham*) system which allowed cooperatives to contract production directly to households on a seasonal basis while maintaining the collectivised farming structure (Beresford 1988; Kerkvliet 1995b; Watts 1998: 459). Under Directive 100, the households provided 70 per cent of the labour and received labour-point credits (Nguyen Huy 1980: 11) while the cooperative did the ploughing, irrigation and drainage and provided seedlings, fertiliser and insecticide (Hy Van Luong 1992: 263). Households were given a two- to five-year contract with the cooperative under which they became responsible for trans- planting, tending and harvesting the crop. Land was allocated to members on the basis of the size and resources of the family and the contractor had to give a certain amount of the harvest back to the cooperative; there was, therefore, an incentive to produce more.

Under this system, rice output grew by over 2.5 million tonnes between 1980 and 1987 and there was a productivity increase of 12 per cent in the north and 16 per cent in the south (Hoang Van Thoi 1999: 48). In some cases, crop yields increased more than 100 per cent (Fforde, 1990: 118). These increases, and the liberating of small-scale farming and other small industries such as crafts and services, increasingly gave farmers capital for other activities (Fforde 1990: 118; Van Arkadie 1993). However, although Directive 100 did represent a fundamental decentralisation of production, state policy between 1982 and 1988 tried to main- tain the cooperative as the nucleus and the central focus in the socio-economic and political organisation of the village (Fforde 1990: 118). Fundamental decisions were still being made centrally, in many cases without considering local prefer- ences and market conditions. The system was problematic in some cases because allocation was based on the amount of labour, which was hard for families with many dependants. As it was also a short-term system, soil productivity deteriorated due to the lack of an incentive for the farmers to maintain fertility (Le Trong Cuc and Sikor 1996).

### Land allocation

Although at first the economy grew, this initial period was followed by economic recession and, in 1985, hyper-inflation, resulting in many redundancies in the public sector (Irvin 1995: 726, 741; Watts 1998: 460). The economic downturn led to new reforms which were announced at the sixth Party Congress in December 1986 in the form of the official policy of *doi moi*, an agenda of agricultural-led growth and economic renovation (Vo Nhan Tri and Booth 1992; Watts 1998: 460). The process of *doi moi* involved the reduction of subsidies to state enterprises and local cooperatives and the recognition of private industry (Forsyth 1997; Watts 1998: 460). Since 1986, cooperatives have been progressively abolished as the basic administrative unit and the country has undergone a transition from a traditional socialist economy with central planning to a market-orientated form of socialism (Beresford and Fraser 1992: 3; Forsyth 1997).

The 1988 Land Law and Decree 10 effectively dismantled the agricultural collectives by allocating land to peasants for 10 to 19 years, supplying credit and developing a marketing system. This encouraged those who were the most productive and persuaded farmers to invest in intensive cultivation (Hy Van Luong 1992: 37; Kolko 1997; Liljestrom *et al.* 1998: 10). Decree 10 replaced the compulsory grain purchase quota with free market pricing systems which meant that farmers no longer had to sell a large part of their output to the state at a low price. The household became the basic unit of production and the cooperative was no longer the monopoly supplier of inputs (Economic Intelligence Unit 1996: 37; Kerkvliet and Seldon 1998: 37; Watts 1998: 454). Land allocation depended on the total number of household members, and rights of inheritance and transfer were recognised (Kolko 1997).

There was some opposition to the reforms at this stage because they were considered 'ambiguous' (Kolko 1997: 90) and would, it was feared, lead to unacceptable disruption, with three quarters of the workforce in agriculture and, to some extent, dependent on the cooperatives for their security. For farmers in the poorer central and northern areas, the cooperatives acted as a safety net and produced a united rural system which provided for the needs of most (Kolko 1997: 89). Over five million veterans and cadres, and more than a million soldiers, benefited from the cooperative system during the late 1980s and the changes also involved the removal of food grain subsidies given to government workers and army veterans.

Decollectivisation advanced rapidly in the Mekong Delta but cautiously in the north where a high population density meant that collectivisation was seen as important for social stability as it reduced the risk of the concentration of land in a few hands (Kolko 1997: 90). In the north and centre of Vietnam, many land-use rights have traditionally been collective in order to diminish the risk of extreme social differentiation (Fforde 1993: 313). In many senses, cooperatives had reverted to these traditional functions. In the south, where these issues are less important due to lower population densities, farmers' access to land after 1989 seems to have returned to the direct possession of user rights; cooperatives were abandoned and farmers paid tax to tax stations rather than to cooperatives as in the north (Fforde 1993: 313).

Under the 1992 Constitution, all land was said to belong to the state but 'land is allocated by the state to organisations and individuals for stable long-term use'

(Vietnam Government 1992). The 1993 Land Law, executing the code set out in the 1992 Constitution fully, recognised private agriculture. This law extended peasants' land use rights to 20 years for annual crops and to 50 years for long-term crops, and it defined the rights of individuals to build on land (Kolko 1997: 90). It allowed for the issuing of land lease certificates for rural and urban land which can be sold, leased, inherited and mortgaged or used as collateral (Irvin 1996: 184). This clearly turned land into a commodity and the ability to trade in land encouraged land to be concentrated in fewer hands. By 1996, the Party had thoroughly altered the whole land system – and, indeed, the nation's social structure.

The four million hectares of land scheduled for reallocation after 1988 has resulted in a significant transformation of rural society. The disbanding of the cooperatives means that the ownership of land is now crucial for farmers. In areas such as the Mekong Delta, tens of thousands of people have mortgaged land and have no land to cultivate. Many peasants have been marginalised and hired labour has once again become a major component of the rural economy and social structure. A marked outcome of the rapid move towards a market economy has been increased inequality (General Statistical Office and United Nations Development Programme Survey 1997–8, reported in Vietnam News Agency 1999). Nevertheless, for many, decollectivisation and increased market access has opened up substantial new production opportunities (Hirsch and Nguyen Viet Thinh 1996: 176). The issues surrounding the process of land allocation are discussed further in Section 2.

## Conclusion

The historic tension between Confucianism and Buddhism described by Jamieson (1993) in his 'yin' 'yang' analogy can be extended to cover the history of environment and development in Vietnam. I argue in this chapter that there is a constant tugging between strong historical traditions and capitalist forces, evident in every era from Chinese colonialism to the modern day. The land issue has been central to the nature of development in Vietnam. The context in which land reform is presently taking place, and shaping the response to environmental change, is coloured by a diverse range of historical and environmental, geopolitical and cultural features that characterise the regions and ethnic groupings of Vietnam.

The process of capitalist penetration has affected the landscape in very divergent ways to colonialism, while post-colonial land reforms have been carried out in radically different contexts. It is still unclear therefore whether the theories discussed above to explain the political upheaval and the persistence of various social organisations under colonial and pre-colonial rule are relevant today in the current period of post-collectivisation. What is clear, however, is that the breakdown of collectivisation cannot be likened to the historical erosion of traditional communal structures. The process of collectivisation, although influenced by communal aspects of Vietnamese culture, was placed within a very specific context of colonial and political struggle. This chapter provides strong evidence that the uniquely Vietnamese discourse on tradition, hierarchy and innovation needs to be explicit when considering resilience and adaptation in the context of present-day social and environmental change.

# Part 2

# The natural resource base

The chapters in this section of the book examine social and environmental change in key natural resource areas for Vietnam: land use, forestry and agriculture in the lowlands and uplands, mangroves, water and coastal resources. They detail how farmers, fishers, local to national government officials and other actors have adapted to the changing political and social environment over the past decade and the implications of this evolution for the environment itself. These chapters, therefore, shed light on the changing resilience and vulnerability of particular groups in the rural sector over this critical period. The issue of right, or of access, to resources provides a focus for a number of these discussions, echoing and illustrating the arguments presented in Chapter 2.

# 5 Property rights, institutions and resource management: coastal resources under *doi moi*

*W. Neil Adger, P. Mick Kelly, Nguyen Huu Ninh and Ngo Cam Thanh*

## Introduction

Changes in property rights can radically alter how individuals and groups cope with external change. In Vietnam, the system of governance and property laws has changed dramatically during the 1990s from central planning towards a market-oriented economic policy. Part of this trend has resulted in the effective privatisation of productive assets such as agricultural land, forests and marine resources. These resources were previously managed through state organised cooperatives. Often, local government overlaid traditional common property rights regimes, resulting in hybrid complex organisations and property regimes (see, for example, Hy Van Luong 1992). As Chapter 4 has demonstrated, prevailing property rights regimes during the 'collectivised' period and their subsequent evolution were, indeed, local compromises in ongoing resource allocation struggles for land and livelihoods.

The distribution of resources is a further important parameter in determining social vulnerability and resilience. The distribution of benefits from common and private resources is important in determining whether common property resources, within the hybrid systems of Vietnam, can be effectively managed. Common property management literature suggests that the impact of rising inequality may have ambiguous impacts on the efficiency of resource use (Baland and Platteau 1997; 1998).

The chapter proceeds by outlining recent theory surrounding common property resource management, including necessary conditions for the persistence of common property regimes and causes of conflict in management. The chapter introduces the political and cultural context of property rights in coastal Vietnam and the implications of recent change. In common with many other parts of Southeast Asia and elsewhere in the world, the conversion of mangroves is causing detrimental impacts to the local environment, as well as to local residents dependent on fishing and related sources of income.

In addition to ownership and control issues within agriculture, there has been continued regional development planning to ensure widespread dispersal of economic benefits. Some of these policies, for example, in the Red River Delta, have resulted in attempts to meet the dual objectives of dispersing population and economic growth, as well as encouraging economic development through the diversification of agriculture into other activities such as aquaculture. This has entailed conversion of coastal wetlands dominated by mangrove species from collectively managed resources into enclosed areas for agriculture and aquaculture that are then allocated on a private basis to either local residents or other concessions. Such conversion undermines the existing common property resource management systems of these areas which have evolved over many decades, if not centuries, and which have generally allowed and encouraged sustainable management of the mangrove resources. The details of a case study of mangrove conversion are outlined with reference to a particular instance in Quang Ninh Province where, in one area 1900 ha of traditionally managed mangrove forest are presently being converted for use in agriculture and aquaculture.

## Property rights, their evolution and sustainability

Property rights refer to the bundle of entitlements defining the specified opportunities for individuals or groups to privileges regarding limitations on the use of a resource. Under capitalist systems, these rights are often vested in individuals while under socialist systems rights tend to reside with the state. But this dichotomy is stylised. In capitalist societies, governments provide public goods and retain property rights to certain resources – national parks, for example. Similarly, socialist states have never completely nationalised all rights to all resources. Some agricultural production, for example, effectively belonged to individual producers under collectivised agriculture in Vietnam. Property rights are universally complex, ranging from rights allocated to individuals through to sophisticated common property regimes and to open access resource use where property rights effectively do not exist (see Bromley 1991; Sandler 1997). Property rights to productive resources in Vietnam are partly vested in the state, partly collective property rights, and increasingly, through the 1990s, rights have been allocated to individuals.

How sustainable are these property rights regimes? The desirability of one type of property rights (private, state or common) to any other on environmental and sustainability grounds is not proven. Ostrom *et al.* (1999) argue for an open mind rather than an ideological approach in analysis of the desirability of alternative property rights. They argue that both government ownership and individual ownership are subject to failure in particular circumstances. They give the example of rangeland management in Siberia, Mongolia and northern China to illustrate these potential problems. Mongolian grassland in recent decades has been much less degraded and remains more productive than that of its neighbours (Sneath 1998). This observation is explained by the appropriateness of the property rights. Siberian grassland is managed in state collectivised farms, while China has moved to a privatised system where individuals have grazing rights. Ostrom *et al.* (1999) argue that the management of Mongolian grassland has facilitated more sustainable resource management than that of Siberia or China – the Mongolian system allowed pastoralists (till more recent changes – Mearns

1996) to continue their traditional group property institutions including movement between seasonal pastures. Thus, collective management based on traditional knowledge (in harmony with evolved cultural constraints on over-exploitation of the environment) can, in certain circumstances, be the most sustainable option.

Common property, in this context, is defined as that property owned, controlled and used by a group of individuals together, and hence contrasts with either private property or state-regulated property. The persistence of common property resource management has been hypothesised to be the result of a complex interaction of the resource, the technologies of enforcement, the relationships between resources and user groups, the features of the user group, and the relationship between users and the state and legal system. Under this formulation, the likelihood of successful management is affected by a number of factors (following Wade 1987; Ostrom 1990; and others).

First, the boundaries of the physical resource should be defined. The more clearly defined, the greater the chances of successful commons management. This is known in economic terms as universality. Second, if the users are resident in the location of the resource then this increases chances of success through reducing enforcement costs. Enforcement and other transaction costs are weighed against the benefits from the resource. Third, the greater the demand for the outputs and the reliance on the resource within a livelihood system the greater the chance of successful common property management. Fourth, the better defined the user group, the greater the chance of success. But conflicts can still occur between small groups of users. The congruence between appropriation and provision rules and local conditions is a critical prerequisite for successful management.

Cultural influences and power relationships determine the enforceability and rights of individuals under common property regimes. Collective management can, however, break down even where well defined and enforced. External environmental shocks or, more frequently, changes in the institutional or economic environment, including privatisation and state intervention, can cause such break-downs. One common theme in observations of common property management is the role of increased market penetration and skewed distribution of resources in causing regime breakdown. As Shanmugaratnam (1996) has observed, management regimes for natural resources:

> come under strain and may even completely break down as inequalities and divergence of interests among resource users increase along with the marketisation of the rural economy. . . . It would seem that sustainable common property resource management is more difficult to achieve in a community with highly uneven, than in one with a relatively better, distribution of private wealth.
>
> (Shanmugaratnam 1996: 166)

Quiggen (1993) argues, however, that common property management is not necessarily 'equitable'. Analysis of common property management regimes often 'carries a baggage of utopianism' (Quiggen 1993: 1135) in this respect, whereby they are assumed to be more equitable. Further, it is argued that greater inequality in some circumstances favours the provision of common public goods (e.g. Bardhan 1993). By contrast, the analysis of Baland and Platteau (1997) shows that, although wealthier individuals inevitably contribute more to collective good provision, it cannot be inferred that inequality encourages greater efficiency. Indeed, this is particularly the case when common property forms only part of a livelihood system such as is common in agriculture in Vietnam (Quiggen 1993).

Much of the theoretical literature on property rights and the management of resources uses stylised models of institutions, focusing on incentives to individuals to cooperate and use resources. Most natural resource management systems, particularly those associated with present-day agriculture in Vietnam, are in fact 'mixed' in terms of their institutional arrangements. In other words, household livelihoods are characterised by income from both private resources as well as from collectively managed resources. The underlying structure of inequality may then be related to the distribution of private property income. It has been argued that pure collective agricultural systems, such as was experienced in northern Vietnam over the past two decades, promote equality compared to 'mixed' systems (Putterman 1983).

The list of necessary conditions for the persistence of regulated or unregulated common property resource management outlined above encompasses both the biophysical and institutional factors and economic aspects of their use. But they are not prescriptive for how to initiate newly forming property regimes. This is the challenge facing Vietnam as it implements Decree 327, the 1993 Land Law and other regulations that move the society towards private property.

The history of land and resource use in Vietnam over the past millennium is dominated by the overarching political structure and by cultural issues. Collective ownership and management of resources has been prevalent since the large-scale settlement and development of agricultural infrastructure in both lowland Vietnam and the higher lands (see Chapter 4 and Bray 1986; Wiegersma 1988; Le Trong Cuc *et al.* 1990). Various explanations for the prevalence of the collective form of ownership and action have been proposed, including the cultural resonance of the Vietnamese to this manner of resource use, the deterministic nature of the physical environment and the technological necessity to share common resources.

The traditional collective property rights of lowland agriculture have carried over into the most recent formulations of collective agriculture and resource use in Vietnam. In the past two decades, resource ownership and control by the state has been tempered with reference to traditional collective action on particular aspects of resource management. District- and commune-level institutions form part of the common property resource management system outside their traditional roles of allocating resources within agriculture and the other marketed economic activities of the localities. Resources falling into this non-traditional management category include access to fisheries, forests in the uplands, and other coastal resources such as mangroves. The role of the local state organisations in this respect is implemented in a number of ways, including the provision of spaces for resource users to make decisions on the common resources and on resources for exclusion.

Property rights surrounding land resources in Vietnam followed the evolutionary path detailed in Chapter 4. Fisheries resources are, however, fundamentally different in nature and have complex management structures associated with them. Some intertidal resources display some characteristics of both static terrestrial resources and fugitive marine resources. Coastal wetlands, for example, include aspects of management of marine or aquatic resources, as well as of systems that operate for land or terrestrial resources. Essentially, wetlands exhibit the nested rights described for many forest and other resource management situations. Adger and Luttrell (2000) argue that specific aspects of wetland characteristics include the multiple-resource nature of the ecosystem, the joint-production nature of many of these resources, and the seasonal alteration of wetland resource use.

In Vietnam, fisheries have undergone abrupt and radical management changes. Traditional management, as surveyed by Ruddle (1998), is based on the shrine system (*van chai*) whereby each local ancestral shrine in each village laid down the rules for fishing. These included rules for acceptable technology and gear to be used, mutual assistance among fishers and disposal of catch and profit sharing. In addition, the shrine dictated conciliation of conflicts and sanctions against those breaking the rules. This shrine-based management varied from locality to locality. The French colonial period had little impact on the governance of fisheries, although the colonial government designated *de jure* nationalisation of all fisheries. In the reunification period after 1975, the coastal areas of north and south came under the system of fisheries governance developed in the north, that of village-level cooperatives. The cooperatives owned all the fishing gear, subsidised by central government, but, with few markets for the catch, the fisheries developed slowly (Ruddle 1998). During this period, the government did not pay much attention to traditional management of fisheries, but, in parallel to other changes in property rights, traditional systems are now re-emerging as resilient management alternatives. In the words of Ruddle:

> because the salient characteristic of traditional management systems in Vietnam is regulation of inter-relationships among fisheries stakeholders, within the framework of the strong moral authority of the community shrine, rather than the governance of fishing and the fishery *per se*, the core of the systems has proven remarkably resilient.
>
> (Ruddle 1998: 3)

Local state organisations are not, at present, exclusively dominant in the role of resource management. Other traditional institutions play a role in common property resource management in the present era of decollectivisation. Indeed, the role of informal 'civil society' within resource use in Vietnam has been increasingly recognised in the debate concerning village state relations (Hy Van Luong 1992; Kerkvliet 1995b). Many reforms in Vietnam in the *doi moi* era have

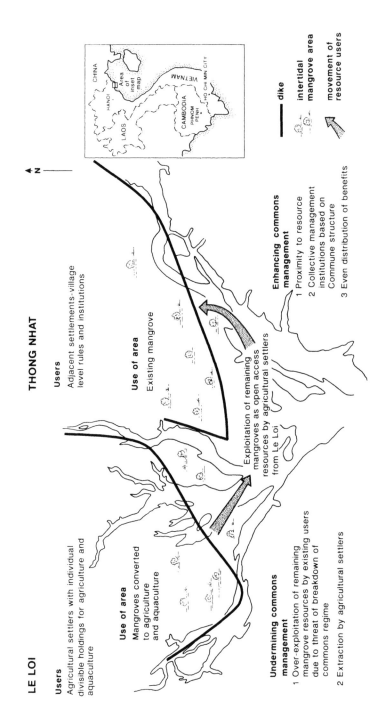

*Figure 5.1* Property rights regimes and conflicts in mangrove conversion in Quang Ninh Province, northern Vietnam

at least in part risen from below rather than been directed from central government (Fforde and de Vylder 1996). Similarly, many production arrangements in agriculture have evolved at the village level, even in the collectivised period, worked out between villagers and local officials (Kerkvliet 1995b). Thus, property rights and access to resources in Vietnam exhibit complex forms: there are examples of state, private and collective resource management and various hybrids of the forms. The following section examines the implications for social resilience of rapid changes in existing property rights in coastal areas.

## Property rights changes in coastal resources

### *Bac Cua Luc wetland*

In Quang Ninh Province, the conversion of mangroves for alternative uses is common, as elsewhere in Vietnam and in other areas of Southeast Asia. This section examines a particular case of property rights change. A government-subsidised 'land reclamation' scheme, in an area called the Bac Cua Luc wetland in Hoanh Bo District, Quang Ninh Province, involves the enclosure of two distinct areas of mangrove wetlands totalling 1900 ha. The first area comprises 1200 ha of mangroves and lies adjacent to the Commune of Le Loi, and the second comprises 700 ha adjacent to Thong Nhat Commune (Figure 5.1). The policy leading to this and other land reclamation schemes is known historically as 'New Economic Zones' (Thrift and Forbes 1986). New Economic Zones promote resettlement and agricultural development in both lowland and highland areas. In coastal areas, resettlement often involves aquaculture development and conversion of mangroves. This is one of the major environmental conflicts in Quang Ninh Province (see also Bach Tan Sinh 1998). The case of Bac Cua Luc discussed in this section demonstrates the potential impacts of conversion of these coastal resources when rights to the mangrove area are overturned and conversion takes place through government subsidy, at the resulting expense of local inhabitants.

In the Bac Cua Luc project, the construction and land reclamation has been undertaken through a grant from central government. The aim is to resettle agricultural households from the densely populated agricultural areas of the Red River Delta to Quang Ninh through allocating the reclaimed agricultural land. In addition, as payment for undertaking construction, one third of the area is retained by the construction agents, and is currently leased by them or sold off to private individuals on a ten-year lease. The leasehold areas for aquaculture range from less than 1 ha to about 100 ha.

### *The economics of reclamation*

Appraisal of the impacts of mangrove conversion on ecosystems and property rights focuses on two related aspects. First, there are economic arguments, couched in efficiency terms, for economic 'development' based on this wetland conversion. Traditional cost benefit analysis demonstrates the efficiency gains or losses in resource use in such a change, though the external impacts of loss of ecosystem functions are often ignored. The second unaccounted for impact of conversion is the impact of disruption of property rights on the institutions of common property management. It is argued in the following section that privatisation in the Ba Cua Luc case undermines the sustainability of common property resource management.

The economic rationale for the conversion of mangroves is based on the argument that the increased economic output from the use of the converted land to agriculture and aquaculture is greater than the costs. The costs arise directly from reclaiming the land and from the opportunity cost of foregone income from present use of the mangrove area in extractive uses such as fishing. In the Bac Cua Luc case, the costs of constructing the dike have been subsidised through the Ministry of Planning and the Ministry of Science Technology and Environment, with capital costs approaching VND 6 billion (US $0.5 million) spread over a six-year period. The productivity of the reclaimed area is based on the future income streams deriving from agricultural and aquaculture output. The cost of under-taking the infrastructure development, that of dikes, sluice gates and the construction of fresh water supplies for agriculture, are also included. Each of these costs and benefit streams occur at different points in time, with the capital costs of construction incurred for a number of years before the agriculture becomes viable. The economic viability, therefore, depends on the implicit rate of discount of future costs and benefits against present costs and benefits.

One of the major driving forces behind the conversion of mangroves is the short-term profit to be made on extensive aquaculture, though benefits often reduce rapidly after only a few harvests. A critical question regarding conversion of mangrove forests to aquaculture concerns, then, the sustainability of the resulting yields of shrimp. In many cases, conversion in Vietnam has resulted in high yields for the first few crops but then decreasing yields due to deterioration of the aquaculture environment through loss of nutrients and soil acidification resulting from insufficient water exchange (Phan Nguyen Hong and Hoang Thi San 1993). These problems are, to some extent, the result of poor design and management and it may be possible with more effective planning to reduce these difficulties in scale (Dierberg and Kiattisimkul 1996). Nevertheless, unsustainable yields characterises the experience of aquaculture in many parts of the world and there are, at present, few examples of successful management. Conversion to agricultural land in Vietnam has encountered similar problems with declining yields as oxidation occurs during long dry periods with strong sunlight and soils degrade (Phan Nguyen Hong and Hoang Thi San 1993). Both aquaculture and agriculture are susceptible to storm impacts, exacerbated by the loss of mangrove, with shrimp farming at considerable risk given its exposed location.

Considering the wider context, the mangrove ecosystem plays an important role in maintaining the well-being of the coastal zone and its inhabitants, providing a range of functions and services (see, for example, Mitsch and Gosselink 1993). Its destruction can have wide-ranging external consequences, including the deterioration of fisheries further afield as spawning grounds are destroyed, the displacement of feeding and breeding grounds for birds, and so on (Barbier 1994). In addition, mangrove forests also provide an effective form of coastal protection, reducing the impact of incident cyclones and lowering dike maintenance costs (see Chapter 9). While tropical cyclone impacts are not a frequent problem in the relatively sheltered area of Bac Cua Luc, severe storms do affect the region (see Chapter 10) and the loss of the mangrove ecosystem will lead to increased impacts in the coastal zone. These functions and services have, it is widely recognised, a positive economic value, even though, as noted above, this value is rarely taken account of in decisions regarding conversion.

Economic analysis of the costs and benefits of conversion of the mangrove forests of Bac Cua Luc demonstrates consequences for social resilience when effects on distinct stakeholder groups, agricultural settlers and present extractive users, are examined (Adger *et al.* 1997). On the basis of data provided by government agencies and a household survey of households in Le Loi and Thong Nhat Communes,[1] there is no economic case for conversion – the costs outweigh the benefits. Moreover, poorer households are more dependent on mangroves and the richer households benefit from conversion. Mangrove conversion exacerbates the maldistribution of resources, already skewed by land allocation (Ngo Vinh Long 1993; Hy Van Luong and Unger 1998; Watts 1998; Adger 1999a). It has been shown that the mangrove resources are important in the overall livelihood system of the coastal communities, representing 13 per cent of total consumption at the household level of marketed and non-marketed economic activity (Adger *et al.* 1997). Clearly, resource dependency can lead to low levels of resilience at the household level: when the resource disappears, the impact on household livelihood security can be significant.

### The impact on mangroves as common pool resources

What of the impact of mangrove loss on the institutions of resource management and social resilience? The resilience of the communities dependent on the Quang Ninh mangroves is affected by their loss. The ability of the community to maintain sustainable common property management of the remaining mangrove and fishing areas is undermined by the changes in property rights and changes in inequality brought about by externally driven enclosure and conversion. The pre-requisites for common property management to be successful discussed earlier in the chapter include a reliance on the resource involved within the livelihood system and a relatively homogenous distribution of benefits within the user group. Thus, resilient common property management is enhanced as users cooperate on the basis of a relatively equitable share of the benefits of use and the critical role of the resource in their livelihood stability. In terms of the stability of income sources and the perceived legitimisation of the imposed property rights changes, resilience of the existing local social systems would appear to be undermined as conversion occurs.

We have suggested that the Bac Cua Luc mangrove conversion scheme is not easily justified in terms of direct economic efficiency and equity arguments.

*Table 5.1* A comparison of resource dependency and aspects of resilience for settlers and mangrove extractive users, Quang Ninh Province

|  | *Present extractive users* | *Agricultural settlers* |
|---|---|---|
| Mangrove dependency | >13 per cent of livelihood sources dependent on mangroves.* | Low dependency on mangroves. Implementing mangrove conversion to aquaculture with higher risk. |
| Assets | Leased agricultural holdings of 20–50-year leases under the 1993 Land Law (mean 0.42 ha).* | Leased agricultural land and aquaculture plots (>5 ha for aquaculture). Also 20–50-year leases. |
| Property rights and social institutions | Household land allocation has led to effective privatisation of mangroves. Breakdown of common pool resource management in remaining mangrove since partial conversion. | Exclusively private holdings. Isolated from existing social institutions of settlements. |

Source: Adapted from Adger *et al.* (1997).

Note

*Imputed income and land areas based on representative household survey (n = 141).

The change in property rights and use in Bac Cua Luc also directly affects the livelihoods of those people who previously utilised the mangrove resources, and additionally has negative impacts on management of the remaining commonly managed resources, particularly those mangrove resources adjacent to Thong Nhat Commune. These aspects of privatisation have been examined through analysis of information regarding directly affected households in nine villages in the two Communes in Figure 5.1. There are stark differences in household characteristics and income between those using the Bac Cua Luc mangrove area and those not using it at present.

Social resilience at the community level is also affected by property rights changes. Research among the households most affected by conversion shows that, due to the loss of part of the mangrove resource, there is enhanced conflict over remaining resources (Table 5.1), leading to non-cooperative exploitation of the mangrove fisheries. In addition, conversion to agriculture and aquaculture in this case increases income inequality within the population, thereby reducing the likelihood of cooperative action within a heterogeneous community.

The high reliance on fishing as an income source for the user group makes sustainable management of Bac Cua Luc mangroves more likely, because, for example, critical dependency on the resource on the part of its users is an incentive for common management to take advantage of economies of scale. This is backed by analysis of the households that are engaged in managing the mangroves, showing that not only did the Bac Cua Luc mangrove ecosystem form a critical part of their income sources but that these households are the most vociferous in expressing the resulting impacts of loss. For example, two household respondents in Le Loi Commune, who judged that the loss of mangrove resources was proving significant to household income and welfare, stated:

> We have stopped fishing now completely. We used to go fishing at Bac Cua Luc from April till November, but never go fishing any more. Where else can we go? To mangroves of other villages?

> Many people benefited from building the dikes, but they were from other communes and districts. Once the dike is completed the whole environment will be worse. Maybe it is better for the economy, but not for this village.

Income in Le Loi is more evenly distributed than in Thong Nhat (Figure 5.2), demonstrated by the distance of the Lorenz curve from the line of equality. Inequality in income as measured by this method is generally lower in Vietnam than in other parts of the world with similar levels of income, and is indeed a feature of the former centrally planned economies based on their relatively even distribution of assets underpinning the income generation. But the key factor examined is the relative difference between the two Communes. Le Loi, along with being more dependent on income sources from the mangrove areas, also has highly even distribution of income and, indeed, even distribution of land area.

Across both Le Loi and Thong Nhat, commercial farming activities, such as flower production for export, and wages and transfers are contributing more to inequality than to their share of income. Extractive uses from Bac Cua Luc, principally fishing activities concentrated there, and agriculture are contributing less to the observed inequality than to their share of income. This is illustrated in Figure 5.3 summarising the contribution to total income analysis of and the contribution to inequality of each source of income (Adger *et al.* 1997). The move towards privatisation of coastal resources in the agricultural economy in this District is, then, demonstrably enhancing inequality in the distribution of income.

For the mangrove areas, the observed increase in inequality increases the chances of dissent among residents on future direction and management of any resources that continue to exist. Conflict among users is particularly evident regarding the conversion of remaining mangrove resources on the grounds that allocation of common fishing rights and zoning becomes more contentious as the resource disappears. In addition, respondents highlight the negative impact on the locality of the majority of the construction work associated with the conversion scheme being undertaken by 'outsiders'. They also stress the negative implications of settlers coming from outside the District, and even from outside the region.

This case demonstrates that social resilience can be eroded when mangroves are lost. The process of conversion removes or constrains the necessary conditions for the persistence of common property management in this case. The 'disturbance'

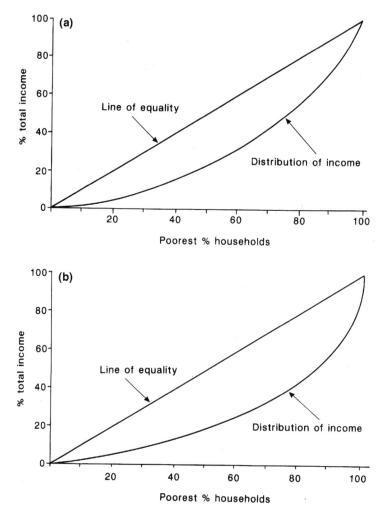

*Figure 5.2* Lorenz curves showing distribution of incomes in (a) Le Loi and (b) Thong Nhat. Based on 1996–1997 household survey, individual basis (n = 141 for full sample; Le Loi n = 89; Thong Nhat n = 52). Distribution of income determined by distance from line of equality

of the institution undermined the social capital of collective management and has resulted in a breakdown of collective action for the remaining resource. This is demonstrated in this study in Vietnam but similar findings have emerged elsewhere – commentators from Central America to the Philippines and Scandinavia have highlighted the negative social and ecological impacts of aquaculture and mariculture (Folke and Kautsky 1992; Kelly 1996). Part of this story from Vietnam is related to the ecological resilience of the system. Aquaculture relies on a narrow

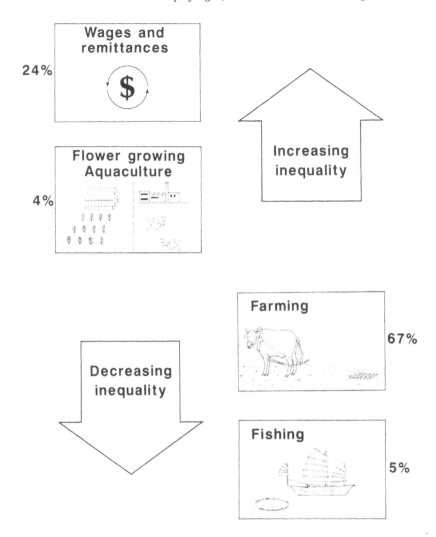

*Figure 5.3* Contribution of income sources to inequality for Le Loi and Thong Nhat Communes. Figures based on sample survey of households – see Adger *et al.* 1997). Percentages are contribution of income source to overall income

range of commercial species prone to pests. In addition, conversion of mangroves to aquaculture ponds actually increases the risk of inundation and coastal flooding. This means that the higher returns, accruing to a smaller number of users, typical of the process of conversion from mangrove to aquaculture occur with less regularity and with a higher variance. Aquaculture entrepreneurs are effectively heavily discounting the future, and often abandon their ponds after less than a decade.

## Conclusions

As mangrove forests have been converted to food production or to commercial aquaculture, so changes in property rights and income distribution highlight changes in vulnerability and resilience. The critical chain of processes can be defined as follows. The move towards effective privatisation of land and other resources enhances the inequality in the distribution of income. The case study from Quang Ninh Province shows that the poorer households rely on the mangrove more for their livelihood than the wealthy, who are more interested in the private commercial utilisation of the coastal resources. The trend in income distribution increases the heterogeneity within the resource user group and lessens the likelihood of cooperative management. Thus, the ability of the community to maintain sustainable common property management of the remaining mangrove and fishing areas is undermined by the changes in *de facto* property rights and income distribution brought about by externally driven enclosure and conversion. At the household level, the poor, dependent, families suffer disproportionately with the loss of the habitat functions and the loss of resilience is self-evident. But, at the community level, there is also a loss of resilience as families compete for the remaining resources, leading to a lack of cooperation in use of the ecosystem. Levels of inequality rise as some households gain in the new commercialism and others disproportionately lose since they have no fallback when the formerly commonly managed resource is denied them. By this means, vulnerability is created and resilience is undermined.

## Note

1 Primary and secondary data were collected in Quang Ninh in 1996 and 1997. Household survey data was collected in the Le Loi and Thong Nhat Communes (Figure 5.1). A randomised sample of 141 households was drawn using commune census and family planning records as the sample frame. The survey details the income sources from 1995, the most recent year when the mangrove resources of Bac Cua Luc were available to the residents of Le Loi Commune (see Adger *et al.* 1997). The survey allows qualitative comparison by the respondents between that year to the situation after the mangrove conversion when much of the economic activity associated with extraction from the mangrove area and its management had ceased.

# 6 Forest land allocation and deforestation processes in the uplands

*Davide Pettenella*

## Introduction

Two thirds of the Vietnamese territory is mountainous, with this area characterised by a large portion of steep slopes and soils subject to erosion, some of which was previously covered by rich tropical forests. Nineteen million hectares, or 58 per cent of the territorial area, is still classified by central planning authorities as forest. This is, however, an administrative classification: only 8.3 million ha, or 25 per cent of the land area, is covered by forests (Table 6.1) while the remainder (10.7 million ha) is, in fact, bare land (Nguyen Van Dang 1997). A large part of the Vietnamese population, estimated to be 9.4 million inhabitants or 13 per cent of the entire population, still depends directly on the use of forest resources and the management of sloping land in mountainous areas (Rambo *et al.* 1995). The future of these mountain dwellers is threatened by the continuous degradation of forest land. Annual deforestation has been evaluated equal to 137 000 ha. The poor and marginal people, mainly members of the 54 ethnic minorities living in Vietnam, are particularly exposed to the negative impacts of this process (Smith 1998).

The macroeconomic and structural reforms initiated in Vietnam in the 1980s with the *doi moi* policy have stabilised the economy and yielded a remarkable level of economic development. Per capita income growth has been in excess of 5 per cent per year (World Bank 1997b) for most of the 1990s, although in 1996 and 1997 there were signs that the economy was beginning to slow down. Despite this overall growth rate, half of the Vietnamese population is still classified as poor and income disparities between the urban and rural populations are widening; average rural income was barely one fifth of the urban income in 1995. The gap is even more serious between GNP per capita in the two rural delta areas and the mountainous regions where the forest land is mainly concentrated.

The success of the recently introduced policy of agricultural land allocation has induced the Vietnamese government to allocate forest lands to individual households (see Chapter 4). Officially, the aim of the forest land allocation policy is to help farmers to solve their food security problems and, as a consequence of the improved living conditions, to gradually abandon unsustainable farming activities. This, it is hoped, will preserve and, where possible, enlarge the forest areas

*Table 6.1* Socio-economic and physical indicators for the forestry sector in Vietnam

| | |
|---|---|
| Land classified as forest | 19 000 000 ha (57.6% of the land area) |
| Forest cover | 8 312 000 ha (25.2% of the land area) |
| • Tropical rain forest | 2 894 000 ha (34.8%) |
| | Annual deforestation: 47 700 ha |
| • Most deciduous forest | 3 382 000 ha (40.7%) |
| | Annual deforestation: 55 700 ha |
| • Dry deciduous forest | 952 000 ha (11.5%) |
| | Annual deforestation: 15 700 ha |
| • Hill and mountain forest | 1 084 000 ha (13%) |
| | Annual deforestation: 17 900 ha |
| Total annual deforestation | 137 000 ha (1.6% reduction of total area) |
| Closed broad-leaved forest | 4 946 000 (59.5%) |
| Forest stock | 1 523 560,000 t (183 t/ha) |
| Forest area/population | 0.11 ha per capita |
| • Average logging intensity | 30 cm/ha |
| • Logged area 81–90 | 26 000 ha (new) + 73 000 ha (previously logged) |
| Plantations | 2 100 000 ha (6.5% of land area) |
| Annual plantations | 70 000 ha |

Sources: Food and Agriculture Organization (1993) and other Food and Agriculture Organization publications.

Note
There is no accordance among different sources of data on the forest resources (see, for example, data from the Cadastral Department and the Forest Protection Department presented in the Vietnam Ministry of Agriculture and Rural Development and Food and Agriculture Organization (MARD–FAO), 1998).

(Nguyen Quang Ha 1993; Vietnam Ministry of Agriculture and Food Industry 1993; Cameron 1994; Ahlback 1995; Poffenberger 1998).

Unlike the situation for agricultural land, there are key issues constraining the reform of the traditional forest land tenure system, issues that lead to negative social and environmental impacts on mountainous areas (Prosterman and Hanstand 1999). This chapter examines the implications of the present ongoing forest land allocation process for forest resource use and the potential social and environmental consequences. First, some background information regarding the experiences of land allocation to individual households in Vietnam is outlined. Second, the constraints and bottlenecks of forest land allocation and their possible impacts on local environmental and social systems are considered. It is argued that the relevant role of externalities in land allocation, the transaction costs involved in property right reform and financial and institutional constraints are having negative impacts on the efficient allocation of forest resources in Vietnam. The chapter draws on the results of field research in Thua Thien Hue and Quang Ninh Provinces in the central and northern highlands, respectively, and on other recent reviews, field research and policy pronouncement regarding the issue of forest land allocation.[1]

## The land allocation policy

Vietnam's land reform programme has changed the structure of the agricultural sector through a process of dismantling the collective system in favour of a system based on individual families with a guaranteed tenure security (Barker 1994). After the revision in 1993 of the 1988 Land Law, the process of agricultural land allocation in the Mekong and Red River Deltas has developed quickly bringing about a spectacular increase in land productivity. In a few years, Vietnam has developed from the position of net importer to that of the second largest exporter of rice in the world.

The most recent regulations under the 1993 Land Law allow paddy land to be allocated for a period of 30 years, based on the number of members in each household. Land classified as forest, with or without tree cover, is allocated either to a variety of institutions or households (as outlined in Table 6.2). In the latter case, land allocation is based not on family size as with agricultural land but on the willingness and the financial ability of the household to reforest the land, if required, and to manage it. If a household receives land where some forest amelioration has already been carried out, the household is obliged to share the future profits deriving from timber exploitation with the former manager, normally the local State forest enterprise (Hayami 1994; Rambo 1995a). Allocations of both the paddy and forest land are, in effect, privatisation: land can be transferred, exchanged, leased and mortgaged, though each of these incurs the payment of a tax.

In the period up to 1997, three quarters of the communes undertaking agricultural land allocation (7500 out of 10 050) and two-thirds of their available agricultural land, equal to 4.2 million ha, has been allocated to 7.5 million households (Vu Van Me 1997). By contrast, 1.0 million ha of forest land (12 per cent of the total forest area, see Table 6.1) has been allocated to 334 000 households, 4.4 millions ha to 327 State enterprises and 0.5 million ha to 1700 cooperatives and other institutions (Vu Van Me 1997; Doan Diem 1997).

Considerable efforts have been expended under the Land Law of 1993 and the Law for the Protection and Development of Forests (19 August 1991) in defining the legal and organisational framework for the forest land allocation process implementation as well as in providing incentives for sustainable forest management, including initiatives under Decree 327 on 'barren lands' (Sikor 1995). The commitment of the policy-makers towards forest land allocation is extremely clear: 'local people are seen as the driving force in forestry development' (Vice-Minister of Agriculture and Rural Development at the last World Forestry Congress – Nguyen Van Dang 1997: 215; see also Nguyen Quang Ha 1993 and Vietnam Ministry of Agriculture and Food Industry 1993). The Forestry Sector Review (Vietnam Ministry of Forestry 1991) and the Tropical Forest Action Plan for Vietnam both highlight forest land allocation to households as a priority in Vietnamese forest policy.

A series of other implicit motivations undoubtedly play an important role in the formal commitment of the Government towards the forest land allocation policy.

*Table 6.2* Main allocation alternatives for forest land under the present regulations

| Forest land allocated | Main features/problems of the property rights system definition |
|---|---|
| To State enterprises and other public agencies | Practically unlimited use of the resources; not always a clear definition of the management objectives (production vs. public non market services) and compensation criteria. Need of an efficient, independent control authority (risk of over-exploitation in autonomous, 'privatised' enterprises). Conflicts with customary users of productive land. |
| To State enterprises or public authorities contracting the land to households or individuals | Forest land can be given to households or individuals (normally former or current employees of the enterprise) that have to take care of it following a management plan. Contract length: 50 years for protection and 'special use' forests; till the end of rotation for productive forests. For protection forests the contractees receive a compensation (Decree 327); compensation measures defined arbitrarily. No clear benefit and cost sharing system between contractor and contractees. |
| To local communities, villages, groups | Forest land traditionally managed in common for shifting cultivation, grazing, collection of non-wood forest products, etc. is allocated to the community. The head of the village/community is formally responsible for the allocation. Contract length: 50 years. Problems may arise if the traditional regulation system is altered by internal or external factors (unsustainable pressure on the resources due to increased number of users, change in management techniques, increased market penetration, etc.). |
| To private companies | Bare land or forests are allocated to large private investors for timber production on the basis of an approved management plan. Large tracts of land (>1000 ha) are allocated by the government. Conflicts with customary users can arise. In the case of conversion of natural forests to fast growing plantations not much attention is given to non market public forest services. |
| To households or individuals | Forest land is allocated for a period of 50 years on the basis of a management plan. Right to exchange, transfer, inherit or mortgage are granted. If the land is bare, householders have to make investments for increasing soil protection and production; if some forest amelioration works have been already carried out, they have an obligation to compensate the former manager. Normal size of the allocated plots does not allow shifting cultivation, therefore households have to change their farming systems. Incentives, compensations and technical assistance are often lacking. |

Central government desires to stabilise and limit three million people estimated to be shifting cultivators (Do Dinh Sam 1994), including nearly all of the ethnic minorities of Vietnam.[2] The implicit objective may be to reduce the demographic pressure on major urban areas and on the two deltas. It is necessary to solve the conflicts in land use between the traditional upland inhabitants and the 3.5 million ethnic Vietnamese (*Kinh*) resettled in the central highlands and the northern midlands, especially in the 1980s, under the Fixed Cultivation and Settlement Programme.

The willingness of government ministries to respond to the external pressure, and to make use of the available financial assistance, from international institutions for the conservation of the residual part of the forest cover is also significant. The influential position of many representatives of ethnic minorities elected in local administrations in mountainous areas (as highlighted in Chapter 7) and the historical support given by ethnic minorities to the northern *Kinh* community during the liberation war, may also play a role in affecting the government's position on forestry policy. Both factors have a positive effect in the way in which the central government is considering problems and needs of the mountainous population.

In explaining the unsuitable use of Vietnamese forest resources, a top-down, simplified explanation of causal factors and solutions is often presented, despite the complexity of local and external factors. Rural poverty is considered the primary factor causing deforestation and this is perceived as being directly associated with shifting cultivation and the unclear definition of property rights on forest land. As observed by many authors in the context of Vietnam (Do Dinh Sam 1994; see also Chapter 7) and elsewhere in Southeast Asia (Brookfield and Padoch 1994; Porter 1995), however, slash and burn agriculture cannot be considered an unsustainable practice *per se*. The concept of shifting cultivation is extremely vague compared to the differentiated social, economical and environmental contexts that can be found in Vietnamese mountainous areas. In most observed cases, even in the presence of traditional slash and burn practices there are multiple proximate causes of deforestation (Baland and Platteau 1996; Lang 1996). Increased population pressure, reduced fallow periods, privatisation of forest State enterprises and the consequent stimulus to commercial logging and related activities such as new road building and secondary logging, increased market integration with market prices favouring some cash crops, lack of credit and technical assistance, and hydro-power all contribute to loss of forests in the Vietnamese uplands, as widely observed elsewhere in Southeast Asia.

Thus, the process of forest land degradation and the possible use of the allocation of forest land to households, villages or enterprises as an instrument of sustainable economic development of mountainous areas must be examined, taking on board this wider framework of causal factors.

## Constraints on forest land allocation as sustainable economic development: two case studies

Clarification of land resource property rights is an important instrument used to internalise the benefits of conservation to those undertaking this process. There is, however, a serious risk that the market will allocate resources inefficiently in an economic sense if the use of natural resources brings about significant negative externalities, the costs of property right assignment and protection is high,

*Table 6.3* Socio-economic characteristics of the case study areas in Thua Thien Hue and Quang Ninh Provinces

|  | *Province* | |
|---|---|---|
|  | *Thua Thien Hue* | *Quang Ninh* |
| *Administrative division* |  |  |
| District | Nam Dong | Tien Yen |
| Commune | Thuong Lo | Phong Du |
| No. of villages | 4 | 16 |
| No. of households | 144 | 480 |
| No. of inhabitants | 919 | 3086 |
| Agricultural land (ha) | 80 | 491 |
| *Field work area* |  |  |
| Forest land (ha) | 839 | 8.6 |
| Allocated land (ha) | 269 | 1.6 |

and compensation measures are not provided (Hayami 1994; Otsuka 1999). In Vietnam, cases of market failure would appear to be pervasive in forest land privatisation. This section reports on case study research designed to address this issue in two densely populated mountainous communes, severely affected by food security problems, where the Ministry of Agriculture and Rural Development (MARD) and the Food and Agriculture Organization (FAO) have been promoting participatory approaches to land allocation in forestry.

The two case study areas are highly heterogeneous in socio-economic and environmental terms (see Table 6.3). Thuong Lo Commune is located in the central plateau (Thua Thien Hue Province, see Figure 6.1) and has a population of 919 in 156 households. *Katu* people represent 95 per cent of the population (41 per cent in the whole District). Population growth is estimated at 3.1 per cent. Almost all the land area (97 per cent) is classified as forest, but only 79 per cent has forest cover. Shifting agriculture based on slash and burn techniques is the traditional cultivation system of the *Katu* people. Till recently, forest land in Thuong Lo was theoretically managed by the State Forest Enterprise (59.2 per cent), the commune (8.3 per cent) and the Bach Ma National Park (32.5 per cent). The Forest Protection Station (i.e. the local forest office which is a division of MARD) was responsible for a general supervision of the three institutions. In practice, however, right of access to the forest land has traditionally been, and continues to be, mainly based on a priority rule: the farmer who first clears the plot of forest land will have the right to cultivate it until the end of the cycle. The cultivation period generally ranges from two to four years, and afterwards the land is left fallow to restore fertility.

In June 1993, a resettlement programme was launched in Thuong Lo, based on a number of components including construction of small dams and wells and provision of home-gardens, promotion of livestock and reforestation of bare lands. In the reforestation programme, farmers were simply informed of the activities to be carried out in the area and the species to be planted (acacia and

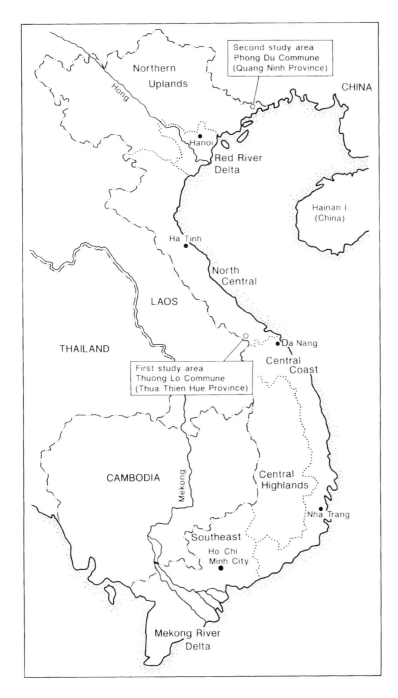

*Figure 6.1* Location of forestry case study areas: Thuong Lo Commune, Thua Thien Hue Province, and Phong Du Commune, Quang Ninh Province

eucalyptus) without any consultation on species selection or without the possibility to check seedling quality. The areas for plantations were allocated to farmers according to the traditional (i.e. pre-collectivisation) ownership rights. One month after planting, forest technicians from the Forest Protection Station examined the resulting reforestation and some farmers were compensated for incurred losses after detracting the seedlings, planning and survey costs (50 000 VND/ha gross compensation or 33 000 VND/ha net compensation). The most fertile areas in Thuong Lo, located in the foothills of the valleys, are now covered in eucalyptus and acacia plantations but with low productivity due to unfamiliarity of house-holders with the silviculture of these species and due to unsuitable and degraded soils. With the support of MARD and FAO, a total of 269 ha of forest land had been allocated to 144 households in Thuong Lo by December 1997.[3] The forest land allocation policy creates problems in land management and investments because it is not, nor could it easily have been, completely coherent with the traditional land use and access systems. The traditional land rights have only partially been respected because the allocation was carried out according to equality criteria (each household received a portion of upland, midland and foothill land). The result seems contradictory. On the one hand, a high land fragmentation results from the allocation process. On the other hand, there is little correlation between household needs and allocated land area using this allocation system, thereby, in welfare economic terms, not maximising overall welfare.

The second case study discussed here is that of Phong Du Commune, located in Tien Yen District, Quang Ninh Province, in the north-eastern part of the country. Despite constant population increase at the District level, the population of Phong Du has decreased significantly between 1994 and 1996. This trend is due to migration programmes organised by the District to nearby Communes or to neighbouring Districts, in response to perceived severe food security problems and acute demographic pressure on limited resources in the Commune. Natural forests represent only 8 per cent of the total Commune area of Phong Du, while 81 per cent is forest bare land. At present there are 3100 inhabitants and population growth of 3.9 per cent a year. Apart from five *Kinh* households, the whole population in Phong Du is represented by ethnic minorities (*Tay, Thanh Phan, Thanh Y, San Chi*) with different migration histories and different endorsements of paddy land per household. Location of the four ethnic groups roughly corresponds to different levels of accessibility. *Tay* villages have settled in the area along the main road since at least eight generations and their relationship with the resources is very similar to those of the *Kinh* households: paddy rice represents the main cultivation and the slash and burn technique has been given up decades ago. By contrast, *Than Y* groups are settled in more remote areas (they arrived around thirty years ago in the commune) and have a strong relationship with the use of forest resources, particularly in the collection of non-wood forest products. The living conditions of the *Than Y* are generally lower than other ethnic groups, shifting cultivation being still an important component of their economy. *Thanh Phan* and *San Chi* groups have similar resource use patterns to the *Than Y* in that

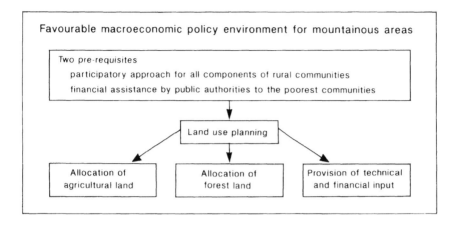

*Figure 6.2* A proposed framework for organising the land use planning and allocation
process

they rely mainly on stabilised farming activities and the scarcity of paddy fields is
offset by intensive management of cash crop plantations.

It would seem that forest land allocation has the potential to enhance livelihood
conditions particularly in those areas where food shortages are most serious. Fruit
tree plantations on forest land are, indeed, perceived by many forestry officials
to be the only means to generate legal income flows and to overcome rice
deficiencies due to the limited paddy field area. Farmers themselves indicate that
tree plantations are a feasible alternative to the heavy task of non-wood forest
products collection, particularly given the scarcity and reduced availability of
many forest resources. Yet there is limited input availability and a lack of technical
knowledge related to plantation forestry such that the forest land allocation policy
loses its potential role and fails to stimulate sustainable alternatives. The
controversial land distribution and the distance of many allocated plots to
settlements also reduce any potential positive impact of alternative land uses.

From discussions and field research in the two case study areas, possible
improvements to the local implementation of the forest land allocation policy have
emerged. Figure 6.2 presents a general framework for the organisation of land use
planning and the land allocation process (Prosterman and Hanstand 1999). In the
following sections, specific issues are discussed.

## The problem of high transaction costs

The process of forest land allocation has been defined through an extremely
complex system of official regulations, as documented in Box 6.1. Problems of
misinterpretation of the land allocation rules are common. The activities
connected with forest land allocation such as the reorganisation of cadastral

*Box 6.1* Legal framework for forest land allocation in Vietnam

**Principal Laws**
Land Law of 1988, revised in 1993
Law for the Protection and Development of Forests (19 August 1991)

**Other relevant decisions, decrees and regulations**
1   Government Decree No. 02/CP, dated 15 January 1994, Concerning Allocation of Forest Land to Organisations, Households and Individuals for Long-term Forestry Purposes
2   Circular Letter No. 06.LN/KL, dated 18 June 1994, by the Ministry of Forestry, on Guiding the Implementation of Decision No. 02/CP
3   Decision No. 202/TTg, dated 2 May 1994, by the Vice Prime Minister, on Contracting Forests in Order to Protect, Regenerate and Plant Forests
4   Decision No. 01/CP, dated 4 January 1995, by the Prime Minister, on Land Allocation of Agricultural, Forestry and Fishery Lands Belonging to State Enterprises
5   Circular No. 158/TB, dated 25 November 1994, by the Vice Chairman of the Government Office, on Some Conclusions of the Prime Minister On Some Important Work in the Forestry Sector
6   Decision No. 25/UB/KHH–DM, dated 14 March 1994, by the Chairman of the State Planning Committee, on the Funding for Planting New Forests, for Regenerating Forests and for Enriching Natural Forests
7   Decision No. 327/CT, dated 15 September 1992, by the Chairman of the Council of Ministers on Policies for the Use of Bare Land, Denuded Hills, Forests, Alluvial Flats and Water Bodies
8   Circular No. 03–UB/NLN, by the State Planning Committee, on Implementation of Decision No. 327/CT with Reference to Appraisal and Approval of Projects and Compilation of Programme 327
9   Circular No. 32 TC/DT, by the Ministry of Finance, on Management and Provision of Credits from the State for Programmes and Projects for Use of Bare Land, Degraded Hills, Forests, Alluvial Flats and Water Bodies
10   Circular No. 300 CV/RD, by the General Department for Land Management, on Allocation of Land in Accordance with Decision No. 327/CT
11   Circular No. 10 KH, by the Ministry of Forestry, Concerning Implementation of Decision No. 327/CT
12   Decree No. 556, dated 15 September 1995, by the Prime Minister, Revising and Supplementing Decision No. 327/CT of 15 September 1992

13 Decision 264/CT, dated 27 July 1992, by the Chairman of the Council of Ministers, on Policies Encouraging Investment for Forestry Development
14 Interministerial Circular No. 11–TT/LB, by the State Planning Committee, the State Bank of Vietnam, the Ministry of Finance and the Ministry of Forestry, on Implementation of Decision No. 264/CT
15 Government Regulation No. 13–CP, dated 2 March 1993, on Agricultural Extension

Source: Desloges and Vu Van Me (1997).

records, demarcation of the borders of the allocated plots, land tenure certificate issuing and the associated activity of land use planning are, in absolute terms, not extremely costly as indicated in Table 6.4. However, considering the low value of the allocated land, the costs are in relative terms quite high both for the public agencies and for the beneficiaries.

Transaction costs often increase significantly due to the presence of traditional and recently established use of forest land by households without legal rights, because of the need to negotiate land allocation between conflicting interests such as State forest enterprises, local authorities and households, and due to the frequent land disputes between hamlets and households with overlapping land ownership. Moreover, in upland areas, 'where criteria for land allocation are less transparent, and village solidarity often less well developed, opportunities for officials to take advantage of their titling process have been much greater' (Rambo 1995a: 8).

### *The need to integrate the forest land allocation process with other development policies and with some institutional reforms*

When farmland is limited and there is an increasing demographic pressure on forest resources, the land allocation process alone is unable to respond to the long-term food security needs of very poor households (Castella *et al.* 1999). Allocated land is frequently on very steep slopes and may have been degraded by past over-exploitation. From a legal point of view, this land is still classified as forest land, and it should be utilised as such, but in most cases the forest cover is extremely limited or absent and the soils degraded, thereby not facilitating sustainable livelihoods for agricultural dependent households. On the basis of the prescriptions reported in the Land Tenure Certificates (the so-called 'red books'), households with allocated land should manage it as forest, planting trees and ameliorating the stand. In the two study areas, no mention has been made of the

*Table 6.4* Transaction costs of land use planning (LUP) and forest land allocation (FLA) in Vietnam

| Activity | VND/ha | Basis of estimate | Source |
|---|---|---|---|
| LUP and FLA | 11–24 000 | Official estimate by MARD of average costs in Vietnam | MARD–FAO (1996) |
| LUP and FLA | 12 000 | Average costs of an experimental activity based on a participatory approach in the Thua Thien Hue and Quang Ninh Provinces | Desloges and Vu Van Me (1997) |
| FLA | 20 000 | Average costs in two communes at Son La and Lai Chau Provinces | Pham Minh Tuan and van der Poel (1998) |
| FLA | 17 500 | Average costs for FLA of 416 000 ha in Thanh Hoa Province (2500 VND for computer application for mapping) | Hoang Cao Trai (1998) |
| FLA | 30–40 000 | Range of costs in three communes Hoa Binh Province | Vu Van Me (1997) |
| FLA | 20–30 000 | Range of costs for FLA of 15 500 ha at Nghe An Province (Rapid Rural Appraisal: 1000–1700 VND; boundaries marking: 1300–2600 VND) | Nguyen Thanh Binh (1998) |
| FLA | 25–35 000 | Range of costs in two communes at Son La and Lai Chau Provinces | Vu Van Me (1997) |
| LUP | 60 000 | Average costs in 78 villages of Lao Cai Province | Jonsson and Nguyen Hai Nam (1998) |
| FLA | 21 000 | Average costs in two communes of Ha Giang Province | |

implementation of agro-forestry systems since this is not formally recognised under current land use planning in Vietnam.

From observations in both case study areas, land and time constraints on poorer households ensure that forest tree planting and management are not a top priority. Forest tree plantations are not perceived as important because they provide long-term benefits but do not meet immediate household needs. Notwithstanding the heavy physical effort required for shifting cultivation in remote forest areas, farmers prefer to rely on traditional agricultural systems rather than novel opportunities offered by the land allocation.

Forest land allocation alone, therefore, cannot be claimed to be enhancing sustainable development. To make such a contribution, it must first be integrated with agricultural land allocation, land use planning and the provision of technical services. Such an integration would require reform, not only of the traditional land regime regulation, but also of public institutions responsible for the extension and credit services. Extension services are still mainly supply oriented and organised for enhancing the productivity of State and cooperative organisations (Nguyen Thuong Luu *et al.* 1995). As a consequence of this institutional arrangement, they are often unable to meet the diversified needs of small owners of forest and agricultural land.

In many of the observed villages of the two case study areas, the plots of forest land allocated to individual households are extremely limited (1.8 ha per household in Thuong Lo, 3.4 ha per household in Phong Du – see Table 6.3). A large part of the forest land is still formally under the management of the State forest enterprises which are reticent to give up their land. Farmers, however, claim ownership rights to forest land because of their long customary possession. Reform of forest enterprises, with a clear definition of its objectives and responsibilities, is, therefore, a fundamental step both for the future development of the forestry sector in Vietnam and in enhancing the contribution of these resources to sustainable development.

## A participatory approach to the land use planning and forest land allocation process

Officially, as indicated in numerous forest policy statements, public authorities are committed to a participatory approach in the forest land allocation process. Guiding principles in land allocation state that: 'the people allocate the land themselves and make the decisions by themselves', and that 'farmers are not forced to plant trees, rather they can decide for themselves the method of cultivation . . .' (Nguyen Cat Giao and Vu Van Me 1998: 79).

Land allocation must be carried out in 'accord with the sentiment first (*hop tinh*)' and then in 'accord with reason (*hop ly*)'. Nevertheless, in the two study areas, participation of householders in land allocation has been largely absent. In most of the villages visited, farmers had only been informed about the forest land allocation process but could not actively contribute to the decisional process (see Box 6.2).

Even in the relatively small studied areas, the settlements and hamlets are characterised by a diversity of socio-economic status and conditions due to a different use of land resources. This diversification is similarly observed in communes inhabited by different ethnic minorities such as in Phong Du Commune as well as in areas populated by the same ethnic group such as Thuong Lo Commune. Even among the households of the same hamlet, the apparent skewed distribution of wealth is often striking. Forest land allocation activities should be, therefore, more responsive to different local needs, an objective which can only be achieved through the adoption of an effective participatory approach. Moreover, in some areas, the newly allocated land used to be a common resource managed for grazing, fuelwood and non-wood forest products collection. A conflict between new individual use entitlement and traditional common access to the land may have very negative social and environmental impacts and provides even greater justification for a participatory approach in land use planning

---

*Box 6.2* An example of a non-participatory process of forest-land allocation

From a farmer's speech, during a village meeting in Phong Du:

> Troubles arise with the Working Group (i.e. the local authorities) when they presented contract papers to the farmers. They refused to sign because they got two to three plots but on papers even five resulted in their hands. The area was bigger than the real one, for every farmer. They just worked on maps without accurate measurements on the field. They didn't take into account our opinions. We have never been involved, we just agreed at the beginning on receiving three plots per household but now the situation on the map is different. Only the hamlet head and his brother signed the contract for the allocation. Nobody else did it. Farmers stated that despite the Working Group menacing, they would accept to sign papers only in case they reflect the real situation. Only 600 ha are available for allocation but 2000 ha result from papers. Another mistake regards land classification on the map. Cinnamon plantations, more than ten years old, are classified as bamboo forest to be protected. This is another reason for not signing contract papers.

---

and forest land allocation as a pre-requisite to promote the sustainable use of forest resources. In addition local indigenous technical knowledge and the traditional pattern of land use are ignored when forest land allocation design and implementation does not encourage the participation of key groups directly affected by the allocation.

### The need to provide compensation

When forest land is conserved it provides a series of positive externalities (such as erosion control and biodiversity conservation), as well as contributing to poverty reduction, stabilisation of the local population to the benefit of the local and national community. Effective open access to forest resources owned by the State has been in the past a major reason not only for the observed forest degradation process, but also for the unequal distribution of the population on the land and equity conflicts (Nguyen Thuong Luu *et al.* 1995). The former land tenure system, based on cooperatives and State enterprises, was in theory more equitable, being in charge of providing assistance to the more vulnerable groups and public services to the community. The financial and human resources on which this system relied for providing these services were, however, exclusively local and therefore often inadequate with respect to the social and economical local problems: 'Even though the old system was called *bao cap*, or "subsidised system" it is clear that the rural sector in general was not a net receiver of subsidies' (Bloch and Oesterberg 1989:7). As observed in the most degraded forest area in the Phong Du Commune, the sustainable use of newly allocated land for long-term forest investments implies a reduction in time needed for alternative, more profitable activities. Planting and tending of trees requires different labour inputs according to the various species utilised. Compensation for plantation establishment and forest amelioration are an essential complement to the process of land allocation.

## Conclusions

Despite Vietnam's impressive economic achievements of the past decade of structural reforms, there are many reasons for concern underlined by recent trends. The World Bank argues, for example, that: 'substantial inefficiencies persist and growth is inward-looking, increasingly capital-intensive and biased in favour of urban dwellers' (World Bank 1997b: iv). The bias towards the urban and capital-intensive sectors is manifest in the fact that paddy land allocation to small private farms in the Mekong and Red River deltas has received much more attention than the problems of property rights re-definition in the uplands. The national focus of land allocation on agricultural land reflects the priority to increase agricultural production and the desire of the government to raise revenue. Under most of the public policy criteria, forest land allocation is perceived as not contributing to, or even hindering, national economic objectives.

In the uplands, the allocation of forestland to the farmers will not in itself reduce deforestation problems. Such issues need to be tackled with new macro-economic policies for the intensification of agriculture (Porter 1995) and the reduction of demographic pressure on the natural resources together with the reform of the State institutions involved in management of the forest areas and in the enforcement of strict requirements for logging and reforestation practices. Shifting financial resources and policymaking attention in favour of mountainous forest areas may result in a reduction in short-term growth of the economy, but is more likely to lead to a more stable, equitable and environmentally sustainable development in the long term. Such reforms, coupled with the reform of the State institutions involved in problems of forest management, could represent an important element able to promote labour-intensive growth, to reduce income disparities and to preserve the long-term productivity of land resources.

## Acknowledgements

I gratefully acknowledge the work carried out by B. Vinceti, A. Gribaudo and T. Gomiero in collecting field data in this research project financed by the FAO to develop an appropriate methodology for participatory land use planning and forest land allocation (Project GCP/VIE/020/ITA). M. Paoletti and P. Palmeri, with the general coordination of P. Faggi from University of Padova (Italy), organised the research project and have given an essential input of ideas to this paper. Members of the Thu Duc University of Agriculture and Forestry (Ho Chi Minh City), the Xuan Mai Forestry College (Hanoi), and a representative of the University of Hue have carried out the field research activity in cooperation with the University of Padova.

## Notes

1 Several national and international initiatives in recent years have analysed or promoted the process of land allocation in forestry and the provision of related services such as extension and credit. See, among others, Bloch and Oesterberg (1989), Neave and Bui Ngoc Quang (1994), Cameron (1995), Hines (1995), Desloges and Vu Van Me (1996; 1997), Fisher (1996), Food and Agriculture Organization (1996), United Nations Development Programme and Food and Agriculture Organization (1996), Vietnam Ministry of Agriculture and Rural Development and Food and Agriculture Organization (1996), Jonsson and Nguyen Hai Nam (1998). A comprehensive review of the different initiatives under the forest land allocation process is presented in Vietnam Ministry of Agriculture and Rural Development and Food and Agriculture Organization (1998).
2 It is difficult to definitively estimate total population numbers for the ethnic minorities. Le Trong Cuc (1988: 6) states that 'the minorities, with a population of only about 1.8 million, live mostly in the uplands and middle lands' while Food and Agriculture Organization and International Institute of Rural Reconstruction (1995: 9) identifies 'ten million people belonging to 54 ethnic groups in the upland of Vietnam'. See Chapter 7.
3 In the period since the area was surveyed for this research, under the MARD–FAO Project in Thuong Lo, 95 ha of forest land have been temporarily allocated to farmers for tree planting.

# 7 Sustainable agriculture in the northern uplands: attitudes, constraints and priorities of ethnic minorities

*Andreas Neef*

## Introduction

In the late 1980s, as Vietnam initiated the process of *doi moi*, the family farm was restored as the principal unit of agricultural production. Within several years, the country became the world's third largest rice exporter and experienced double digit economic growth rates (Pingali and Vo Tong Xuan 1992; World Bank 1995a; Pingali *et al.* 1997). However, the economic gap between urban and rural areas and between high-potential lowlands and marginal upland areas has widened since that time (Kerkvliet and Porter 1995), with many upland regions still suffering from temporary and seasonal food insecurity (Rambo and Le Trong Cuc 1996a). Moreover, environmental concerns about fragile mountainous regions are increasing.

According to the World Bank (1995a), 47 per cent of the northern mountain regions of Vietnam are highly susceptible to deterioration or erosion. Vietnam has one of the lowest land areas per capita in the world, and expansion of agriculture into forest areas and marginal uplands is a continuous process. In the northern mountain region, 60–65 per cent of the total area is presently classified as 'barren land'. Hence, a major share of present development efforts in northern Vietnam is dedicated to the restoration of heavily degraded land and the introduction of more sustainable land use systems in the uplands. These areas are the main homelands of ethnic minorities (World Bank 1995a; Rambo and Le Trong Cuc 1996a; Sikor and O'Rourke 1996).

This chapter examines the prospects for the policies and programmes targeted at the upland agricultural regions by drawing on the perceptions of ethnic minorities as to the potential for and constraints on sustainable agriculture. In the upland areas of northern Vietnam, the management practices of the ethnic minorities are a key part of the human landscape. The analysis is based on primary field research in Son La and Bac Kan Provinces and on a review of other case studies on ethnic minorities in the uplands. The chapter does not detail all the development issues facing the ethnic minorities in upland areas. The intention is to call into question common assumptions and pre-conceptions regarding the agricultural practices of ethnic minority farmers in Vietnam and to draw attention to the various constraints affecting minority people with regard to both resource

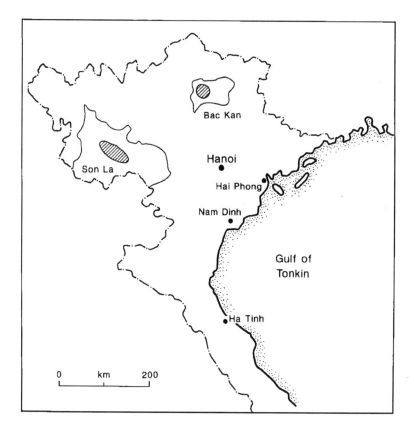

*Figure 7.1* Location of case study areas: Son La and Bac Kan Provinces

and capital endowments, the endowments which ultimately determine the resilience and adaptive options of these communities.

For the past 30 years, government policies towards many ethnic minorities have followed a single centralised approach, implicitly based on the concept that ethnic minority farmers constitute a threat to both natural resources and national security. The perceived threat led to policies to encourage settlement in 'fixed' villages and the adoption of permanent agricultural practices. Case study and policy research, however, indicates that traditional and newly evolving agricultural practices of minority farmers in the uplands cannot simply be classified as 'sustainable' or 'unsustainable,' as some official sources would have them portrayed. Minority farmers throughout the mountainous regions of mainland Southeast Asia have proven to be highly flexible and resilient in adapting to a variety of agro-ecological conditions and economic settings.

Minority people in northern Vietnam are, nevertheless, constrained by numerous external and policy factors. The present land allocation processes in

*Table 7.1* Population of the main ethnic minority
groups in Vietnam, 1995

| Group | Population (million) |
|---|---|
| Tay | 1.41 |
| Thai | 1.26 |
| Muong | 1.09 |
| Kho Me (Khmer) | 1.03 |
| Hoa | 0.88 |
| Nung | 0.81 |
| Hmong (Meo) | 0.67 |
| Mien | 0.58 |
| Other ethnic groups | 2.15 |
| **Total** | **9.88** |
| As % of national population | 13.6 |

Source: Kampe (1997).

the uplands often lack transparency and appear to disadvantage some ethnic minority groups (see Chapter 6). Upland farmers in northern Vietnam face considerable difficulties in gaining access to formal credit, extension services and markets, important constraints both in the integration of perennial crops and livestock and in making existing farming systems more sustainable. Perceptions by ethnic minority farmers of priorities indicate that these farmers are directly engaged in diversification of cropping systems and integration of fruit trees and livestock to raise incomes and ensure food security. But this evolution towards resilient and flexible livelihood systems requires a favourable policy and appropriate institutional arrangements if they are to be fully realised.

## Ethnic minorities in Vietnam: demography and government policies

### Demography, resettlement and forest cover change

There are approximately ten million ethnic minority people living in Vietnam. Ethnic minorities comprise as many as 53 different groups, mainly located in the northern mountainous regions and the central highlands (Table 7.1). In some provinces, such as Son La Province, ethnic 'minorities' even constitute the majority of the population. The socio-economic status of ethnic minorities is, on average, much lower than that of the Vietnamese *Kinh* (Rambo 1995b) and has variously been explained in relation to the remoteness of many upland regions, the poor conditions of the physical infrastructure and the low agro-ecological potential of these areas.

Although there are many individuals within government structures at the commune, district and province levels in the mountainous areas of Vietnam who

*Table 7.2* Change in forest cover in selected regions of Vietnam

| Region | Forest cover in 1943 (% of land area) | Forest cover in 1991 (% of land area) |
|---|---|---|
| Northern mountains | 95 | 17 |
| Northern midlands | 55 | 29 |
| Central highlands | 93 | 60 |
| Total country | 67 | 29 |

Source: World Bank (1995a).

belong to ethnic minority groups, there is widespread recognition that top-down rural development planning by these levels of the state have been unsympathetic towards ethnic minorities over the past 30 years. Government policies towards ethnic minorities have been mainly led by national security and perceived environmental concerns as well by attempts to eliminate production and trade of illegal narcotics. Particular ethnic minorities have been demonised as threats to national security, major producers of illegal opium and the agents of destruction of natural resources such as upland forests. In reality, opium production is mainly concentrated among those ethnic minorities living in altitudes greater than 800 m above sea level.[1] Moreover, the destruction of natural resources by 'inappropriate' agricultural practices such as 'slash and burn' agriculture is not proven. While the upland forest regions have been the location of the fastest rates of forest clearance in this period (Table 7.2), many commentators consider the main culprit for the extremely high rate of forest destruction in the uplands of northern Vietnam as being immigration into these areas, and engagement in upland agriculture, by farmers or other individuals from lowland areas who are not familiar with local agro-ecological conditions (Rambo *et al.* 1995; Poffenberger 1998).

In the 1960s, roughly one million people were encouraged to migrate from the densely populated Red River Delta to the northern midlands and highlands. Under the government's Fixed Cultivation and Settlement Programme, 2.4 million people were resettled in the period 1981–90, mainly migrating from the Red River Delta to the central highlands (World Bank 1995a). One of the objectives of this programme has been to 'sedentarise' ethnic minorities and thus encourage permanent cultivation and socio-economic development. The construction of the Hoa Binh Dam, which now supplies energy for a major part of the country, forced the relocation of nearly 60 000 minority families who lost their irrigated paddies and dry fields and were driven to settle in higher altitudes on extremely steep slopes above the dam where they clear around 2000 hectares of forest annually. Erosion of the upland slopes in this area has increased to such a degree that the projected life-span of the dam is estimated to be reduced from 300 years to only a few decades because of the increased sedimentation associated with run-off and erosion (Poffenberger 1998).

## Policy responses

Property rights regimes have a significant impact on, as well as a central role in, the use and management of natural resources. Many environmental problems, such as soil degradation and forest depletion, can be characterised as a result of incomplete, inconsistent or non-enforced property rights (Bromley and Cernea 1989; Wachter 1992). Hardin (1968) has argued that, where property regimes are non-existent, natural resources are prone to overexploitation because the costs of negative externalities such as pollution of water or overgrazing of pastures are borne by the community as a whole but the potential benefits accrue to the individual. Coase (1960) argued that the absence of clearly defined property rights inevitably leads to degradation of soils and other natural resources. The general interpretation of these theories from the 1960s onwards was that collectively owned property was the culprit of forest destruction, land degradation and water pollution and that private property was crucial to sustain natural resources. However, as discussed in Chapter 5, growing evidence amassed by economists, anthropologists and political scientists suggests that sustaining environmental resources does not primarily depend on whether the property rights regime is based on states, communities or individuals. It is more important that property rights are congruent with their ecological and social context (Bromley 1991; Hanna and Munasinghe 1995; Adger *et al.* 1997). Examination of these issues and a careful examination of the robustness of institutions for management, as argued in Chapter 5 is the key to understanding the resource management strategies of the ethnic minorities and the potential impact of externally imposed property rights on their continued sustainability.

Recent land allocation policy in Vietnam has evolved through three main reforms, all directed towards greater individualised land ownership and control. First, Directive 100 in 1981 introduced an initial stage in devolution of responsibility for production to farm households. Second, Resolution 10 in 1988 led to large-scale decollectivisation in most parts of the country (Tran Thi Van Anh and Nguyen Manh Huan 1995). The third stage, initiated in 1993, has allocated land use rights to farm households, providing long-term security for agricultural activities (Dao The Tuan 1995; Rambo *et al.* 1995). The 1993 Land Law, based on the principles of the 1992 Constitution, guarantees to farmers long-term use rights for a period of 20 years (for annual crops and aquaculture) to 50 years (for forest and perennial crops). In the first five years of implementation of this law, the 'Property Reports' by which these leases are confirmed – one of the 'Red Book Certificates' – have been distributed mainly in urban centres and in the main lowland rice growing areas. Ethnic minorities in the uplands appear to be last in line to obtain formal land use rights (Wandel 1997). Research in particular upland areas by Bergeret (1995), Rambo and Le Trong Cuc (1996a), Hirsch and Nguyen Viet Thinh (1996) and Mellac (1997) all indicate that land allocation processes in the uplands often lack transparency and provoke inequalities among ethnic minority groups and between individuals.

In parallel with this land allocation process, a major programme has been undertaken for upland development under the umbrella of the Vietnamese land use policy. This is Decree 327 'Restoration of Barren Land', initiated in 1992. The objectives pursued by this programme are described by Sikor (1995) as converting unoccupied land, denuded hilly areas and degraded forests to a productive use. The success of Decree 327 has been limited due to lack of participation by local people, diversion of funds for other purposes and poor planning (Sikor 1998; World Bank 1995a). The slow process of land allocation in the uplands is another major inhibiting factor (see above). Jamieson *et al.* (1998) argue that the programme was refocused by Decree 556 issued in 1995. Decree 556 specified the rural household as the key actor in the restoration of barren land and protection of natural resources. Rural households, however, face major constraints in implementing these objectives.

## Ethnic minorities and sustainable agriculture

### Concepts and definitions of sustainable agriculture

Both the popularity and elusiveness of sustainability as a guiding principle for agricultural and rural development are demonstrated by the plethora of definitions of 'sustainable agriculture' and 'sustainable development' in recent decades. An often-cited contribution to this discussion on sustainability is that of the Brundtland Commission, emphasising the intergenerational dimension of sustainability: 'Sustainable development is development that meets the needs of the present without compromising the ability of future generations to meet their own needs' (World Commission on Environment and Development 1987). For agriculture, the power of the concept of sustainability is evident in the orientation of both agricultural research and rural development in developing countries. The Consultative Group for International Agricultural Research (CGIAR) suggests that: 'sustainable agriculture should involve the successful management of resources for agriculture to satisfy changing human needs while maintaining or enhancing the quality of the environment and conserving natural resources' (CGIAR document quoted in Chantalakhana 1995).

The international debate on 'sustainable development', newly stimulated by the United Nations Conference for Environment and Development of 1992, raised the profile of this issue in agriculture. 'Sustainable Agriculture and Rural Development' under Agenda 21, among other issues, emphasises: agricultural policy review; participation and human resource development; diversification of farm and non-farm employment; land resource planning; land conservation and rehabilitation; water for sustainable food production and rural development; and the conservation and sustainable utilisation of plant and animal genetic resources.

There is no doubt regarding the relevance and efficacy of these guiding principles for rural development in Vietnam. Choosing the right political setting for upland agriculture and involving minority people in the process of land use planning and conservation of biodiversity while respecting indigenous knowledge will be crucial for achieving sustainable agriculture. The feasibility and resilience of any approach will also depend on property rights and the institutions of governance. Farming systems advocated by various bodies for sustainable agriculture in the uplands include integrated farming,[2] agroforestry systems, and various practices of soil and water conservation. Many authors emphasise the importance of rural linkages for sustaining agricultural productivity and a robust rural development process (Timmer 1993; Kerkvliet and Porter 1995; Fleischhauer and Eger 1998). This is supported by evidence from Thailand (see, for example, Mellor 1995). The importance of participation in achieving sustainable land use is highlighted by Pretty (1998) who argues that policies which aim to create conditions for development based on locally available resources and indigenous knowledge must establish dialogue and interaction between the various actors in land management: 'Sustainable land management should not, therefore, be seen as a set of practices to be fixed in time and space. It implies the capacity to adapt and change as external and internal conditions change' (Pretty 1998).

## Attitudes, constraints and priorities of ethnic minorities as to sustainable agriculture

### Traditional and newly evolving agricultural practices in the uplands

Much of the literature describing the traditional agricultural practices of ethnic minorities in the uplands of Vietnam categorise ethnic groups as being *conservers* and *non-conservers* of natural resources. The *Black Thai* minority is reported to practice a traditional form of 'conservation agriculture', growing paddy rice primarily on terraced fields. Other ethnic groups such as the *Hmong* are considered to be a threat to the forest as they practice 'slash and burn' (shifting cultivation), with relatively short fallow periods resulting in both forest depletion and soil degradation.

This distinction is false, both because of the difficulty in appraising sustainability and because of the social and farming systems differentiation within ethnic groups. As highlighted above, shifting cultivation cannot be regarded as an unsustainable practice *per se*; it may well be sustainable at relatively low population densities, such as below 50–70 persons per km[2] (World Bank 1995a) depending on soil properties and crops grown.[3] Indeed, van Keer *et al.* (1996) demonstrate that slash and burn agriculture can have both beneficial and negative effects on the environment. The pioneering work of Kunstadter and Chapman (1978) on shifting cultivation in Northern Thailand revealed that this practice can be at least as effective as irrigated agriculture in terms of labour productivity (cited by Rerkasem and Rerkasem 1998). In most regions of northern Vietnam, however, rural population densities nowadays exceed by far the carrying capacity of the fragile natural resources (Rambo 1995b).

In reality, agricultural practices are not necessarily determined by ethnic origin as perceived by both researchers and by government institutions, but rather are differentiated by agro-ecological conditions, the socio-economic environment and

*Table 7.3* Land use systems and land tenure status of different minority groups in
Song Da watershed, Son La Province, northwestern Vietnam (per cent)

| | Valley based systems | | Upland based systems | |
| --- | --- | --- | --- | --- |
| | *Paddy based system* | *Diversified system (paddy/others)* | *Medium altitudes (<800 masl)* | *High altitudes (>800 masl)* |
| *Ethnic group* | | | | |
| Thai | 93 | 73 | 48 | 1 |
| Hmong | 0 | 0 | 0 | 99 |
| Other ethnic groups | 7 | 27 | 52 | 0 |
| *Land tenure status* | | | | |
| Red Book Certificate or permitted tenure | 48 | 66 | 25 | 5 |
| Without permits | 51 | 34 | 73 | 95 |
| Rented | 1 | 2 | 2 | 0 |

Source: Social Forestry Development Project, Song Da watershed 1994.

demographic factors (Rerkasem and Rerkasem 1998). In Son La Province, the
*Thai* minority group, believed to have migrated to this area hundreds of years ago,
was able to choose the most favourable places for agriculture (Poffenberger 1998).
They mainly occupy the scarce places in the valley where paddy cultivation is
possible (see Table 7.3). The *Hmong*, who arrived later, were driven to the highly
erosive uplands where they rely primarily on annual food crops and opium
cultivation (Social Forestry Development Project 1994). Case studies of agricultural practices at the community level in Son La Province suggest, however, that
indigenous soil conservation practices are common among *Hmong* minority people
(Poffenberger 1998). Rambo and Le Trong Cuc (1996b) report that *Hmong*
farmers in the northern mountain regions, who are said to dislike wetland rice
cultivation, have in fact developed paddy fields by learning from *Dao* farmers
(quoted from Rerkasem and Rerkasem 1998). In the highlands of Lao Cai
Province, *Hmong* and *Dao* farmers have converted whole hillsides into impressive
landscapes of terraced fields.

Drawing on the case of minority farmers in southern China, who are generally
regarded as pure shifting cultivators, Brookfield and Padoch (1994) emphasise the
diversity of agricultural practices, involving more than 100 different agroforestry
combinations. Evidence from Thailand suggests that minority people are highly
flexible in adapting their agricultural practices to a changing economic environment. *Hmong* farmers in Chiang Mai Province of northern Thailand, for
example, are presently the main producers of highly profitable cash crops like
fruits, flowers and vegetables (Thailand Development Research Institute 1994).

Farmers in northern Vietnam's uplands are currently trying to intensify
agricultural production by increasing livestock numbers. The major share of
formal credit granted to farmers in Yen Chau district, Son La Province, and in
Ba Bê district, Bac Kan Province, is used for investments in animal husbandry

(Schenk 1998; personal observations 1999). However, formal credit institutions often request animals as collateral (see below). Hence, farmers are caught in a vicious circle: without credit they cannot buy animals and without animals they cannot obtain formal credit. These constraints on the intensification of animal husbandry has major implications for the further promotion of the highly sustainable VAC system. VAC is the Vietnamese acronym for combining gardening '*vuon*', aquaculture '*ao*', and animal husbandry '*chuong*'. This system has significantly contributed to the increase in farm income of lowland farmers but has not been significantly adopted by minority farmers in the higher altitudes of the uplands until now.

## Land allocation practices in the uplands and farmers' responses

There is increasing evidence that minority groups in Vietnam are disadvantaged with regard to access to permanent land use rights, for example, with the lack of progress on land allocation in the uplands after the 1993 Land Law. But even between different ethnic groups considerable inequalities can be observed arising from the land allocation process itself. In Son La Province, the implementation of the land allocation process has been initiated in the valleys rather than in the uplands. Thus, the *Black Thai* who occupy most of the paddy rice area, as shown in Table 7.3, have been favoured by the land allocation process, whereas land tenure of the majority of the *Hmong* remains without any legal status to their land.

Similar processes can be observed in Cho Don District, Bac Kan Province, where long-established *Tay* families who hold the political and administrative power control both the redistribution of lowland paddy fields as well as the allocation of sloping lands and forest areas. Thus, the *Kinh* and especially the *Dao* minority people are directly disadvantaged by being forced to give up their extensive agricultural practices with long fallow cycles due to lack of land resources. For the *Dao*, the shift from slash and burn agriculture to permanent agriculture has led to immense problems of adjustment in farming systems, and emigration is observed particularly with younger people leaving the villages as they are not able to adapt to the new state (Mellac 1997).

Flexible adaptation is also observed in these communities. In the same area, local scientists promote agroforestry systems[4] against erosion in sloping land. Although *Dao* minority farmers do not perceive soil erosion to be a major problem in this area, they adopt the new system to demonstrate their willingness to practice 'conservation farming'. They do not, however, use the stalks and leaves of the bushes to improve soil fertility and stop erosion in the uplands, as recommended by extension workers, but carry the mulch to their lowland paddy fields to improve rice yields (Gutekunst 1998). This unconventional adaptation of a new technology to specific needs is often regarded as a failure by scientists and extension agents. It may, however, be more sustainable within the *Dao*'s evolving farming systems. In summarising experiences from montane mainland of Southeast Asia, Rerkasem and Rerkasem (1998: 1328) conclude that 'preventing soil erosion has never been found to be among the farmer's primary land use objectives'. This is confirmed by interviews with key persons in various villages of Yen Chau District, Son La Province, as well as Ba Bê District and Cho Don District, Bac Kan Province, carried out by the author from 1997–9. Soil and water conservation practices often serve more obvious functions. Farmers in Cho Don District, Bac Kan Province, stated that they adopt soil conservation to secure their long-term land use rights. In some cases, the adoption of soil conservation measures seems to be a precondition to obtaining Red Book Certificates (Gutekunst 1998).

At this early stage of distribution of land use titles in the uplands, it is difficult to assess whether secured and individual land use titles stimulate long-term investments in the land. Some conflicts are emerging due to the allocation of forest land to individual households (see Chapter 6). The headman of the *Dao* village of Ban Kuan raised the problem of ruminant keeping on individualised forest land because free grazing of ruminants may cause conflicts between neighbours, implying that even under individualised tenure arrangements an effective communal system of regulation is still essential for sustainable land management (see also Adger *et al.* 1997).

### Access to formal credit

Access to rural credit is considered to be an important prerequisite for adopting new technologies and sustainable agricultural practices (Kerkvliet 1995a). In the context of *doi moi,* the Vietnam Bank for Agriculture was created in 1990 to provide banking services to rural people. According to a World Bank estimate, however, 75 per cent of rural credit needs were still met by informal credit sources in 1995. As the majority of the rural poor was not reached by the Vietnam Bank for Agriculture, the Vietnam Bank for the Poor was founded in mid-1995 with the mandate to provide microfinance for the poorest farm households in remote areas (Nguyen Xuan Nguyen and Nachuk 1998).

Results of a survey of 80 *Black Thai* farm households in four villages of Yen Chau District, Son La Province, suggest that the two formal institutions, the Bank for Agriculture and the Bank for the Poor, only satisfy 50 per cent of the demand for credit. Land use certificates had no significant impact on access to formal credit as land transfers are still uncommon due to bureaucratic restrictions and reluctance of rural households to sell their land use rights. Richer households, being able to use their assets such as concrete houses, animals or fruit trees as collateral, had significantly better access to formal credit than poorer households (Schenk *et al.* 1999). The *Hmong* being the poorest people in Yen Chau District seem to be totally excluded from access to formal credit. Their transaction costs are higher as compared to the *Thai* minority because they live in remote areas, often do not speak the official language and do not have the necessary collateral (Schenk 1998). Most of the rural women also mainly rely on informal credit sources backed up by evidence from Tran Thi Que (1998) who found that women are extremely disadvantaged as to access to formal credit in two communes in the northern Provinces.

## Access to markets and extension services

Experience from Thailand and elsewhere suggests that erosion control measures and more sustainable agricultural practices are more easily adopted on fields with highly productive crops such as irrigated vegetables or fruits (Thailand Development Research Institute 1994; Rerkasem and Rerkasem 1998). Given the low levels of rural infrastructure and access to markets and extension services throughout rural upland Vietnam, in contrast to the lowland intensive agricultural areas, the development of highly integrated production systems which are both ecologically sound and economically profitable is limited. Poorly maintained roads inhibit the production and marketing of perishable crops such as fruits. Cultivation of opium with a high value-to-weight ratio will remain attractive as long as transport costs are high. *Hmong* farmers in Hoa Binh are reported to earn as much cash from one kilogram of opium as from twenty tons of cabbage (Rambo 1995b). Access to extension is also directly related to the development of rural infrastructure as extension agents often target farmers in more easily accessible villages.

## Sustainability and the rural labour force

In general, sustainable agricultural practices can be characterised as being more labour intensive in the upland smallholder sector. Since many male household heads in rural areas are typically engaged in off-farm activities or migrate to urban centres, gender differentiation in labour inputs and the role of women are crucial in the realisation of sustainable agricultural practices. In rural Vietnam as a whole, a relatively high share of farms are managed by women. According to the 1989 population census, 1.9 million women became widows at an early age in the rural areas of Vietnam. There is at the present time, however, scant knowledge of the intra-household labour allocation among most ethnic minorities. A case study of Page and Thanh (1993) suggests that, among *Thai* households in Son La Province, it is the women who take most of the agricultural decisions for the household including new investments. In *Hmong* society, which is patriarchal and male dominated, women are excluded from important investment decisions and are tied to reproductive tasks and household work.

Research is required to identify opportunities and the constraints facing women-headed households in adopting conservation measures. Such research would provide valuable recommendations to extension services who so far target their activities primarily towards male farmers. Even in male-headed households, the role of women is still key: for example, throughout rural Vietnam women are observed to be in charge of livestock. Since the promotion of integrated livestock-based systems is a key issue for sustainability in the highlands, the different gender roles in the diffusion of sustainable practices is of the highest significance.

A major constraint for the implementation of labour-demanding sustainable agricultural practices in the highlands of northern Vietnam is the rising number of drug addicts among ethnic minority people.[5] As these drug addicts are mostly men, more workload is put on women who already contribute more than 70 per cent to field work in Vietnam's highlands (Tran Thi Van Anh and Le Ngoc Hung 1997).

## Conclusions

Government policies towards ethnic minorities in northern Vietnam represent a considerable externally imposed driving force with regard to the agricultural and socio-economic environment of many ethnic minority people. Perceptions of most of the traditional agricultural practices of these minorities by higher level government institutions remain: they constitute a threat to the natural resource base, particularly the forest cover in the uplands. The pressure for fixed settlement and for abandoning shifting cultivation practices in favour of permanent agriculture is often too rapid to allow sustainable adaptation by farming communities. Institutional assistance for such adaptation is usually inadequate.

Major constraints for ethnic minority people include: lack of permanent and secure land use rights; limited access to formal credit; and lack of access to markets and extension services. Long-term tenure security is an important prerequisite for the adoption of sustainable agricultural practices. When land use rights are not secure, it is observed that the motivation for farmers to adopt conservation measures promoted by development projects is not their economic profitability or perceived conservation benefits, but rather a strategy to claim long-term land use rights in an unstable environment. In these circumstances, effective allocation of land to individual households through Property Reports and Red Book Certificates in the mountainous areas of Vietnam could have beneficial effects for minority farmers by promoting long-term investments in profitable soil conservation practices.

Formal credit institutions in rural areas of northern Vietnam do not match the demands of the poorest households. Some ethnic groups such as the *Hmong* are extremely disadvantaged with regard to access to formal credit because of the remoteness of their settlements, because of language and communication difficulties with formal institutions, and by their lack of collateral. Integration of perennial crops and livestock in existing farming systems could potentially enhance sustainable production of the upland ethnic minorities but such integration is constrained by access to credit. Based on ideas discussed in Chapter 6 in the context of the forest land allocation process, the transaction costs for obtaining formal credit need to be considerably decreased for ethnic minorities.

Integrated farming systems with fruit trees and livestock depend on access to local and regional markets. As long as physical infrastructure is poor, minority farmers are forced to rely on annual food crops for home consumption or on cultivation of less perishable and high value crops such as the poppy. Sustainable agriculture in mountainous regions is more difficult and demands greater knowledge and skills than agriculture in high-potential lowland areas. Extension agents therefore require a more holistic view of existing farming systems in the uplands, should be better trained and should be given incentives to visit even remote rural areas.

Sloping agriculture technologies promoted by both government agencies and development projects are still not available for any agro-ecological conditions, are lacking economic profitability on the farm level and are often not flexible enough to cope with market dynamics. Unless sustainable agricultural practices that are able both to protect natural resources and increase rural incomes can be successfully implemented in mountainous areas, ethnic minorities in these regions will be threatened by even greater poverty and food insecurity than today or will return to illegal opium production.

## Notes

1 In Vietnam, cultivation of the poppy (*Papaver somniferum*) is only possible at an altitude greater than 800 m above sea level.
2 Integrated farming generally involves the integration of multi-cropping (e.g with fruit trees) and livestock production.
3 Rambo (1995b) considers only 20 people per km$^2$ as the threshold for sustainability.
4 Most of the agroforestry systems currently promoted in Bac Kan Province follow the patterns of Sloping Agricultural Land Technology (SALT) created by the Mindanao Baptist Rural Life Center in the Philippines.
5 According to Food and Agriculture Organization estimates, 35 per cent of the local opium producers are drug addicts.

# 8 Environmental change, ecosystem degradation and the value of wetland rehabilitation in the Mekong Delta

*Duong Van Ni, Roger Safford and Edward Maltby*

## Introduction

The Mekong, one of the world's great rivers, flows from its source on the Tibetan Plateau through, or along the borders of, Myanmar (Burma), Laos, Thailand and Cambodia before reaching the sea via a delta covering five million hectares, 80 per cent of which lies in Vietnam (Le Dien Duc 1989). The rich and relatively accessible natural resources of the Mekong Delta have made it one of the prime target areas for economic (principally agricultural) development in Vietnam during recent decades (Netherlands Engineering Consultants 1993; Vietnam General Statistical Office 1996, 1998). This development has involved land, water and human resources, and so inevitably has altered the delta's environment.

Two main challenges associated with this intense development now face environmental managers in the Mekong Delta. First, disturbance of certain soil types as a result of land development has caused acidification and pollution of surface water (Hanhart *et al.* 1997; Le Quang Minh *et al.* 1997, 1998). Second, large areas previously occupied by wetland ecosystems have been converted to alternative uses (Le Dien Duc 1989; Nguyen Van Nhan 1997), resulting in the loss of goods and services provided by those ecosystems (Maltby *et al.* 1996). One means of combating these problems of environmental degradation and associated widespread poverty is through enhancing sustainable income by diversification of income sources, thereby reducing trends towards land consolidation and reduction of farming populations (see Ellis 1998, on livelihood diversification). Rehabilitation of the delta's wetlands has been suggested as a means of providing opportunities for income diversification, as well as for improving environmental quality and restoring biological diversity (Le Dien Duc 1991; Maltby *et al.* 1996). Some wetland has been rehabilitated, but the potential benefits have not yet been fully realised (Le Dien Duc 1993; Beilfuss and Barzen 1994).

In this chapter, we focus on these problems of acidification and wetland loss, and identify a progression linking environmental degradation, pressure on natural ecosystems, and poverty. We present the first results of experiments that test whether, and if so how, wetland rehabilitation can enhance the quality of water in the acid soil areas typical of the Mekong Delta. We are expanding these experiments to test the practicability, economic viability and sustainability of a wetland-and-rice farming system designed to capitalise on the multiple benefits of wetlands in a demonstration plot, and we discuss briefly the significance of early findings from operating this system.

## The Mekong Delta

### Geography and early development

Nguyen Huu Chiem (1993) has divided the Mekong Delta into five land-form units. These are floodplains, which include natural levees up to one metre high along the two main arms of the river; a coastal complex of sand ridges, flats and mangroves; a broad depression covering most of the Ca Mau peninsula, which does not receive floodwater from the Mekong; an old alluvial terrace in the far northeast; and mountainous 'islands' of altitude up to 710 m which were never wetland.

Floodplains and the broad depression of Ca Mau cover most of the delta (Figure 8.1) (Tran Kim Thach 1980; Nguyen Huu Chiem 1993). These low-lying lands are referred to, henceforth, as the depression areas. On most of these, sediments rich in pyrite ($FeS_2$) were deposited under a tidal, brackish-water mangrove swamp. Fertile alluvial sediments were deposited on top of these sediments (Nguyen Huu Chiem 1993). Mekong floodwaters seasonally overtop the natural levees. Without artificial drainage, the lower-lying depressions further from the river were permanently submerged as large areas retained ponded water, the level of which gradually lowered as the dry season progressed (Beilfuss and Barzen 1994). Such areas were very rich in biological resources (Le Dien Duc 1991), with plant communities dominated by wetland forest and diverse grasslands.

The historical development of the Mekong Delta has been dominated by paddy rice cultivation, with associated creation of canals and new villages. Access to the extensive wetlands of the depression areas was originally very difficult, and these environments remained long undisturbed. The few inhabitants lived on the riverbanks where natural resources were abundant and a thick layer of alluvial sediment was available for agriculture. From the mid-nineteenth century, under French colonial rule, however, canals were extended into the depression areas (Brocheux 1995). The canals served to connect central provinces and districts, but also made the depression areas accessible to settlers, who created more villages and hamlets on natural raised 'islands' and along canal banks. Canals are now the major transport routes of the delta, and expansion of the canal system continues to the present.

### Land use changes since the 1960s

The canals and associated activities dramatically changed the face of the delta. Previously inaccessible and uninhabited areas were settled, and surface water drained quickly from the depressions. The average period of flooding in the depressions decreased from 12 months to between four and six months (Hanhart

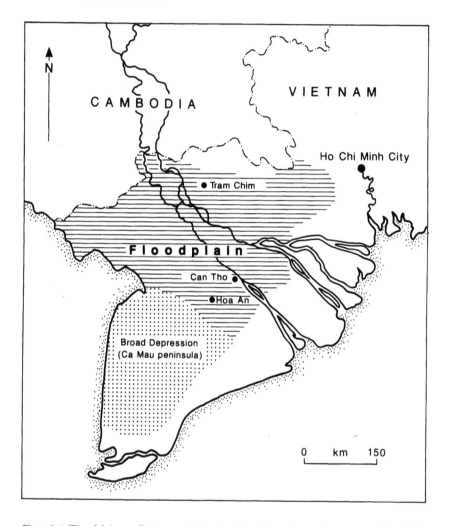

*Figure 8.1* The Mekong Delta and the two low-lying depression areas with different
        geomorphology and hydrology, labelled 'floodplain' and 'broad depression'
Source: Depression areas identified by Chiem (1993).

and Duong Van Ni 1993), and only the lowest areas remained submerged all year
round.

Where undrained depression areas were used for cultivation, the traditional
method was the *Mua* system, using a transplanted, long-stemmed rice variety in
the shallowest areas only. Rice was planted at the onset of the rainy season and
harvested after the recession of the flood (see Nguyen Huu Chiem 1994 for more
detail). Since the introduction of short-duration dwarf rice varieties in the delta in

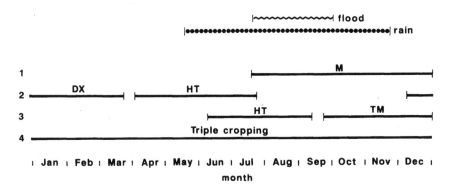

*Figure 8.2* Rice cropping systems calendar of the Mekong Delta
Key: M: *Mua* cropping pattern; DX–HT: *Dong-Xuan–He-Thu* cropping pattern; HT–TM: *He-Thu* – Middle *Mua* cropping pattern.

the late 1960s, rice cropping systems have changed completely over most of the delta. A representation of these alternative cropping calendars is shown in Figure 8.2. Farmers began to cultivate an early rainy season (*He-Thu*) crop harvested before the flood. Where fresh water is available from the irrigation system, farmers are also able to cultivate a dry season or winter–spring (*Dong-Xuan*) crop planted after the recession of the flood and harvested near the end of the dry season. In areas with intrusion of saline water into the canals, it is impossible to cultivate an irrigated dry season crop. Where the rainy season is long enough, however, farmers cultivate an early rainy season crop (*He-Thu*) using early ploughing and dry seeding techniques, followed by a long-stemmed short duration variety (TM, or Middle *Mua*) harvested after the recession of the flood. On the levees, with unrestricted drainage and irrigation, triple cropping is possible.

These recent changes from single to multiple rice cropping were made possible by the new canals, but driven also by interrelated socio-economic changes. The major socio-economic changes are population increase partly through immigration, the national drive towards increased rice production through tax incentives, available credit from banks and the policy of encouraging landless farmers to 'tame the wild lands', and the rise of the free market.

### The Mekong Delta in the late 1990s

The human population of the Mekong Delta in 1996 was 16.9 million (Vietnam Department of Agriculture, Forestry and Fishery 1996), growing at around 2.2 per cent per annum (data for 1990 from Netherlands Engineering Consultants 1993). Three main types of environment now exist in the delta. First, towns and cities (defined as areas with population density over 600/km²) cover 10 per cent of the land area and represent 30 per cent of the population. Second, agricultural land including villages and hamlets with economies dependent on agriculture, covers 83 per cent of the land area and represents approximately 70 per cent of the population (Vietnam Department of Agriculture, Forestry and Fishery 1996). Rice dominates agricultural land use, although upland crops are also important locally. The third environment, natural or semi-natural wetland ecosystems, now covers less than 7 per cent of the delta, mainly in the depression areas (Nguyen Van Nhan 1997). The resident population of the depression is insignificant although many rural dwellers utilise their resources. About one third of the area of wetland ecosystems is swamp forest dominated by the tree *Melaleuca cajuputi*, a papery-barked species in the family Myrtaceae, whose distribution extends from Southeast Asia to Australia (Blake 1968). Most *Melaleuca* is either replanted or naturally regenerated forest managed by private forestry companies or Provincial authorities, or is effectively open access land.

## Environmental problems

Reclamation of the depression areas of the Mekong Delta for agriculture has been critical in allowing a rapid increase in national rice production – a 4.6 per cent annual growth rate in 1995 (Vietnam General Statistical Office 1996) – and promoting economic growth (Le Quang Tri *et al.* 1993; Vo-Tong Xuan 1993). The large socio-economic and environmental changes have created environmental problems. Saline intrusion and soil acidification have increased (Netherlands Engineering Consultants 1993), storm or flood damage were very severe in 1996 and 1997 (personal observation, all authors) and natural ecosystem functions (including biodiversity support) have been lost. Overcoming these problems is vital for the area's future. This account focuses on soil acidification and loss of ecosystem functioning, as these problems can be addressed most directly by the local development initiatives discussed in the next section.

## Soil acidification and its immediate consequences

The lengthening of the dry period in the depression areas results in increased evapo-transpiration. In these circumstances, the ground water level in the depressions often drops below the soil surface. Aeration of subsoil layers allows oxidation of the sulphur in the pyrite-containing sediment, and acidity is released (Dent 1992; van Breemen 1993). The process may be simplified as follows:

$$FeS_{2(s)} + 7/2\ O_{2(aq., g)} + H_2O_{(l)} \rightarrow Fe^{2+}_{(aq.)} + 2SO_4^{2-}_{(aq.)} + 2H^+(aq.)$$

These soils are referred to as acid sulphate soils and cover almost half of the Mekong Delta (Vietnam State Commission for Sciences 1991). Acid sulphate soils are often considered to be a major constraint on agricultural intensification (Dent 1992). Under severely oxidising conditions with high acidity, jarosite $(KFe_3(SO_4)_2(OH)_6)$ forms, which has a characteristic yellow colour often seen in material dredged from ditches and canals. Among many associated reactions, the hydrogen ions formed in the oxidation reactions react with the clay mineral aluminium, liberating dissolved aluminium (aqueous $Al^{3+}$), highly toxic to plants (van Breemen 1980).

The excavation of canals two to three metres deep brings fresh pyrite-containing material to the surface. The resulting dikes are capable of acidifying the canal and surface water for many years, depending on the drainage capacity of the canal (To Phuc Tuong *et al.* 1989). During the first few years, pH values of 2.5 to 3.5 in the canal water are not unusual. Acidification of canal water reaches a yearly peak at the onset of the rainy season (June to July) resulting in dramatic fish deaths downstream. The dikes also contribute strongly to acidification of the surface water of adjacent land at the same time.

These processes had little effect on the early settlements on the levees and in undrained depression areas, for a number of reasons. First, the pyrite-rich sediments on the levees were deeply overlaid with alluvium, and thus topsoil could be drained without danger of pyrite oxidation (van Breemen and Pons 1978). Second, year-round flooding of shallower alluvial sediments prevented aerial oxidation of pyrite in the depression areas. The traditional *Mua* rice-cropping system, for example, did not require drainage. Acidification was, therefore, most severe on drained depression areas.

Many new settlers using canals to penetrate depression areas were from higher areas, often far from the Mekong Delta, and were ignorant of the dangers of acid sulphate soils. They burned the original vegetation during the dry season and planted rice, often unsuccessfully. Rice seeds were killed by high toxin concentrations in surface water and topsoil solution. When the seedlings survived, the yields were generally very poor, usually below 0.5 tons $ha^{-1}$. Settlers abandoned their land to settle elsewhere, and the original forest and diverse grassland ecosystems did not recover on the degraded soil. Instead, the acid-tolerant sedge *Eleocharis dulcis* became dominant over wide areas. This contrasts with the high species richness often found in some other grassland types in the delta. For example, in a recent, single-site survey, 95 species of grass and sedge were found (Tran Triet *et al.* in press).

Overall, significant land use change in the depression areas as a result of agricultural expansion led to a rapid increase in soil acidification. Release of acidity and associated toxins has affected wide areas, and this has become one of the major challenges for land management in the Mekong Delta.

## Loss of biological diversity and ecosystem functions

The undrained wetlands of the depression areas were important habitat for flora and fauna (Le Dien Duc 1989). A few of the most biologically diverse and productive natural freshwater ecosystems in the Mekong Delta are strict nature reserves, covering a total of around 20 000 ha (Le Dien Duc 1989; Safford *et al.* 1998). Habitat alteration and intensive exploitation of remaining wetland have resulted in population declines and extinctions of numerous species that cannot exist year-round on paddy fields or are intolerant of intensive exploitation. Surveys

*Table 8.1* Farm household income by sources for Hoa An, Cantho Province, and the
Mekong Delta region

|  | *Mekong Delta** | *Hoa An* |
|---|---|---|
| Income per household (VND 000) | 6804 | 4235 |
| Income per person (VND 000) | 1265 | 706 |
| Source of income |  |  |
| On-farm resources (%) | 41 | 40 |
| Off-farm resources (%) | 33 | 25 |
| Off-farm wage labour (%) | 23 | 28 |
| Natural resource exploitation (%) | 3 | 7 |

Source: Duong Van Ni and Vo-Tong Xuan (1998).

Note
*Estimates from Dao Cong Tien (unpublished data).

since 1995 by Safford and colleagues (Safford *et al.* 1998; Tran Triet *et al.*
in press; unpublished data) have failed to locate several rare waterbird species,
such as Giant Ibis *Pseudibis gigantea,* Milky Stork *Mycteria cinerea* and Greater
Adjutant *Leptoptilos dubius.* These species were all previously reported in the delta
(Le Dien Duc 1989). Many other species have become rare and localised, and
these include directly harvestable taxa such as several fish species (Duong Van Ni,
personal observation). At the same time, reduction in habitat area (Nguyen Van
Nhan 1997) has caused declines in valuable species such as wild rice *Oryza rufipogon,*
so that genotypes that might have proven valuable in selective breeding have
almost certainly been lost. Finally, services provided by the wetlands and their
biodiversity (Safford and Maltby 1997), such as water purification and floodwater
storage, have been reduced.

### Environmental change, poverty and ecosystem degradation

Many land-owning farmers have proven unable to adapt over recent decades to
the socio-economic and environmental changes that have affected the Mekong
Delta. Debts built up and a proportion of farmers were forced to sell their land in
order to pay them off. Landless farmers now constitute 15–20 per cent of the
delta's population (Duong Van Ni and Vo-Tong Xuan 1998). There are three main
livelihood options for landless households. The first is to remain and work for those
farmers who still have land. Average income from such farm labour in the Mekong
Delta is around 30 per cent of the national average (Duong Van Ni, personal
observation). Farmers dependent on farm wage labour, therefore, typically suffer
extreme poverty. The second option is to migrate to the towns and cities, but this
is often a risky strategy, as shown by high urban unemployment rates (Duong Van
Ni, personal observation). Finally, many households turn to the harvest of natural
resources in order to supplement their incomes, or in some cases to provide their
main income source.

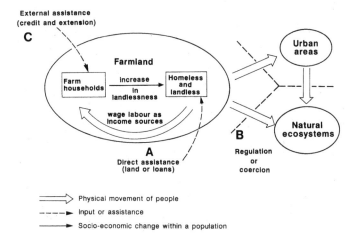

*Figure 8.3* Conceptualising the relations between natural resources and rural inhabitants of the Mekong Delta

Source: Adapted from Duong Van Ni and Vo-Tong Xuan (1998).

Data from a farming community at Hoa An, a typical acid sulphate soil area in Cantho Province, illustrate the dependence on farming and the problems of poverty, the need to enter the labour market and the pressure to exploit natural resources (Table 8.1). Around 7 per cent of income equivalent results from extraction of products from wetland areas, and one quarter from off-farm wage labour. Also, at Hoa An, knock-on effects, such as competition between labourers for the limited available work, have increased social conflicts within the community (Duong Van Ni and Vo-Tong Xuan 1998). This sequence of events results in ever-increasing pressure on natural ecosystems, environmental degradation and poverty as represented in Figure 8.3. Further damage to wetlands results when failed migrants to towns and cities return, landless, to the depression areas to harvest natural resources.

## Potential management strategies

For continued economic development to be achieved with minimal environmental degradation, the negative progression described above and in Figure 8.3 must be broken. A number of strategies for intervention have commonly been used. The first is the provision of direct assistance such as land or loans to landless people in order to avoid the need to move to towns or exploit natural resources (link A in Figure 8.3). Such an approach may fail if the ultimate causes of the changes are not also addressed. Households that had recently been landless may lose their land or money again, and indeed, this often happens more rapidly than in the first instance (Duong Van Ni, personal observation). The second approach (B in

Figure 8.3) is to use laws or force to protect natural resources, or to prevent migration to towns and cities. Regulation and exclusion often cause resentment among poor households, and such regulations are often subverted or at least challenged. The third approach (C in Figure 8.3) is to attempt to avoid the need for farmers to sell their land, by providing education, technical assistance and short-term financial assistance.

Income raising and diversification strategies for income sources are central to this final form of intervention. Successful activities on acid soils in various parts of the delta include rice–shrimp and rice–crab systems, fishponds, various food-crops and traditional wetland plants raised as crops (Vo-Tong Xuan 1993). All of these activities can be combined with livestock rearing. Three techniques have been successful at Hoa An (Duong Van Ni and Vo-Tong Xuan 1998). First, in rice–fish systems, fish reduce the impact of insect pests on rice. They can also be harvested from the paddies and, hence, provide valuable protein. Second, rearing livestock, particularly ducks and pigs, has been widely adopted, using family resources efficiently and hence reducing wage labour competition. The third technique has drawn upon *Melaleuca* trees. These are one of the most important resources that have supported the socio-economic development of the Mekong Delta's human communities for many generations (Le Dien Duc 1991; Maltby *et al.* 1996). *Melaleuca* wetlands have also become economically important elsewhere in Southeast Asia (Wyatt-Smith 1963), particularly to rural communities.

A wide variety of direct values of *Melaleuca* is documented (Brinkman and Vo-Tong Xuan 1991; Safford and Maltby 1997), such as wood, honey, essential oil, fodder, medicines and edible species among associated plants. It has also been suggested that *Melaleuca* can provide major benefits in environmental quality, in particular in improving water quality in the acid sulphate soil areas that dominate the areas of the Mekong Delta where environmental problems are most severe. Several factors, however, have made it difficult to capitalise upon these benefits. Primarily, the hurdles stem from the need of households for rapid returns from investment. These rapid returns are inherently difficult to obtain from tree crops. A lack of investment for rural development in remote areas, and the difficulty of demonstrating the true value of the benefits, both qualitatively and quantitatively, compounds the problem of rapid return. The improvement in water quality goes unnoticed, and such benefits are invariably ignored in economic analyses locally. Understanding the ecosystem processes that affect water quality, and applying this understanding to develop optimal land use systems, is central to overcoming these technical and practical problems.

The remainder of this chapter describes a series of experiments and farm trials with three main aims. The first is to understand the processes that affect surface water quality in *Melaleuca* ecosystems. The second aim is to demonstrate water quality benefits that can be provided for farmers living on acid sulphate soils by a simple model which can prevent acid water run-off and reduce environmental pollution. The third aim is to evaluate the ability of the community to adopt such agronomic practices. Tentative conclusions based on initial results of this demonstration plot are presented.

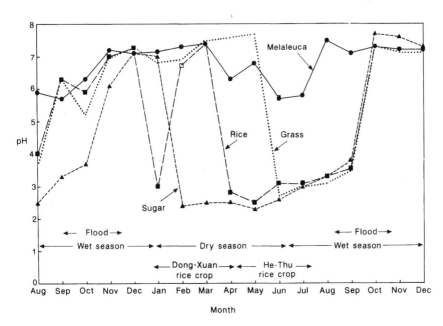

*Figure 8.4* Seasonal variation in acidity of surface water in relation to present land uses on acid sulphate soils at Hoa An, Mekong Delta

## *Melaleuca* management for development and environmental quality

### *Background*

In the early 1990s, surface water quality on acid sulphate soils under different forms of land use in the Mekong Delta was monitored routinely. The results were unexpected. Acidity in the surface water on a flooded *Melaleuca* plot remained low: pH was always above 5, while values in all neighbouring croplands dropped below 3, as shown in Figure 8.4. These data indicate a difference of great significance to water resource users. For example, at pH<4, a one unit decrease in pH is typically associated with a ten-fold increase in the concentration of dissolved aluminium (van Breemen 1993; confirmed in Vietnam by Hanhart and Duong Van Ni 1993).

### Melaleuca *ecosystem processes*

In order to understand the low acidity on the flooded *Melaleuca* plot, a series of experiments is being carried out at Hoa An, investigating the components and processes that may affect the quality of surface water in *Melaleuca*. The rainy season extends from June to December with 90 per cent falling in four months (August to November). Every year, surface water accumulates from rain, run-off

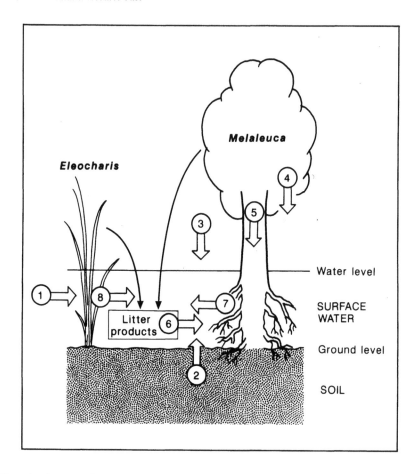

*Figure 8.5* Sources of dissolved substances in surface water in a *Melaleuca* ecosystem in the
Mekong Delta
Key: 1 Run-off; 2 Ground water; 3 Rain; 4 Rain through canopy; 5 Trunk flows;
6 Litter decomposition products; 7 Living *Melaleuca* roots; 8 Living *Eleocharis*.

(floodwater) and ground water. These water sources each carry different dissolved
substances, whose mixing and dilution may chemically affect the surface water
quality (sources 1, 2 and 3 in Figure 8.5).

The changes in water quality are caused by processes in the *Melaleuca* ecosystem.
Rain and rainwater passing through the canopy may dissolve some substances and
add them to water on the ground (processes 4 and 5 in Figure 8.5). The most
abundant associated plant species in *Melaleuca* ecosystems, especially while the
*Melaleuca* is still young, is the acid-tolerant sedge *Eleocharis dulcis*. When
the *Melaleuca* matures, *Eleocharis* is shaded out and its litter buried in the litter from
the *Melaleuca*; this produces a very thick litter layer, to which other plant species

also contribute. Decomposition products from this layer add to the surface water, as shown in process 6, Figure 8.5. The roots of various plants growing on acidic soils release organic acids, which bind strongly to ions such as aluminium and iron in surface water (Jones *et al.* 1996), rendering them non-toxic (processes 7 and 8, Figure 8.5).

The results of the experiments conducted to date lead to a number of conclusions. First, precipitation dripping through the *Melaleuca* forest (down trunks and through the canopy) does not affect the pH of surface water, apart from minor dilution effects. Second, the leaf litter in *Melaleuca* ecosystems significantly affects the quality of water, demonstrated by decreased acidity and lower mobile iron and aluminium concentrations. Thus, litter fall in the *Melaleuca* forest enhances surface water quality, and may play a major role in improving quality of water resources. Living plant matter may also have a role to play in water quality; this is currently being tested.

## Application to land management

Applications of the results of these fundamental experiments have been investigated in 1996–8 by a field trial or demonstration at a scale useful for farmers (15 ha). Farmers participated in the trial, and field methods were partly based on their experience. A 6-ha plot of five-year old *Melaleuca* on severely acid soil was impounded adjacent to a 9-ha plot used for rice production (as shown in Figure 8.6). The *Melaleuca* was used as a reservoir for floodwater which was used for irrigation of dry season rice crops; enough water was stored to allow the system to remain closed during the rice growing season. Acidified water drained off the rice paddies was diverted into the *Melaleuca* land, rather than into canals or other water sources (Figure 8.6). The *Melaleuca* thus became a sink for the acid water. The system allows farmers to benefit in the first few years from rice and also products associated with the *Melaleuca* land such as fish, honey bees, wood and wild vegetables. *Melaleuca* grows well on acid soil (Brinkman and Vo-Tong Xuan 1991) and can serve as collateral for bank loans to finance start-up of new farming techniques.

The areas of *Melaleuca* and rice, and sizes of dikes, ditches, and gates were determined by the water budgets for rice and *Melaleuca*. Operating costs, including yearly expenses (such as farm labour) on rice land and *Melaleuca* land, were reported for comparison of benefits between *Melaleuca* and rice crops. Farmers' experiences and suggestions were also noted in order to simplify the practical procedure.

The results have positive implications. The demonstration plot exhibited important improvements in the quality of water drained off the rice paddies. This is shown by reduced acidity, and decreased aluminium and iron concentrations. Diversification allowed farmers to raise incomes quickly; farmers who were previously landless obtained their first profits from rice, fish and livestock within three years, even before the *Melaleuca* started to yield income from wood (which begins with thinnings after the third year). The farmers were willing and able to

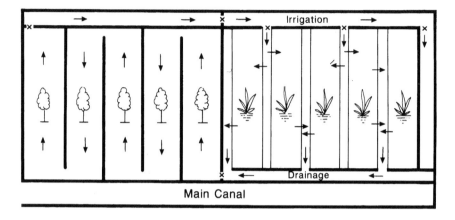

*Figure 8.6* Demonstration and experimental plot of the new rice-*Melaleuca* farming system at Hoa An, Mekong Delta

adopt such practices. The main factors limiting their ability to do so were related to poverty and lack of resources. Households were limited in land availability and capital to develop the water management system. Institutions, including farmers' organisations, played a crucial role in the model's development, through key inputs by individuals, and by ensuring community participation. Social benefits of adoption of *Melaleuca* integrated systems can also be surmised. Greater food security and more stable year-round labour requirement of a diverse farming system, for example, freed children to attend school. However, long-term effects on the economy and environmental quality are certain to be determined also by the macro-economics of the environment and by outside support such as infrastructure development, training and extension.

### Future developments and applications

Our research on *Melaleuca* ecosystem processes and their application to community-based natural resource management is continuing. The results so far, though, strongly reinforce the Vietnamese government's programme of rehabilitating and maintaining *Melaleuca* on acid sulphate soil areas at a larger scale. The farming model including *Melaleuca* use is well suited to implementation in an extension system or rural development project.

A second trial encouraging farming diversification is taking place in the Plain of Reeds, in the vicinity of Tram Chim National Reserve (Duong Van Ni 1997).

As at Hoa An, rice cultivation continues but is complemented by other land uses including *Melaleuca* plots. The reserve is surrounded, and its biodiversity threatened, by communities who suffer severe poverty and adverse environmental conditions caused largely by acid sulphate soils. By linking the development initiatives to management plans for the reserve, it is hoped that the reserve's biodiversity will be maintained while the living standards and resilience of local resource users are enhanced.

## Acknowledgements

This work was carried out as part of the Darwin *Melaleuca* Wetlands Project, which is funded by a grant from the Darwin Initiative for the Survival of Species, administered by the Department of Environment, Transport and the Regions (UK government). The project is implemented by the Royal Holloway Institute for Environmental Research (Royal Holloway University of London), Cantho University and other collaborating organisations. Additional support has been provided by British Embassy in Vietnam, IUCN – The World Conservation Union, Royal Holloway University of London, the British Council (Vietnam), and British Airways Assisting Conservation. Particularly valuable assistance has been provided by the following staff in Vietnam: Professors Vo-Tong Xuan and Tran Thuong Tuan, Mr Vo Lam and Mr Tran Duy Phat.

# 9 Mangrove conservation and restoration for enhanced resilience

*Nguyen Hoang Tri, Phan Nguyen Hong,*
*W. Neil Adger and P. Mick Kelly*

## Introduction

This chapter considers one aspect of ecosystem management widely practised in coastal Vietnam, namely, mangrove restoration, undertaken to enhance and restore local environments as well as to benefit local populations. It is argued, in the context of this volume, that mangrove restoration enhances the resilience of ecosystems at the landscape scale through enhancing species and functional diversity of suitable coastal regions. In addition, such practices enhance the resilience and adaptive potential of the coastal population.

Here, we discuss the economic benefits of mangrove restoration to local populations and quantify them for case study areas in the Red River Delta region. Data from three districts in Nam Dinh Province are used to demonstrate that reinstating mangroves can be a desirable activity both from the local perspective and in buffering coastal regions from the impacts of coastal flooding (see Chapter 10). Thus, the direct economic benefits and the environmental benefits of enhanced coastal protection both enhance the resilience and the adaptability of these coastal populations. It would appear that mangrove planting is a prime example of a 'win–win' activity in the case of present-day coastal landscapes and communities in improving the livelihood of local resource users as well as enhancing sea defences. Kelly *et al.* (1994) argue that, in the context of long-term environmental change, facilitating adaptation must involve identifying such 'win–win' situations in which action to reduce future risk also minimises vulnerability in the present day. The ownership, control and property rights surrounding such activities, as discussed in Chapter 5, remains the key constraint to the sustainability of mangrove use in this context.

## Trends in mangrove ecosystem conversion

The term 'mangrove forests' refers to those areas which include some species of mangrove tree. These forests are commonly found in the intertidal tropical and sub-tropical coastal wetlands. Despite being amongst the most productive ecosystems in the world, the global area of mangrove forests has been declining to less than half the former area from the beginning of the twentieth century (Field

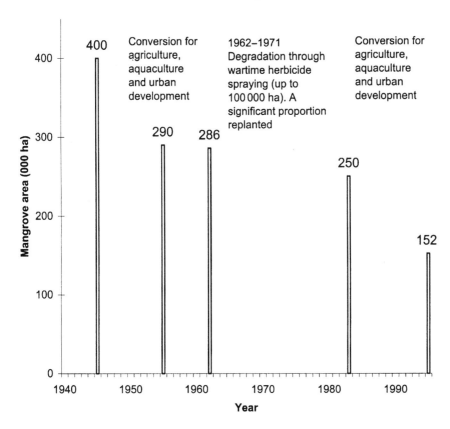

*Figure 9.1* Total mangrove area in Vietnam, 1945–1995
Source: Based on Phan Nguyen Hong (1994) and various sources including Mauraud (1943) and Rollett (1963).

*et al.* 1998). These trends are due to conversion for agriculture, forestry and urban uses and the extraction of timber for fuel and other uses (Farnsworth and Ellison 1997). In Vietnam, large areas of mangroves have been converted to agriculture and, in particular, to shrimp aquaculture, causing ecological disturbance and enhancing instability in the coastal physical environment (Phan Nguyen Hong and Hoang Thi San 1993).

The total mangrove area of Vietnam has been in decline in the second half of this century, according to historical estimates of the total area as shown in Figure 9.1. Before the two Indochina wars, mangrove forests were distributed extensively in coastal and estuarine areas. They were estimated to have a total area of around 400 000 ha in 1943 (Maurand 1943), with over 250 000 ha in the southern parts of the country, concentrated in the Ca Mau peninsula (150 000 ha) (Phan Nguyen Hong 1994). Conversion has occurred throughout the past half

*Table 9.1* Estimated timber loss from direct degradation by herbicides and subsequent productivity for selected sites in southern Vietnam

| Location | Potential production loss over 1971–1995 (000 m³) | |
|---|---|---|
| | Marketable timber | Other usable wood |
| Can Gio, Ho Chi Minh City | 1206 | 705 |
| Mekong river mouths | 1721 | 1208 |
| Ca Mau Peninsula | 4268 | 741 |

Source: Phan Nguyen Hong *et al.* (1997).

century and an estimate by Rollett (1963) puts the area at less than 300 000 ha in the early 1960s.

From 1962 to 1971, nearly 40 per cent of the mangrove area in southern and central Vietnam, or almost 160 000 ha, was destroyed as a result of chemical warfare, including over half of the mangrove in the Ca Mau peninsula (Phan Nguyen Hong 1994). Estimates of the impact on productive areas suggest that more than four million cubic metres of usable timber (Column 1 in Table 9.1) was destroyed during this period, with severe impacts on the fauna and flora in these areas. In addition, the legacy of dioxins remains even today, not only in the higher fauna but also in human populations in these areas. A study by the Hatfield Group (1999), reported in Cooney (1999), argues that dioxins persisting in the environment from the spraying of Agent Orange have moved into the food chain and that people living in heavily sprayed areas experience higher rates of deformities, early cancers and other medical conditions.

But, as shown in the estimates of total mangrove area in Figure 9.1, a significant proportion of the area degraded by herbicides was subsequently replanted, such that a mid-1980s estimate of the total area suggests that around 250 000 ha of mangrove remain (Phan Nguyen Hong 1994). Despite this restoration effort, major mangrove forests all along the coast of Vietnam, including those rehabilitated, have now been converted again for use in agriculture, aquaculture and human settlement. The best estimate of the remaining area of mangroves is shown by province in Table 9.2, which also shows that nearly one quarter of the coastline of Vietnam is now protected by sea dikes. Where agricultural land has been reclaimed from mangrove, it is often necessary to build significant sea dikes to protect the land from storm surges induced by tropical storms where once the land would have been buffered by the mangrove forests.

Mangrove conversion represents an important ecological, as well as economic, issue. The functions and services provided by mangrove areas are diverse and have been well documented and appraised (Lugo and Snedaker 1974; Mitsch and Gosselink 1993; Ewel *et al.* 1998b; Field *et al.* 1998). Mangrove ecosystems are also diverse in their species, yet little is really known about the role of biodiversity *per se* in maintaining the services they provide. It has been shown that the functions and services provided by the mangrove ecosystem have positive economic value,

*Table 9.2* Coast and sea defence length and mangrove area for the coastal provinces
of Vietnam

| Coastal provinces | Protective dikes | | | | Mangrove (ha) |
|---|---|---|---|---|---|
| | Coastline (km) | River dikes (km) | Sea dike (km) | Total | |
| *North* | | | | | |
| Quang Ninh | 377 | 66 | 64 | 130 | 13 294 |
| Hai Phong | 110 | 48 | 67 | 115 | 3 382 |
| Thai Binh | 50 | 135 | 135 | 270 | 4 200 |
| Nam Dinh | 65 | 15 | 76 | 91 | 4 000 |
| Ninh Binh | 17 | 6 | 24 | 30 | 400 |
| Thanh Hoa | 83 | 59 | 41 | 100 | 700 |
| Nghe An | 83 | 120 | 33 | 153 | 600 |
| Ha Tinh | 130 | 304 | 17 | 321 | 645 |
| *Central* | | | | | |
| Quang Binh | 115 | 92 | 4 | 96 | 100 |
| Quang Tri | 65 | 94 | 132 | 226 | 0 |
| Thua Thien Hue | 105 | 30 | 126 | 156 | 10 |
| Quang Nam Da Nang | 189 | 68 | 31 | 99 | 30 |
| Quang Ngai | 122 | 71 | 65 | 136 | 205 |
| Binh Dinh | 206 | 92 | 28 | 120 | 10 |
| Phu Yen | 204 | 39 | 6 | 45 | 54 |
| Khanh Hoa | 422 | 0 | 12 | 12 | 33 |
| Ninh Thuan | 115 | 0 | 0 | 0 | 9 |
| Binh Thuan | 216 | 0 | 0 | 0 | 9 |
| Ba Ria-Vung Tau | 104 | 0 | 9 | 8 | 300 |
| *South* | | | | | |
| Ho Chi Minh | 51 | 0 | 1 | 1 | 40 000 |
| Tien Giang | 40 | 0 | 43 | 43 | 4 000 |
| Kien Giang | 154 | 0 | 146 | 146 | 2 532 |
| Ben Tre | 103 | 0 | 33 | 33 | 6 126 |
| Tra Vinh | 68 | 0 | 48 | 48 | 5 591 |
| Soc Trang | 75 | 0 | 14 | 14 | 3 058 |
| Bac Lieu – Ca Mau | 435 | 0 | 239 | 239 | 65 779 |
| **Total** | **3 704** | **1 239** | **1 395** | **2 633** | **155 067** |

Source: Reported in Nguyen Hoang Tri *et al.* (1997) based on various sources.

though this is often ignored in the ongoing process of mangrove conversion
(Barbier 1993; Ruitenbeek 1994; Daily 1997; Barbier and Strand 1998). In
Vietnam, conversion often occurs because traditional common management of
the resource is overridden and functions and services are undervalued, as discussed
in Chapter 5 (see also Bailey and Pomeroy 1996; Adger and Luttrell 2000).
Necessary first steps in promoting sustainable utilisation of such resources are,
therefore, identification of relevant functions and services and the incorporation
of these into the land allocation process alongside the encouragement of
appropriate property rights, whether communal or private.

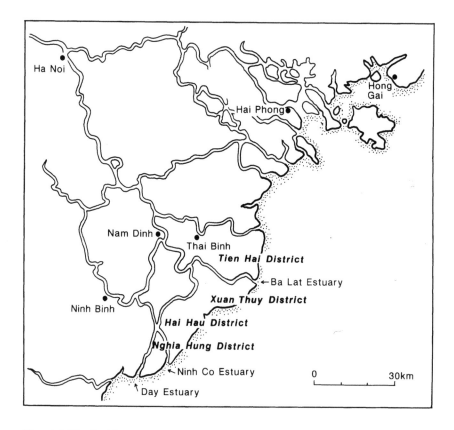

*Figure 9.2* The Red River Delta and mangrove research locations

Two related environmental characteristics of mangrove habitats facilitate their restoration. First, substrate and physical factors only loosely explain the natural zonation of mangrove trees (Ewel *et al.* 1998a). In other words, there is some flexibility in the conditions in which mangroves can be planted. Second, disturbed mangrove ecosystems lead to rapid vegetation re-establishment and substrate stabilisation (Field *et al.* 1998). These two factors suggest that the physical niches in which mangroves can enhance landscape resilience are widespread. This is particularly so given the circumstances in Vietnam where much planting is aimed at restoring former mangrove areas.

The remainder of this chapter documents the economic rationale behind mangrove restoration in the case of three coastal districts of Nam Dinh Province in the southern Red River Delta area. In these areas, mangrove rehabilitation is often subsidised by international development agencies such as the World Food Programme, Oxfam and Save the Children Fund as part of rural income generating projects, based largely on an assumed benefit to local communities. This analysis provides some quantification of the economic and environmental

*Table 9.3* Estimated total mangrove area in Nam Dinh Province

| District | Present mangrove areas (ha) | Land estimated to be available for planting (ha) |
|---|---|---|
| Xuan Thuy | 3 000 | 7 640 |
| Hai Hau | 200 | 641 |
| Nghia Hung | 5 200 | 9 826 |
| **Province total** | **8 400** | **18 107** |

Source: Nam Dinh Province data.

parameters of restoration in the context of resilience and adaptability to environmental and social change at different scales.

## Mangrove restoration in Nam Dinh Province

### *The remaining mangroves of Nam Dinh Province*

Nam Dinh Province, located in the southwest of the Red River Delta in northern Vietnam (Figure 9.2), includes three coastal administrative units, Xuan Thuy, Hai Hau and Nghia Hung Districts,[1] and has a sea dike system to protect people, houses, infrastructure and crops. Freshwater reserves help mitigate against the impacts of saline intrusion, flood, storm and sea water rise. The total population of the three coastal districts of Xuan Thuy, Hai Hau and Nghia Hung is almost 0.5 million, with a population density of more than 1000 per km$^2$ which is typical of the densely populated areas of the Red River Delta plain. The total area of the three coastal districts is approximately 72 000 ha.

Within Nam Dinh Province, the impacts of severe storms, as elsewhere, are generally concentrated in the coastal districts (see examples in Chapters 2 and 10). The economy of these districts is primarily dependent on agriculture with paddy cultivation, aquaculture and salt making being the major agricultural activities. Each of these activities is susceptible to, and differentially affected by, typhoon impacts. The impact of typhoons on coastal Vietnam is significant (Chapter 10). Over 400 000 hectares of crops were lost in the coastal provinces of Vietnam as a result of tropical cyclone impacts over the 10-year period 1977–86 (Trinh Van Thu 1991). Estimates of the magnitude of impacts in Nam Dinh Province from floods and typhoons for the 20 years between 1973 and 1992 show that there were more than 990 injured people, including fatalities, in total and over VND 470 billion damage (1993 constant prices) as a result of severe storms. Protecting vulnerable coastal areas from typhoon impacts is, therefore, of high social and economic importance.

Given the prevailing circumstances in the coastal districts of Nam Dinh Province, and similar regions up and down the coast of Vietnam, it is clear that mangrove restoration can have a variety of benefits where the topography of the coastal shelf and other social, physical and ecological factors are appropriate. In such situations, mangrove restoration can provide income where households

*Table 9.4* Spatial distribution and nature of the economic benefits of mangrove
restoration and conservation

| Type of service | Location of goods and services | |
|---|---|---|
| | On-site | Off-site |
| Marketed goods and services | *Timber*, tannin | *Honey from bee-keeping*, fishing |
| Non-marketed benefits | Fish nursery function, *medicinal plants*,* crisis foodstuff, fodder, wildlife habitat | *Storm protection function:* • *avoidance of cost of maintaining sea defences and dikes* • avoidance of impact on agriculture and infrastructure |

Source: Adapted from Dixon and Lal (1997).

Notes
Goods and services in *italics* are quantified in this study.
*Medicinal plants utilised are surveyed and classified, but no economic value is estimated (see text).

are often severely constrained in cash income sources, as well as bringing about
environmental benefits in terms of productive assets and reducing the impact of
coastal storm surges.

### Mangrove restoration: economic analysis

Some economic values of the goods and services resulting from mangrove
restoration can be assessed by observation of existing markets, but some of the
benefits and services are indirect. The crucial aspects of value for local decision-
making, and for the differential impacts of environmental change, are whether
these benefits stem from direct or indirect use. The major goods and services which
accrue from mangroves in this area are outlined in Table 9.4. Table 9.4 highlights
an important distinction between the direct benefits of using mangroves, which
are almost always extractive in nature and located within the mangrove areas, and
those which are largely off-site and indirect. It is often more difficult to quantify
these indirect benefits because of the dynamic nature and ecological complexity of
the relationship between the productive output and the mangrove forest. The fish
nursery function of mangroves, for example, is well established (Primavera 1998).
Yet it is difficult to attribute the value of final commercial or subsistence fish catch
to this single function among many of the whole system (see Barbier and Strand
1998).

It should be noted that some economic benefits of the mangrove resource will
increase in value over time, while others will remain constant or will decline. For
example, as agricultural development intensifies, the potential economic losses
from storm surges increase, so the value of the coastal protection function of the
mangroves will rise accordingly. Exogenous environmental change associated with
global climate change may increase the frequency and intensity of storms. As
reported in Chapter 3, the net impacts of global warming on the typhoon regime

of Southeast Asia are uncertain, but the frequency of flooding will change due to heightened sea level, even in the absence of changes in typhoon frequency. Hence, for this reason too, the value of the coastal protection function of the mangroves is likely to rise over time.

The first step in the appraisal framework is to delineate the resource issue. In this case, the analysis allows examination of the efficiency of using land, labour and capital resources to rehabilitate or restore mangroves in the coastal areas of Vietnam. The second step is the identification of a set of costs and benefits. The allocation of effects into costs and benefits involves determining what is the current situation, and focusing in partial analysis on the values of the marginal changes. The economic cost benefit analysis of mangrove rehabilitation schemes in this case is of the form:

$$NPV = \sum_{t=1}^{\gamma} \frac{B_t^T + B_t^{NT} + B_t^P - C_t}{(1 + r)^t}$$

NPV = net present value (VND per ha)
$B_t^T$ = net value of the timber products in year t (VND per ha)
$B_t^{NT}$ = net value of the non-timber products in year t (VND per ha)
$B_t^P$ = value of the protection of the sea defences in year t (VND per ha)
$C_t$ = costs of planting, maintenance and thinning of mangrove stand in year t (VND per ha)
r = rate of discount
$\gamma$ = time horizon (20-year rotation).

Estimates of the data sources and methods for carrying out the quantification and valuation of costs and benefits in establishing the rehabilitated mangrove stands are summarised in Table 9.5.

### Direct costs and benefits of restoration

The costs of establishing the rehabilitated mangrove stands are estimated primarily on the basis of the cost of labour for the activities described. The survey research was carried out in 1994 with the cost for a work day in that year being typically 2.5 kg of rice or VND 5500. Planting of one hectare of mangroves required 95 work days or VND 522 000, as shown in Table 9.5. The estimates are averaged across the three districts, with variations in costs dependent on where the seedlings were obtained. The planting and handling fees for seedlings obtained from forests in the area under rehabilitation are not significant compared to costs for collecting, handling and transportation for other areas which increase depending on the distance from the seedling source site to the planting site. The seed mortality rate between time of collection and time of planting adds an additional cost factor.

For some mangrove species, such as *Sonneratia* sp, *Avicennia* sp, *Aegiceras* sp and others, planting directly onto mud flats is unsuccessful due to the exposure to

*Table 9.5* Benefits and costs of mangrove rehabilitation in Nam Dinh and Thai Binh
Provinces and methods for their valuation

| Impact or asset valued | Method and assumptions for valuation | Timing of costs and benefits |
| --- | --- | --- |
| *Benefits* | | |
| Timber | Market data: thinning (VND 180 per tree); extraction of mature trees (VND 5000) | Thinning and extraction from year 6 with 3-year rotation |
| Fish | Market data: mean price of VND 12500 per kg; yield 50 kg per ha | Fishing benefits from year 2 after planting |
| Honey | Market data: potential yield estimated at 0.21 kg per ha | Honey collected from year 5 after planting |
| Sea dike maintenance costs avoided | Morphological model: costs avoided = f (stand width, age, mean wavelength) | Benefits rising from year 1 |
| *Costs* | | |
| Planting, capital and recurrent costs | Market and labour allocation data: costs of seedlings and capital (VND 440000 per ha); workdays valued at local wage in rice equivalent (VND 5500 per day) | Planting costs at year 1; thinning from year 6 on 3-year rotation |

Note
US$1 = VND 11000 at time research conducted.

strong wind and wave forces which wash away the seedlings. The cost of raising such species in a nursery and transplanting them at eight months old is relatively high, with fees for maintaining the nursery, care, protection and transportation adding to overall expenditure. The costs of establishing a stand, including planting, gapping and protection, occur mainly in the first year. Maintenance from the second year on incurs an estimated annual expenditure of VND 82500 per hectare. The cost of thinning occurs in years 6, 9, 12, 15, 20 and 25.

The benefits from wood and fuelwood sources from the processes of periodic thinning and extraction are derived from observations in local markets, and are shown in Table 9.5. The timber benefits represent wood for poles and fuelwood. The benefits from direct fishing sources are estimated as on-site revenues only. Fishing activities in the three districts are undertaken through the use of simple fishing nets, simple tools or even by hand. Aquatic products include fish, crabs, shrimps and shellfish. The average yield is estimated at approximately 50 kg per hectare within mature mangrove stands annually for all types of aquatic products. The average unit price in 1994 was around VND 12650 per kg averaged across the products. There is some evidence that present exploitation of mangrove aquatic products in the Red River Delta, in general, may be leading to declines in stocks of species. The yield estimates are considered conservative for the districts surveyed.

Honey from bee-keeping is derived from the flowers of a number of mangrove species, though the season spans a limited number of months. The honey from mangroves is obtained during the first flowering season of *Kandelia candel* from January to March and from July to September for other mangrove species and the second flowering season of *Kandelia candel*. The potential yield from this bee-honey source was estimated to be an annual minimum of 0.21 kg per hectare. Honey production is possible from five years after planting, though some species of mangrove can flower after three to four years, and even after one and a half years, from planting.

In addition to these direct uses, mangroves produce other subsistence and indirect services, as shown in Table 9.4. In particular, the diverse species within a mangrove forest are used extensively for medicinal purposes. For the established mangrove area of Xuan Thuy, the major utilised flora are shown in Table 9.6 along with other subsistence uses. This is based on extensive surveys in Xuan Thuy District and on the identification of these uses more generally by Phan Nguyen Hong and Hoang Thi San (1993).

Leaves, bark or roots from over 30 species of plant are commonly used in medicines in the District, with other parts of the plants being used for tannin, manure and directly as food in times of food shortage. The market value or value of substitutes of these direct consumptive uses of medicinal plants have not been quantified and not included in the cost benefit analysis presented here. It has been demonstrated in many regions of the world, however, that medicinal uses of plants, and their associated cultural significance, is often a primary reason for conservation of forest areas. Such motivations are not readily captured in economic analysis (e.g. Brown 1995; Crook and Clapp 1998). The utilisation of mangrove areas for subsistence and non-market uses adds weight to the argument for restoration and conservation.

### *The mangrove forest as coastal protection*

The planting of mangroves on the seaward side of the extensive sea dike system provides the benefit of avoiding the cost of maintaining of these defences. Such maintenance takes place on an annual basis in most of the coastal districts of Vietnam. As discussed in Chapter 2, maintenance primarily took place in the period up to the 1990s through the obligatory labour of district inhabitants organised by the district committees. These commitments drew a heavy burden on labour-scarce households and were a source of conflict regarding the inter-district allocation of labour contracts (Adger 2000a). Maintenance is now increasingly financed through local land taxes.

The evaluation of the role of mangroves in protecting sea dikes is estimated from expenditure on their maintenance and repair in comparison with a case where no mangroves exist, with the control situation assumed to have similar morphological characteristics. In general terms, the greater the area of mangrove, the greater the benefit in terms of avoided maintenance costs. Establishing a precise set of relationships in order to estimate the benefits is not, however,

Table 9.6 Useful species within the mangrove area of Xuan Thuy protected Ramsar site

| Species | Timber, fuelwood and charcoal | Tannin | Green manure | Food | Fodder | Bee-keeping | Medicinal plant | Medicinal use |
|---|---|---|---|---|---|---|---|---|
| *Acanthus ebracteatus* | | | | | | | + | Boiled bark and roots used for cold symptoms, skin allergies. Bark in malaria treatment and back pain |
| *Acanthus ilicifolius* | | | | | | | + | Leaves for relief of swelling, rheumatic pain, neuralgia |
| *Achyranthes aspera* | | | | | | | + | |
| *Acrostichum aureum* | | + | | | | | + | Leaves as poison antidote and applied to boils |
| *Aegiceras corniculatum* | | | | | | + | + | |
| *Avicennia marina* | | | | | | + | + | Bark used in contraceptives, leaves on abscesses |
| *Azeratum conyzoides* | | | + | | | | + | |
| *Bidens pilosa* | | | | + | | | + | |
| *Bruguiera gymnorhiza* | + | + | + | | | + | + | |
| *Casuarina equisetfolia* | + | + | + | | | | | |
| *Chenopodium album* | | | + | | | | | |
| *Clerodendron inerme* | | | | | + | | + | Leaves used in treatment of jaundice, dried roots for wounds and cold |
| *Crinum asiaticum* | | | | | | | + | Tonic, laxative and expectorant |
| *Cynodon dactylon* | | | + | | + | | + | |
| *Cyperus malaccensis* | | | | | | | | |
| *Cyperus tegetiformis* | | | | | | | | |
| *Cyperus stolonferus* | | | | | | | | |
| *Datura fastuosa* | | | | | | | + | |
| *Derris trifoliata* | | | + | | | | + | Leaves as laxative. Roots used in treatment of malnutrition |

| Species | | | | | | Uses |
|---|---|---|---|---|---|---|
| *Eclipta alba* | + | | | | | |
| *Eupatorium oderatum* | + | | | + | + | |
| *Excoecaria agallocha* | | + | + | + | + | Bark for leprosy. Leaves for ulcers and epilepsy |
| *Heliotropium indicum* | + | | | + | | |
| *Hibiscus tiliaceus* | | | | | | |
| *Ipomoea pes-caprae* | + | + | + | + | + | |
| *Kandelia candel* | + | + | + | + | + | |
| *Lumnitzera racemosa* | | | | | | |
| *Pluchea pteropoda* | + | + | + | + | + | Treatment of scalds and burns |
| *Rhizophora stylosa* | | | | + | + | Bark as astringent, anti-diarrhoea, for scalds and burns |
| *Scirpus kimsonensis* | | + | | | | |
| *Sesuvium portulacastrum* | + | + | | | | |
| *Sonneratia caseolaris* | + | + | | + | + | |
| *Sporolobus virginicus* | | | + | | | |
| *Thespesia populnea* | + | + | | + | + | Scabies |
| *Wedelia calendulacea* | | | | | + | |

Source: Based on Nguyen Hoang Tri *et al.* (1997), Phan Nguyen Hong and Hoang Thi San (1993).

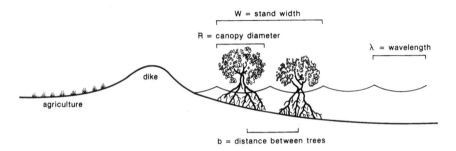

*Figure 9.3* Profile of rehabilitated mangrove stands showing parameters for estimation of avoided maintenance costs

a straightforward matter as the mechanisms by which mangroves protect the adjacent dike are complex. Mangrove stands provide a physical barrier, resulting in drag effects and the dissipation of wave energy. They also stabilise the sea floor, trapping sediment, and can affect the angle of slope of the sea bottom and again the dissipation of wave energy.

Studies in southern China have resulted in an empirical relationship through which the benefit, in terms of avoided cost ($B_t{}^P$), can be expressed as a function of the width of the mangrove stand as a proportion of the average wavelength of the ocean waves that the stand is exposed to and various parameters related to the age of stand (mangrove size and density) expressed as a buffer factor. The key parameters are illustrated in Figure 9.3. The relationship was developed and tested in mangrove stands in southern China and has been calibrated in Vietnam through simulation (Tran The Vinh 1995). We have used a simplified version of this relationship in estimating indirect use value in this study. Here, the buffer factor, $\alpha$, is given by:

$$\alpha = \frac{2\pi R^2}{1.73 b^2}$$

where $R$ is the mean radius of the canopy of an individual tree (m), which increases with age, and $b$ is the typical distance between trees (m), which generally increases with time. As the stand matures, $\alpha$ increases from a minimum of around 0.1 to close to 1.0 as the stand presents a more and more effective obstacle.

For the Nam Dinh example, the model was calibrated using survey data on the annual costs of maintenance of sea dikes in each of the three coastal districts and data on mangrove productivity (growth in terms of mean annual increment, height, and canopy density) for *Rhizophora apiculata* (Aksornkoae 1993). The model was tested for its sensitivity to various parameters including the costs of main-tenance in the districts and the design of the protection schemes in terms of the width of the stand in front of the sea dikes. Observations indicate that a mature stand will avert 25 to 30 per cent of the costs of dike maintenance assuming a

*Table 9.7* Costs and benefits of direct and indirect use values of mangrove restoration compared

| Discount rate | Direct benefits (PV million VND per ha) | Indirect benefits (PV million VND per ha) | Costs (PV million VND per ha) | Overall benefit/cost ratio |
|---|---|---|---|---|
| 3 | 18.26 | 1.40 | 3.45 | 5.69 |
| 6 | 12.08 | 1.04 | 2.51 | 5.22 |
| 10 | 7.72 | 0.75 | 1.82 | 4.65 |

Source: Nguyen Hoang Tri *et al.* (1998).

Notes
Stand width = 100 m; incident wavelength = 75 m.
US$1 = VND11000. Benefit/cost ratio = Net Present Value Total Benefits/Net Present Value Costs.

stand width at least comparable to the characteristic wavelength of the incident waves. Nguyen Hoang Tri *et al.* (1998) use the model described above and show that, beyond a certain point, increasing stand width results in diminishing gains in protection. Typical wavelengths would be between 25 and 75 m and the model suggests stand width should be of the order of 50 to 100 m.

The model of maintenance costs avoided was used for the three coastal districts to derive the indirect benefit of mangrove rehabilitation. The results presented here must be considered an approximation of the benefits associated with reduced maintenance costs for a number of reasons: the model may be over-estimating the benefits when the stand is not fully developed or the width of the stand is much less than the incident wavelength. Nevertheless, uncertainties in this area may not be critical for two reasons. First, the direct benefits from the use of the resources are considerably more significant than this indirect use value. Second, as also discussed below, the value estimated here is only part of the true storm protection value, which must also include broader damage avoidance benefits, and is, therefore, a lower bound figure.

The baseline costs of maintenance are incurred by the District Committees which keep detailed records of work days and expenditure on annual maintenance showing of the number of person-days a year spent on dike maintenance. As the results represent the average situation, the impact of the most severe storm surges on both the cost of maintenance and repair of dikes is not accounted for.

### Comparing the costs and benefits of restoration

The full results of the cost benefit analysis are presented in Table 9.7. This cost benefit analysis is of a partial nature, comparing establishment and extraction costs with the direct benefits from extracted marketable products and with the indirect benefits of avoided maintenance of the sea dike system. It is assumed that present-day conditions continue to prevail with respect to storm frequency, and so on. The results show a benefit to cost ratio in the range of four to five for a range of discount rates. The low relative changes in benefit cost ratios illustrates that

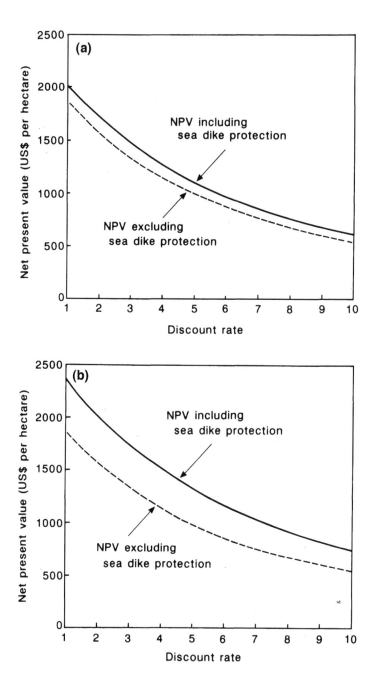

*Figure 9.4* Net present value of mangrove rehabilitation, including value of sea dike protection, for two cases: (a) stand width = 100 m; incident wavelength = 75 m; (b) stand width = 33.3 m; incident wavelength = 25 m

most of the costs, as well as the benefits of rehabilitation, occur within a relatively short time frame, with even the reduced maintenance cost beginning to accrue within a few years of initial planting.

Choosing the rate of discount is considered by many economists to be somewhat arbitrary and dependent on whether the project to be appraised is being undertaken in the public or private domains. A range of real discount rates from 1 to 20 has been used in many circumstances, but rates at the lower end of this range tend to reflect the time preferences implicitly applied by governments in investments on behalf of society (see Markandya and Pearce 1991). The results presented in Figure 9.4 appear to be robust to the discount rates adopted as a sensitivity test.

Figure 9.4 illustrates that the direct benefits from mangrove rehabilitation are more significant in economic terms than the indirect benefits associated with sea dike protection over a range of realistic parameter values. As might be expected, the greater the stand width, the more important the direct benefits in comparison to the avoidance of maintenance costs (Figure 9.4a). Yet even at the lower end of the range of realistic stand widths, offering the greatest return per hectare given suitable conditions, the direct benefits dominate (Figure 9.4b).

As shown in Table 9.4, the sea dike protection estimates do not include the benefits of reduced repair after serious storm damage, nor the potential losses of agricultural produce when flooding occurs. Flooding associated with severe tropical storms can lead to large economic losses, as well as to loss of life, and a reduced probability of flooding associated with the protection from the mangrove itself would be an additional indirect benefit. This benefit has not been estimated to date, though the impact of historic storms can be discerned by examining aggregate agricultural production from district archival records.

Figure 9.5 shows total rice production in one of the case study districts, Xuan Thuy, from 1981 to 1995. The radical increase in agricultural production over the period coincides with the liberalisation of agricultural production practices and distribution of leaseholds to individual households, beginning with the 'output contract' system in the early 1980s (see Chapter 2). But the impact of the major storms which crossed the coast close to this district over the period can be seen to have some impact on agricultural production, at least in 1986 and in 1994. Since there is, however, little evidence of the impact of the 1992 storms on agricultural output, it is clear, following the discussions in Chapter 2, that the timing of extreme events, as well as institutional and other factors, directly affects vulnerability in both the physical and social senses (see also Chapter 10). In this case, the timing with respect to the cropping cycle is critical. The impact of the presence or absence of mangroves in its role as protecting agriculture in coastal areas cannot be directly deduced from this data.

In any event, it is clear from the results of the economic analysis summarised in Figure 9.4 that the direct benefits from mangrove rehabilitation mean that this activity is economically desirable, as evidenced by the positive net present values at all discount rates considered. The increase in net present value associated with mangrove planting resulting from including dike maintenance savings would

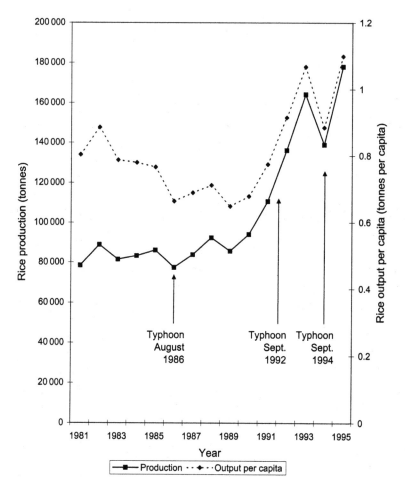

*Figure 9.5* Total rice production and production per capita for Xuan Thuy District and timing of major storms, 1981–1995

Source: Data from Xuan Thuy District archives.

promote the desirability of planting. The results presented in Table 9.5 show that this indirect joint-product benefit of mangrove rehabilitation is significant in further strengthening the economic case for such action in these locations.

## Conclusions

The broader lesson which can be drawn from this analysis concerns approaches to the problematic issue of adaptive responses to long-term environmental change. Decision-makers face difficult decisions in assigning priorities when faced with an uncertain future in a resource-limited present. We would argue that this difficulty can be minimised by adopting a precautionary approach which focuses attention on present-day or near-future benefits which will accrue regardless of the nature and magnitude of the impact of environmental change. See Kelly (2000) for a related discussion of sustainable responses to the climate problem.

In the example presented in this chapter, mangrove restoration, a practice which is widespread and is based on available and appropriate technology, provides immediate economic benefits to residents of the adjacent settlements, those most vulnerable to storm impacts, while reducing the potential for storm damage over the near- and long-term. One alternative strategy, building higher or stronger sea dikes, may be needed in some situations even today, but technological lock-in to 'hard' coastal engineering can be to the neglect of the critical issues of the resilience of the physical and ecological system and the social resilience of coastal populations.

## Note

1 The district names current at the time the research was undertaken are used in this chapter.

# 10 Responding to El Niño and La Niña: averting tropical cyclone impacts

*P. Mick Kelly, Hoang Minh Hien and Tran Viet Lien*

> The wild-raging storm sweeps the whole earth now,
> running adrift the drunken fisherman's boat.
> From all four quarters, clouds thicken and blacken,
> waves surge like the report of beaten drums,
> everything washed out by slashing rain, gust-driven,
> beneath the shuddering menace of this thunder.
> Afterwards the dust settles, the sky grows calm,
> and the moonlit river lengthens out. What time of night is this?
>
> From 'The Four Hills of Existence'
> by Emperor Tran Nhan-tong (1258–1308)[1]

## Introduction

Participants at the workshop *The Impact of El Niño and La Niña on Southeast Asia*, held in Hanoi in February 2000, warned that

> El Niño and La Niña events are the most important cause of climate extremes lasting from a season to a year or more in the tropics and subtropics. Responding effectively to the challenge posed by these events is essential if sustainable development is to be secured.
>
> (Kelly *et al.* 2000: 15)

With immediate benefits, a pro-active effort to avert the worst consequences of El Niño and La Niña represents a 'win–win' strategy with regard to reducing vulnerability to longer-term environmental and socio-economic threats and enhancing social resilience.

El Niño and La Niña events mark the opposite phases of what is known as the El Niño Southern Oscillation (ENSO) phenomenon. El Niño events are associated with warming in the eastern equatorial Pacific Ocean, while La Niña events are associated with cooling. The ENSO phenomenon, though, is the result of more than oceanic warming and cooling. It is the result of a complex set of inter-relationships between the atmosphere and oceans of the tropics and subtropics. See Box 10.1. The process creates marked variability in regional climates with

*Box 10.1* The mechanics of the El Niño Southern Oscillation (ENSO) phenomenon

The ENSO phenomenon has two distinct phases (Rasmusson and Wallace 1983; Diaz and Kiladis 1992).

When El Niño conditions prevail, sea surface temperature is higher than normal over the eastern Pacific Ocean as the South Pacific High weakens, the trade winds drop in strength and the upwelling over the equatorial Pacific associated with the equatorward flow off the South American coast decreases. It is this upwelling which under normal conditions depresses local sea surface temperatures. The increase in sea surface temperature results in greater evaporation rates and heating of the lower atmosphere which, in turn, strengthens convection and vertical motion. Feedback acts to mainta in this phase of the cycle as increased convective activity further weakens the trade wind circulation.

Over the eastern tropical Pacific and neighbouring land areas, the increase in convection creates favourable conditions for a marked increase in rainfall. Combined with the effect on fisheries productivity of the changes in the neighbouring ocean, the resulting disruption in the coastal nations of South America represents one of the prime human impacts of the ENSO phenomenon (Glantz *et al.* 1991; Glantz 1996).

The disturbance in the eastern Pacific spreads round the tropics and subtropics, affecting the monsoon flows and extending, at times, into the temperate zone as the global atmospheric circulation adjusts. Over the western Pacific and the eastern Indian Ocean, the major zone of convection shifts eastwards. Drought is characteristic of El Niño conditions over Southeast Asia as convection is reduced and the monsoons are affected (Ropelewski and Halpert 1987).

The change in the Pacific circulation is referred to as the Southern Oscillation, conveniently measured by the difference in mean sea-level pressure between Darwin and Tahiti (Können *et al.* 1998).

During a La Niña event, conditions are broadly reversed, though this phase of the cycle may not be as clearly defined. Each phase lasts one to two years with an interval of between three and seven years between successive El Niño or La Niña events.

significant impacts on the well-being of the inhabitants and natural ecosystems of those areas (Glantz *et al.* 1991; Glantz 1996; World Meteorological Organization 1999).

The impact of the most recent El Niño event on Southeast Asia in 1997–98 has been well-documented; lives were lost and economies set back as drought was aggravated by haze and smog with the pollution created by forest fires trapped by stable, atmospheric conditions. Drought in southern and central Vietnam seriously affected the coffee and rice crop during spring 1998 (BBC World Service, 17 April 1998). In the south, the country recorded the highest temperature, 40.6°C, for 80 years (Malaysian National News Agency, 2 April 1998), forest fires became a serious cause for concern (Xinhua News Agency, 15 April 1998) and close to half a million people were reported to be short of food (*South China Morning Post*, 13 May 1998). Losses were estimated at over US $400 million (Japan Economic Newswire, 30 May 1998). During the opposite, La Niña, phase, crops are again at risk; drought in central Vietnam during the early months of 1998 was followed by heavy rainfall and flooding during the autumn as the most recent event developed (BBC World Service, 24 October 1998; Reuters, 5 November 1998).

While the large-scale consequences of the ENSO phenomenon are well-established, much remains to be done to define effects on particular regions. The overall impact may be mixed, with both El Niño and La Niña bringing positive and negative consequences to any area. In May 1999, for example, the press agency Reuters reported that La Niña continued to affect rainfall patterns over Vietnam, with early rains creating favourable conditions for bumper rice and coffee crops. But La Niña is also associated with more frequent typhoons on the shores of Vietnam (Nishimori and Yoshino 1990; McGregor 1994; Saunders *et al.* 2000); five major storms hit the country during the closing months of 1998.

Even when impacts can be clearly defined, developing a response capability that can deal with the regular alternation between the different phases of the ENSO phenomenon, thereby providing a basis for more effective management of resources, is a challenge that few nations have addressed. For many parts of the tropics and subtropics, the regular cycle between El Niño and La Niña conditions is a key component of the regional climate, as important a feature to plan for as average conditions. Nevertheless, while the existence of linkages between climate extremes over different parts of the tropics has been recognised since the late nineteenth century (Nicholls 1992), it is only in recent years that understanding of the underlying mechanism has become sufficient to support a predictive capacity (Barnston *et al.* 1994; Ghil and Jiang 1998) and the application of these forecasts is still at the experimental stage. See Kelly *et al.* (2000) for recommendations regarding means of furthering the response to El Niño and La Niña and the potential use of ENSO forecasts.

The aim of this chapter is to examine the lessons that can be derived from strategies developed to reduce the impact of natural hazards related to El Niño and La Niña in Vietnam. We single out the threat of tropical cyclone landfall; the ENSO phenomenon is a key factor in determining interannual variability in the characteristics of this hazard. The location of Vietnam, adjacent to the most vigorous zone of cyclone formation in the world, makes it particularly vulnerable to storm impacts. The surge of water generated by a tropical cyclone, particularly if combined with a high tide, and vigorous waves may cause saltwater flooding. Heavy rainfall may cause flash floods and persistent freshwater flooding. Strong winds may cause damage to crops, boats, buildings, roads, communications, and so on. Successive strikes in rapid succession represent the greatest threat as recovery may not be possible before the next storm makes landfall (Imamura and Dang Van To 1997).

While it is only recently that the variable occurrence from year to year of cyclones making landfall on the Vietnamese coast has been recognised as a direct product of the ENSO phenomenon (Nguyen Huu Ninh *et al.* 2000), there is much that can be learnt regarding the broader response to El Niño and La Niña from Vietnam's long experience of dealing with the cyclone hazard. Despite the high frequency of cyclone landfalls along the 3200 km coastline of Vietnam, an effective civil protection system, evolving over the centuries, has significantly reduced the vulnerability of coastal populations. The system has been described as 'one of the most well-developed institutional, political and social structures in the world for mitigating water disasters'.[2] Here, we first define the link between storm characteristics and the ENSO phenomenon and then review the historic development of protection against storm impacts identifying key strands. The present system for flood and storm control is described and strengths and weaknesses are assessed on the basis of documentary sources, fieldwork (largely focused interviews) and the personal experience of two of us (HMH and TVL). Finally, we draw conclusions regarding the broader response to El Niño and La Niña and means by which resilience to environmental change in general can be enhanced.

## The impact of the ENSO phenomenon on tropical cyclone frequency

In order to determine the role played by the ENSO phenomenon in shaping the cyclone climatology of coastal Vietnam, use is made of a database maintained by the United States National Oceanic and Atmospheric Administration (NOAA) and the United States Navy (United States National Oceanic and Atmospheric Administration 1988; United States Navy and National Oceanic and Atmospheric Administration 1994 and updates).[3] The analysis is based on information concerning track (in terms of location at six-hourly intervals) and maximum wind speed (knots). The data for the western Pacific are considered reasonably reliable from the 1960s onwards, though some information is missing from this data set with respect to storms forming in the East Sea (South China Sea) in the ten years or so following the reunification of Vietnam in 1975.

We first plot the seasonal cycle in the frequency with which the most severe storms, typhoons, approach Vietnam for sets of El Niño and La Niña events. Typhoons in this context are defined as those tropical cyclones with maximum wind speed greater than 64 knots (33 ms$^{-1}$). Figure 10.1 shows the strong contrast between the seasonal cycles for the El Niño and La Niña years. There is little difference in frequency at the start of the cyclone season from May to September but, as the ENSO disturbance develops during the closing months of the year, a notable difference in typhoon numbers is found in October and November with frequencies around three times higher when La Niña conditions prevail, a highly

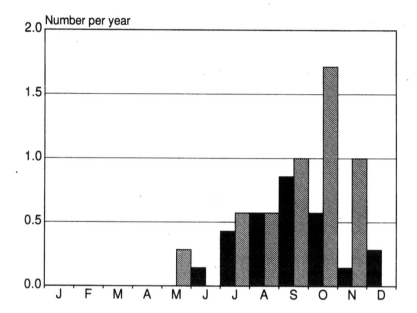

*Figure 10.1* Seasonal cycle in typhoon frequency (average number per year in each month) in the area 7.5–22.5°N, 100–115°E, for seven El Niño (dark shading) and seven La Niña (light shading) events. The years used in the compositing were: El Niño, 1957, 1963, 1972, 1976, 1982, 1986, 1991; and La Niña, 1954, 1956, 1964, 1970, 1973, 1984, 1988

significant increase in risk. These results confirm the main findings of Nishimori and Yoshino (1990) and McGregor (1994).

Figure 10.2 shows the tracks of typhoons occurring during the October–November period during the sets of El Niño and La Niña events. The contrast is striking, with the two dominant cyclone tracks in this area – one due westwards across the Philippines towards Vietnam and the other curving away to the northeast – alternating in strength. When La Niña conditions prevail, storms tend to form closer to Southeast Asia and track due west, with frequencies greatly enhanced over Vietnam and the neighbouring East Sea. During El Niño events, storms form further east and curve away from Southeast Asia eventually heading northeastwards. This alternation in the strength of the dominant storm tracks is consistent with the broader disturbance in the atmosphere over the tropical Pacific Ocean (Chan 1985; 1995; McGregor 1994).[4] It should be noted that, in both sets of years, some storms do develop west of the Philippines, close to Vietnam. These storms are particularly difficult to predict and leave little time for warning and preparation, as illustrated below.

We can draw one immediate conclusion from this analysis with regard to management strategies. The fact that the changes in storm frequency associated

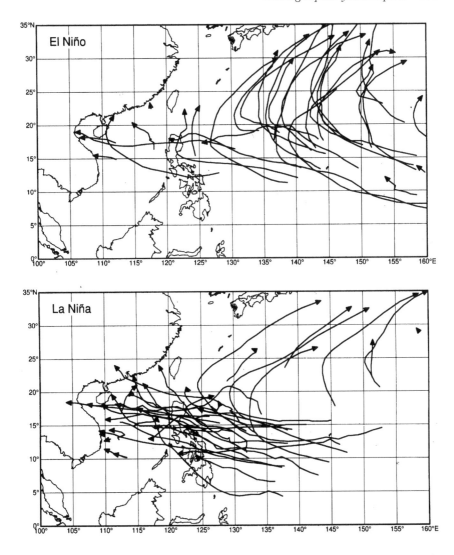

*Figure 10.2* Typhoon tracks during seven El Niño and seven La Niña events for the months of October and November. Years as in Figure 10.1

with the ENSO phenomenon occur late in the storm season, towards the end of the year, suggests that a limited predictive capacity might be based solely on the emergence of early signs that El Niño or La Niña conditions are developing midway through the year. Resources may then, with due caution, be deployed accordingly. Model-based forecasts are available from the international community and empirical precursors are readily available in the large-scale pattern of sea surface temperature over the tropical Pacific, supported by evidence of any local changes in climate occurring early in the development phase.[5]

The changes in typhoon frequency associated with the ENSO phenomenon are marked. Yet seasonal frequency alone does not determine the net impact. It is also necessary to consider other characteristics such as storm intensity, track, predictability (that is, available warning time) and recovery time (that is, time between strikes). During 1992, the first storm of the season made landfall in Yen Hung District in the south of Quang Ninh Province, just north of the Red River Delta, on 29 June (Figure 10.3). According to local informants, it damaged 300 m of sea dike on Ha Nam island, parts of which lie 3 m below sea level. Had the tide not turned, one further hour would have seen vulnerable sections of the dike destroyed, placing the island's population of 55 000 at serious risk. The second storm of the season hit the same area on 13 July, just two weeks later; at least ten days is usually required to repair damaged dikes and longer to drain floodwater from the island. Catastrophic saltwater flooding was avoided as repairs were made in time. Nevertheless, the loss in agricultural production, due, for the most part, to rain-induced flooding, was considerable and no taxes were paid that year. Clearly, the cumulative effect of two storms in close succession can be considerably greater than the sum of two isolated impacts.

The chart of storm tracks for 1992 (Figure 10.3) demonstrates the difficulties in preparing for a hazard that can strike any part of Vietnam's coastline, apparently at random. In any particular year, long stretches of the coastline may not experience storm landfall, yet in any one area multiple strikes can occur within a very short period of time. A further complication lies in the difficulty in forecasting far enough in advance the precise location of landfall in the case of each individual storm. Note the 'confused' track of the first October storm in 1992 (Figure 10.3). With the spatial scale of maximum impact often of the order of tens of kilometres, a high degree of accuracy is desirable if spurious warnings are not to damage credibility and adversely affect the deployment of resources.

The rare nature of an extreme event can play a very direct role in determining the scale of the impact. In the El Niño year of 1997, only three tropical cyclones had an impact upon Vietnam, entirely consistent with the overall average of three a year for the set of seven El Niño events analysed above. But the last storm was Tropical Storm Linda[6] in early November and this caused terrible damage in southern Vietnam (Duong Lien Chau 2000). Linda was a very unusual storm. It developed in the East Sea, close to Vietnam, at a very low latitude (near 8°N) and the track remained south of 10°N before the storm struck the southernmost tip of Vietnam (Figure 10.4). The maximum wind speed reached was 42 ms$^{-1}$ and the lowest pressure was 989 mb. It was the first time that many, if not most, people in the Mekong River Delta had experienced a severe tropical storm. According to the assessment of the Vietnamese government, the damages associated with Linda were the greatest this century for the southern part of Vietnam. There were many casualties; 778 people were declared dead and 2132 remain missing, thousands of fishing boats were sunk, over 300 000 houses were ruined and around 22 000 hectares of rice paddy destroyed; economic losses approached US$600 million.[7]

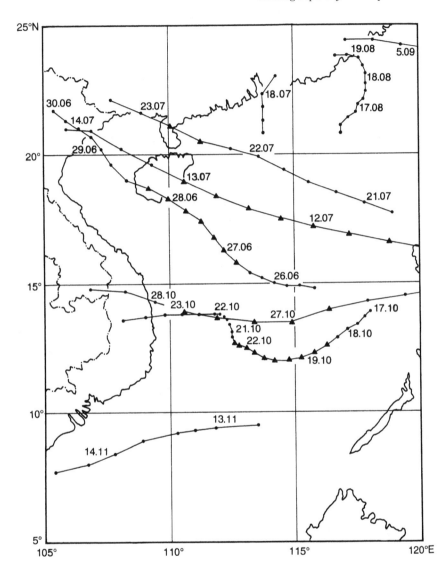

*Figure 10.3* Tropical storm tracks during 1992. The dots and triangles indicate position at six-hourly intervals; the triangles indicate points at which the storm reached typhoon force; the date is indicated by the day/month annotation at 00 UT

There are two important lessons that can be drawn from the experience of cyclone Linda in the context of this present discussion. First, whatever the benefit of a seasonal storm prediction, it is an accurate forecast of each individual event that is of most value. Second, while storm numbers may be lower in El Niño years,

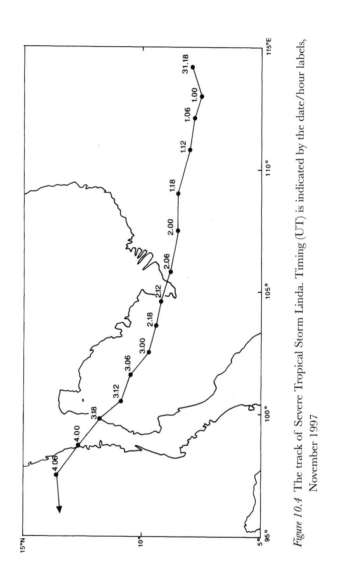

*Figure 10.4* The track of Severe Tropical Storm Linda. Timing (UT) is indicated by the date/hour labels, November 1997

the potential for surprises – for example, storms developing rapidly close to the coast – is at least as great as under La Niña conditions. Even if an accurate seasonal forecast capability were developed, the possibility of 'surprises' means that it must not be permitted to lead to complacency. Effective communication to users of the meaning, value and safe application of longer-term forecasts is essential (Kelly *et al.* 2000). It is a critical step in managing perceptions of risk (Kasperson 1992).

## Protection against storm impacts

### Introduction

To understand the nature of the present-day storm protection system in Vietnam, it is necessary to review its historical development. This account draws heavily on Phan Khanh (1984). Along the coast, sea dikes have long provided the first line of defence (Benson 1997: 64). Dike-building in Vietnam for water-retention in rice agriculture and to withstand river flooding goes back to the Bronze Age (Phan Khanh 1984: 55). The original justification for the construction of coastal dikes was usually the creation of new agricultural land, rather than storm protection *per se*. The Kings of the Tran Dynasty in the thirteenth and fourteenth centuries strongly promoted encroachment on the sea (Phan Khanh 1984: 58; Bray 1986: 94), encouraging their princes to build maritime dikes (though these were relatively small in scale). As agriculture developed in these new lands, the dikes assumed the role of protecting life and livelihood. As a defensive measure, the main need for coastal protection has always been in the north of Vietnam where there is the triple threat of high tides, storm surges and strong wind-induced waves (Vietnam Ministry of Water Resources *et al.* 1994: 51).

Dikes represent the main physical component of Vietnam's coastal defence system. It was during the Tran Dynasty that the other, human, component began to develop rapidly with the emergence of institutions to ensure that dikes were constructed as needed and then adequately maintained. Initially, river flooding was again the main concern. The population as a whole was mobilised and, as noted in Chapter 4, the King (or Emperor) took ultimate responsibility for the protection of agriculture from natural hazards and other threats. The people had the right to revolt if the King failed in this responsibility (Le Thanh Khoi 1955). 'When the people are victims of floods, the King must see to it that they must be rescued and dikes must be repaired, there is no greater duty,' said Prime Minister Tran Khac Chung in the year 1315, advising King Tran Minh Tong of the significance of a tour of inspection (Phan Khanh 1984: 57).

Large-scale construction of sea dikes occurred under the Le Dynasty in the fifteenth to seventeenth centuries, and evidence of Hong Duc dike built under King Le Thanh Tong in the mid-fifteenth century can still be found in Ha Nam Ninh (Phan Khanh 1984: 60). King Le Thanh Tong formalised the process of dike inspection and repair, appointing a mandarin in charge of the country's dikes. In fact, mandarins with responsibility for the dike system were appointed at all levels of the administration, right down to district level. Another recurring theme is evident during this period when dike construction and maintenance can be seen as an immediate priority following liberation. Occupation by the Chinese in the early fifteenth century was accompanied by destruction of the dike system and, as Le Loi advanced with his troops, he immediately ordered dike restoration, such was the importance for agricultural production (Phan Khanh 1984: 59).

The Le Kings developed a series of laws and regulations to ensure annual maintenance of the dike system, formalising the moral duty of their predecessors and creating a regime of responsibilities, rewards and penalties. In 1664, King Le Huyen Tong decreed that the dike system should be inspected at the end of the rainy season and that all construction and maintenance should be completed within five months, before the start of the next rainy season. By the mid-eighteenth century, the regime had extended throughout society as this contemporary regulation at commune level demonstrates:

> On the 5th drum roll on announcement of work for dyke protection, the village inhabitants must assemble to struggle against floods. Anyone who slips away and causes delay to public affairs will be fined three ligatures as a lesson to others.
>
> *Convention of the External Village of Co Linh,*
> *Thanh Hoa Province* (quoted in Phan Khanh 1984: 64)

The early nineteenth century, following the disruption of civil war, was a time of extensive dike building in the north, with a coherent, large-scale dike system developing as river dikes and sea dikes were extended and merged under the Nguyen Dynasty. The sea-dike system now extended to 240 km. There was considerable technical debate at this time regarding the most effective forms of flood protection, and even the value of the dikes themselves was seriously questioned (a debate that has continued almost to the present-day). Dike dimensions were specified and penalties were increased. The destruction of a dike was punishable by death and commune authorities held responsible for dike failure suffered corporal punishment (Phan Khanh, 1984: 69).

Tempered by experience, there had emerged over the centuries a communal view of coastal protection as a process that all strata of society had a duty to engage with, as a continual process intended to anticipate impacts rather than simply react to them. This perspective was backed by a complex set of institutions to ensure compliance. The balance shifted, following the colonial take-over during the closing decades of the nineteenth century, from civil protection as moral duty towards civil protection forced by circumstance. The dike system was extended and strengthened during the colonial period, but, for the most part, in response to catastrophic failure of the system and consequent social unrest rather than any sense of moral responsibility. Nevertheless, it was during this period that serious attention was given to scientific monitoring and study of the tropical cyclones of the region (cf. Bruzon and Carton 1929), providing the basis for modern predictive techniques. During the later years of colonial rule, greater involvement in dike construction occurred but focused more on reducing river flooding than coastal protection.

One aspect of dike maintenance that had significantly reduced the effectiveness of the system for many centuries was inherited by the French. In the event that a dike was breached, the standard response was to fill the gap, raise the height, and strengthen the weakest sections. Often, though, the weakness of a dike was the result of poor materials or workmanship, either at the time of construction or during previous over-hasty repair. This problem required a more radical response, but such a response rarely occurred. It is now clear, for example, that redesign of a dike is often far more effective than simply adding height in increasing levels of protection, though it took until the early 1970s, in the aftermath of a major flood, for this to replace 'conventional wisdom' (Vietnam Ministry of Water Resources *et al.* 1994: 43).

Dike maintenance and the restoration of agricultural production again became a government priority after Ho Chi Minh declared independence in 1945, a priority reinforced by the impact of catastrophic flooding in the Red River Delta that year when as many as two million died as a result of the flood and subsequent famine (Nguyen The Anh 1985). In 1946, the main elements of the modern system for responding to tropical cyclones were established, resting on the foundations of centuries of experience. Dike construction continued and, by the end of the twentieth century, the network of sea dikes extended 3000 km in length (Benson 1997: 64).

As Phan Khanh (1984: 58) has observed, the centralisation of authority and the solidity of the societal structure in thirteenth and fourteenth century Vietnam played no small part in the early development of a large-scale dike system. Indeed, in what might be termed 'co-evolutionary' interplay or feedback (cf. Adger 1999c), the degree of centralisation and the strong societal structure, with its roots in the rural village economy (Chapter 4), may have partially resulted from the need for cohesion both within and across all administrative and spatial levels in a resource-dependent civilisation confronted by diverse hazards. In any event, there is no doubt that the struggle against invasion, in the most general sense, has played a substantial role in the development of Vietnamese civilisation, shaping relations between rich and poor, *inter alia*, and promoting Scott's 'moral economy' (Scott 1976; see also Chapter 4):

> The struggles against foreign invasion and that against natural calamities . . . have, for thousands of years, constituted the two fundamental indicators of the historic vocation of the Vietnamese people, their two most beautiful traditions.
>
> (Phan Khanh 1984: 80)

## The modern storm response system

The modern system for protecting the coastal lands against storm-induced flooding and related impacts has been tried and tested over a 50-year period, not only against the onslaught of natural hazards but also against attempts to breach the dikes during times of conflict (Lacoste 1973). What we will refer to, for the sake of convenience, as the 'storm protection system' is intended to provide protection against a range of natural hazards but we focus here on issues related to the tropical cyclone, the system's primary concern. The storm protection system has continued to evolve during the most recent decades, with the 1994 *Strategy and Action Plan for Mitigating Water Disasters* (Vietnam Ministry of Water Resources *et al.* 1994) providing the current framework for emergency response. The 1994 Strategy adopts a three-pronged approach based on forecasting and warning systems, preparedness and mitigation, and emergency relief, emphasising non-structural solutions on a short- and medium-term basis. Development of the system has also occurred as measures have been adopted to compensate for limited resources by improving the efficiency of existing structural and non-structural components (Benson 1997: 59). The building of storm-resistant structures has been encouraged. Efforts have been made to change agricultural practice to avoid close correspondence between key dates in the cropping calendar and the likelihood of storm landfall. Mangrove planting and rehabilitation has been promoted to provide additional protection to the coastline (see Chapter 9).

The following account of the institutional and temporal structure of the present-day storm protection system and the constraints and challenges this system faces draws on: first, various documentary sources (most notably, Trinh Van Thu, 1991; Vietnam Ministry of Water Resources *et al.* 1994; 1995; United Nations Disaster Management Team and Vietnam Disaster Management Unit 1995; Benson, 1997; Nguyen Huu Ninh *et al.* 2000); second, field research examining the local (community level) to national operation of the formal and informal aspects of the system conducted between April 1996 and February 2000, based on focused interviews[8] with key actors and stakeholders including

(i)   staff of the Vietnamese Hydrometeorological Service, in Hanoi and Quang Ninh Province, the Disaster Management Unit, Hanoi, and the Provincial Department of Science, Technology and Environment for Quang Ninh;
(ii)  members of the Provincial Committee for Flood and Storm Control, Quang Ninh Province, and the People's Committee and the Committee for Flood and Storm Control, Yen Hung District, Quang Ninh Province;
(iii) district and commune officials and members of the community, Yen Hung District, Quang Ninh Province; and, finally,
(iv)  the personal experience of two of the authors of this chapter (HMH and TVL).

At the community level, much of the information draws on accounts of the circumstances of Ha Nam island in Yen Hung District, Quang Ninh Province, in northern Vietnam. Formed by the accretion of sediments over recent millennia, the island was first settled and protected by dikes some 700 years ago. It is now home to over 55 000 people, growing rice and vegetables irrigated by water transferred by canal from a reservoir 15 km distant. Much of the island lies below sea level – by as much as 3 m in places, as noted earlier – and it is only made habitable by the 40 km dike, up to 5.5 m high, that encircles it. It is known as the 'floating boat'.

*Institutional structure*

For the sake of convenience, the storm protection system can be divided into three main components: monitoring and forecasting cyclone activity; coordination; and implementation of disaster prevention and preparedness measures before, during and after the storm season.

*Monitoring and forecasting* is the task of the Hydrometeorological Service (HMS), with the primary responsibility in the main offices at the national level in Hanoi and secondary responsibility at the regional and provincial level. The HMS is responsible for monitoring and forecasting tropical cyclone characteristics (position, movement, severity, likely landfall, and so on) and for issuing warnings to the general public and the relevant authorities. This work is undertaken at the national level, in Hanoi, by the National Center for Hydrometeorological Forecasting (NCHF), and at the regional (city and provincial) level by the local HMS offices. The NCHF is divided into two divisions or branches, responsible for meteorological forecasting and hydrological forecasting. Tropical storm forecasting is the responsibility of the meteorological division, though the hydrological division may become involved in the event of, for example, river flooding. With a total staff of around 150, the NCHF is further divided into sectors responsible for, for example, administration (including issuing warnings to the general public), communications (issuing warnings to regional HMS offices), research and development (information and methods), satellites and computing (data provision and analysis), short-term forecasting and medium- and long-term forecasting.

*Coordination* is the primary role of the Committees for Flood and Storm Control and Disaster Preparedness (CFSC), which exist at the national, city and provincial, and district level and within relevant ministries and their branches. Each commune has a CFSC Officer supported by the Commune People's Committee. The Committees for Flood and Storm Control are responsible for monitoring the state of coastal defences, pre-storm season preparations (including maintenance of the physical defences and resource allocation), disaster prevention and preparedness instructions as a storm approaches, action as a storm makes landfall, coordination of relief at the relevant level in the aftermath of a storm, collation of impact information and post-season assessment of effectiveness.

The Central CFSC, which has ministerial status, was established in May 1990, succeeding the Central Committee for Dyke Maintenance. There is a permanent Secretariat for the Central CFSC with a small staff, run by the Department of Dyke Management and Flood Control (DDMFC) under the Ministry for Agriculture and Rural Development. Construction, maintenance and strengthening

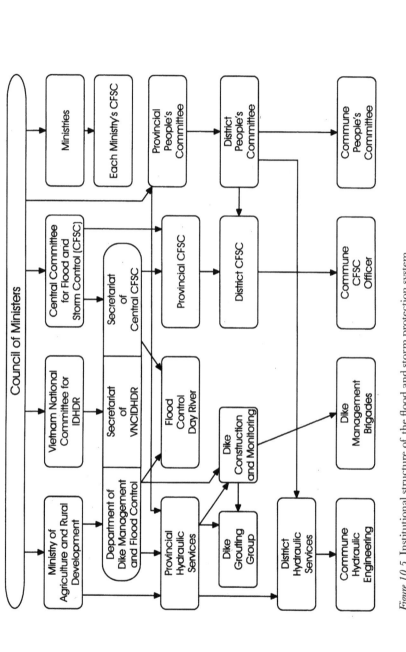

*Figure 10.5* Institutional structure of the flood and storm protection system

Source: Disaster Management Unit, http://www.undp.vn/dmu/dm-in-vietnam/en/institutional-framework.htm.

of dikes is the responsibility of the DDMFC. The Central CFSC is primarily responsible for the emergency response to floods and storms. It has responsibility to monitor and control information, develop annual plans for flood and storm protection and advise the government on the issuing of legal documents. The Central CFSC is chaired at vice-minister level and includes representatives from a wide range of relevant ministries as well as the Chair of the HMS.[9] Longer-term planning is the responsibility of the National Committee for the International Decade of Natural Disaster Reduction (NCIDNDR), set up in January 1991. The Secretariat of the NCIDNDR is also run by the DDMFC. The institutional relationships between the components of the system at the national level and the reach through parallel structures at the provincial and district levels to the commune level are shown in Figure 10.5. Funding for the activities of these organisations and the measures they instigate comes from three sources: government, local communities and international assistance. The Flood and Storm Preparedness Fund, contributed by city and provincial administrations, supplements central government funding.

The Disaster Management Unit (DMU) was established in 1994, with support from the United Nations Development Programme (UNDP), to assist the work of the CCFSC and the NCIDNDR. Its aim is 'to join together over 1000 years of Vietnamese flood protection culture with twenty-first century Western technology to better protect the entire population of Vietnam against the annual natural disasters that ravage the country'. Technology provided through the DMU includes a computer-based emergency warning and disaster damage reporting system, a GIS-based information system of disaster management and relief needs data, a web-based public information system, and decision-making strategies for distributing emergency relief and recovery resources. The DMU also provides disaster awareness and response training at the national, provincial, district and commune levels and is responsible for distributing emergency relief channelled through UNDP.

*Implementation*   of disaster prevention and preparedness activities is the responsibility of government departments and organisations at all levels, the media and the remainder of Vietnamese society. Figure 10.6 shows the flow of warnings and instructions from the national to the commune level. The need for coordination between numerous ministries and between administrative levels in order to facilitate a multi-sectoral response has driven the formulation of this complex chain of communication. Given the spatial scale of the maximum impact of an individual storm and the nature of the Vietnamese administrative system, primary responsibility for many defensive measures and actions is taken at the district and commune level with support available from other levels as considered necessary.

*Temporal structure*

The temporal structure of the storm protection system has five elements: preparation, before the storm season begins; warning, before a storm makes

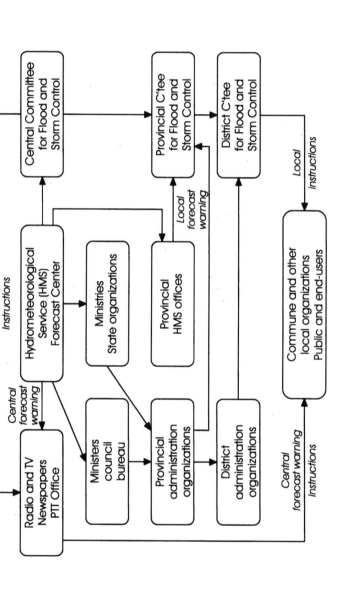

*Figure 10.6* The flow of warnings and instructions as a storm approaches the coast of Vietnam
Source: Updated from Trinh Van Thu (1991).

landfall; response, during landfall; recovery, in the immediate aftermath; and evaluation and improvement, after the event.

Before the cyclone season begins, the state of protection in each area is reviewed on the basis of local reports and any necessary remedial action such as dike maintenance is taken. The local CFSCs will check the state of the dikes and allocate labour to maintain weak points. The Central CFSC receives reports from the provinces and districts on maintenance requirements and distributes resources accordingly. Every 20 to 30 km of dike is the responsibility of a single management team. Dikes are checked on a 24-hour basis throughout the storm season by dike monitors, each responsible for about 1 km of dike, who live permanently by the dike. Extending the process of regular inspection established centuries earlier, this system has been reported to have resulted in significant improvements in dike condition (Vietnam Ministry of Water Resources *et al.* 1994: 61). Unless a substantial breach occurs, communes take primary responsibility for the maintenance of dikes in their area once constructed. Repair materials and other supplies are stocked close to points considered to be particularly at risk.

As a storm approaches Vietnam, the timing of the issuance of cyclone warnings, of the meetings of the various CFSCs, and of implementation of disaster prevention and preparedness activities is largely controlled by the proximity of the storm to the coastline. Warnings begin to be issued as a storm passes the Philippines and continue at regular intervals until the storm makes landfall. During this period (typically four to five days), forecast detail and the level of activity of the CFSCs increases. Most of the actual monitoring and forecasting that is undertaken at this stage as a tropical cyclone approaches Vietnam take place in the NCHF. In recent years, regional centres have taken an increasing role in disseminating and customising forecast information, filling a perceived gap in the structure. Local HMS offices at the city and provincial level receive regular information from the NCHF and regional meteorological centres and will modify and add to this data on the basis of any further information and analysis available locally. As well as issuing warnings to the general public, HMS offices at all levels will be in close contact with the relevant authorities through the CFSCs. The Central CFSC meets as the storm approaches, coordinates the response of the various authorities and issues central disaster preparedness and preparation instructions. Meetings are called by the Central CFSC Chair in consultation with the HMS. Generally, the committee will meet 24 to 48 hours before a storm crosses the coastline although if there are indications that the impact may be serious then the committee will meet earlier. The DMU regularly updates its computer-based information system – the DMU Disaster Communications System (DMUnet) – aiming to provide communications between, in particular, the national and provincial levels.

When a tropical cyclone is east of the Philippines (see Figure 10.3 for geographical orientation), the NCHF will monitor storm behaviour largely on the basis of assessments and forecasts made by other meteorological services. Key information comes from China, Japan and the United States (Guam). The Guam assessment was considered the most useful during the mid- to late 1990s as it was the only one to contain a long-term (three-day) forecast at that time. Tele-communications links to Bangkok and elsewhere are used at this stage and a link to the Internet is available. Three staff work continuously on storm monitoring. At this stage, direction of track is an important indicator; if the angle the track makes with latitude 15°N is small then landfall on the Vietnamese coastline is possible. If the cyclone crosses over south of the central Philippines then landfall on the Vietnamese coastline is very likely.

As the cyclone passes over the Philippines, the NCHF begins to make internal forecasts. These forecasts are based on statistical and dynamical methods, developed by the Center, though not on modelling of any form. When the storm crosses 120°E, the HMS begins to issue storm warnings, unless it is clear that the storm will not strike Vietnam within the next five days. When the storm is located between 120 and 115°E, the warning is known as the 'Far Storm Warning' and this is issued and updated at six-hourly intervals. The warning contains details of the name, position and strength of the storm and a 24-hour forecast. The public are kept informed through the media (radio and television), via the telecom-munications office, and the Secretariat of the Central CFSC is regularly notified of developments. The Secretariat, in turn, informs the CFSCs in the ministries and provinces along with the Ministry for Agriculture and Rural Development and the Office of the Government (see Figure 10.6) At this stage, additional staff begin to be allocated to storm forecasting rising from three to eventually ten to twelve people (in three shifts) as the storm approaches the coast.

When the storm crosses 115°E, the 'Near Storm Warning' is issued at three-hourly intervals. This warning contains additional information including the likely area of impact on the coastline. When the storm is 500 km from the coast in the case of strong cyclone (maximum wind greater than 17 ms$^{-1}$) or 300 km in the case of a weaker cyclone, the 'Critical Storm Warning' is issued, again at three-hourly intervals, containing more detailed information regarding the likely impact area. At this stage, landfall is expected in about 24 hours' time. Warnings continue to be issued to the public and to the CFSCs. The provincial and city authorities in likely impact areas will contact the NCHF at hourly intervals by telephone for updated information. Radio broadcasts covering storm location and other information occur every hour enabling all interested parties to plot the storm track.

During the period when the storm is west of the Philippines, the local HMS offices at the city and provincial level will be issuing their own assessments to the provincial and thence district CFSCs, based on the national forecasts. Communication between the national and local offices is by fax, modem, teletype and telephone. The local offices also receive international assessments when equipment is functioning. As the storm approaches, the geographical focus of the warnings will be refined, narrowing in scale, until it is clear at what point on the coastline the storm will strike. Information and instructions may be sent by car from the provincial to the district level at this stage and telephone, fax and teletype are used for regular communications. One to two days before the storm makes landfall, materials will be sent to vulnerable areas ready for protection activities and the local population will be mobilised, providing both supplies and labour.

As a storm makes landfall, the actual defensive measures in areas of risk are implemented by formal teams of workers organised at the commune level, managed by the District CFSC, with assistance from the army and the local community. The chair of the District CFSC will go to any area at risk to oversee the maintenance of the defences. Over-topping of dikes by waves raised by the surge in sea level and under-cutting then destruction of the dikes by wave action are the principal threats, alongside the damage caused by high winds and heavy rainfall. Parallel instructions are issued through a number of tracks from the national level to the district level (cf. Figures 10.5 and 10.6). This multiple-track approach is considered an efficient means of ensuring that all considerations are met at every administrative and spatial level; the built-in duplication or redundancy permits some degree of compensation if any communication channel fails. It is at the local level, where channels of communication are limited, that the most serious communications problems arise. Telephone lines, for example, may be destroyed in the high winds and command and control then becomes problematic. Each commune has at least one telephone but if this is out of action then communication between communes can only take place by messenger, though gunshots may be used to pass alarms from one village to the next. Apart from ensuring the dikes hold, other measures may have to be taken. Property may require additional protection against flooding, livestock may have to be moved and, in the worst circumstances, evacuation will take place. All coastal areas have a full evacuation plan with emergency boats provided to each household in particularly vulnerable zones.

Following the storm, aid and assistance is provided to the worst-hit areas by the district, provincial and national authorities and international cooperation may be sought. Repair of damaged dikes is a matter of high priority and is normally achieved within a week or so. Depending on time of year, rice may be replanted, taking advantage of rapid-growing varieties. The ultimate responsibility for recovery lies with the District and Provincial CFSCs, with support in terms of finance, labour or materials (such as rice, clothes and schoolbooks) from elsewhere in the province and from neighbouring provinces. Financial assistance may be received from the national level and tax concessions can be granted. The Central CFSC sends a mission to the area affected if relief is required and then coordinates the response of central government (through the State Planning Committee and Ministry of Finance) and, as appropriate, international assistance (through the Ministry of Foreign Affairs and, on the NGO side, the Union of Peace, Solidarity and Friendship Organizations). The Vietnam Red Cross and the Committee for Coordination and Reception of Foreign Aid of the Ministry of Finance are responsible for delivering international relief, alongside the DMU which, as well as collating damage reports, is responsible for distributing UNDP assistance. The continuing responsibility of administrative 'neighbours' as the first line of support resonates with historical practice and the duty to help the less fortunate. It is also reflected in the expectation of international assistance, on which there can be considerable dependence in the case of the most serious events.

Monitoring and evaluation continues to be a fundamental part of the system, with reports prepared after each storm and an extensive process of review. At the end of the cyclone season, the storm reports are assessed and lessons are drawn from the year's experience. This evaluation process takes place on all levels, from the district to the national, and will frequently involve exchange of experience between different administrative areas.

## The way forward

### *Strengths, weaknesses and opportunities for improvement*

Before considering the manner in which Vietnam's storm protection system might evolve in the future by identifying areas in which improvements might be warranted, it is important to underline its main strengths. While failure does occur, the system has undoubtedly been responsible for considerable savings in terms of both human life and economic well-being. We would cite four main strengths. First, the system has developed over many centuries and represents the culmination of a lengthy period of learning through experience. Second, the system is the result of an evolving pact between the elite, the institutions and the people of Vietnam and, as such, has widespread support, commitment and legitimacy. Third, the battle against 'invasion' by the tropical cyclone is as ingrained in the Vietnamese psyche as the struggle against human invaders (Lacoste 1973; Phan Khanh 1984). Finally, the system's structure, though complex, ensures that all levels and sectors of the administration and, indeed, much of Vietnamese society are firmly linked into the process. It contains a degree of duplication and redundancy, increasing the chances that information will be conveyed in the event of failure of any one channel. At a deeper level, its institutional structure controls the social attenuation or amplification of risk (Kasperson 1992; Pidgeon 1999), ensuring, to the extent possible, that the nature and scale of the societal response to each hazard is appropriate. In many respects, the approach to flood and storm control meets the criteria of a 'total flood warning system' (Keys 1997):

> A total flood warning system integrates flood prediction, the assessment of likely flood effects, the dissemination of warning information, the response of agencies and the public in the threatened community, and review and improvement. These components must operate together for sound flood warning performance to be achieved.
>
> (Emergency Management Australia 1995: 5)

The system would likely score very highly in most characteristics and criteria of the Criteria-Development Matrix (CDM) methodology for assessing warning dissemination systems developed by Parker *et al.* (1994) and used, for example, in Mauritius (Parker 1998).[10] Perspectives on total flood warning systems and the CDM methodology have informed this assessment of the Vietnamese system for flood and storm control which draws on documentary sources, interviews with key informants and personal experience (see introduction to previous section for further details).

Despite its strengths, the system for flood and storm control in Vietnam is subject to constraints and weaknesses. Informants at the district level cite lack of finance and credit (notably, extended credit from the government that would support large-scale improvement of local defensive structures rather than the existing, year-by-year funding that can only support piecemeal repair), the need to improve monitoring of dike condition, analysis of dike design and hence construction, limited access to 'new' coastal protection techniques and technologies, inefficient deployment of resources and distribution of relief, the lack of modern technology (such as computers and mobile telephones) and the desirability of better forecasts and communications. Stronger economic growth locally was also seen as necessary to support improvement of the system. Similar issues emerged at the national level, alongside the importance of public training and education, more effective deployment of resources, the need to 'localise' disaster prevention and preparedness measures (spatial scale issues), improved understanding of impacts and risks, better communication and institutional cooperation and the significance of environmental protection (especially forests to reduce flash floods and mangroves to protect dikes and the coastal zone).

Clearly, the limits on the financial resources and technical facilities that are available present a major problem (Benson 1997: 80). The most obvious effect is on the level of physical protection that can be provided by maintaining, strengthening and improving existing coastal defences.[11] Lack of funds limits the purchase of new technology to improve forecast capacity and strengthen communication links and the training of personnel. Having identified resource constraints as a problem, though, there is considerable scope for efficiency improvements within the existing system. Indeed, it may well be that improvements in the societal capacity to respond to warnings, for example, could prove more effective and less expensive than approaches based on costly technology and structures (cf. Burton *et al.* 1993: 92), albeit, perhaps, more challenging to implement.

Despite the comprehensive institutional framework that has developed – or, some would argue, because of the complexity of this framework – there is scope for better cooperation between the various elements of the structure and more efficient organisation and equitable deployment of resources throughout the system. At all levels, information flow can be interrupted or distorted, jeopardising the integrity of the system. As Burton *et al.* (1993: 158) note, in discussing collective adjustment to hazards, 'it is not surprising . . . that competitive relations develop around overlaps or vacuums of role and responsibility.' To select just one example, the increased involvement of international agencies in the system has, at times, resulted in very clear tensions between these groups and the various national institutions as the system adjusts to greater foreign involvement. Some informants expressed the feeling that overseas 'experts' have limited respect for national achievements and expertise and do not understand local conditions. It is noticeable that the strategies and plans fostered by international agencies focus on the national and regional level, but relatively little effort appears to have been made to become involved at the local level where the storm impact actually occurs. There is a perception that funding has been allocated preferentially to the newer, internationally-inspired elements of the system, for example, for computer hardware, with the original Vietnamese institutions remaining impoverished. While these views are, of course, subjective, there is no doubt that a pro-active effort to build understanding, trust and respect is needed.

There remains considerable scope for enhanced understanding of the storm risk and its impacts, ranging from improved forecast capability, through more effective coastal protection techniques, to more comprehensive assessment of damage costs. One reported weakness of the storm protection system in the early 1990s was that the forecasts and warnings were not considered credible (Vietnam Ministry of Water Resources *et al.* 1994: 86). To improve the forecasts, and their credibility, a priority must be faster data acquisition and processing as the predictions are based on data that may be as much as six hours out-of-date (data are only received once every six hours from the international meteorological network). Increasing the frequency of forecast updates from one every six to one every three hours is considered highly desirable; this too is dependent on improved data collection and communication. Storms which form in the East Sea present a particular problem, generally allowing around two, as opposed to at least four, days' warning. From a technological point of view, improved radar coverage[12] along the coast would aid local monitoring of many cyclone characteristics, including rainfall patterns, there is considerable potential for greater use of high-resolution satellite imagery, and the development of computer-based forecast models (to reduce reliance on overseas predictions) would clearly be advantageous. As far as impacts are concerned, there has long been an obligation for communes to report losses and damage to central government via the district and province level, but there is some concern regarding the quality of the resulting data. The Disaster Management Unit has developed comprehensive procedures for damage assessment at the provincial level (Vietnam Disaster Management Unit 1997), although there is, according to one informant, a reluctance to relinquish the older methods.

Improved public understanding of the risks posed by cyclone landfall and of the protective measures that can be deployed to reduce those risks can play a major role in improving the performance of the storm protection system (Trinh Van Thu 1991; Benson 1997: 69). The impact of cyclone Linda was exacerbated by the fact that the local population had little experience of cyclone landfalls and did not heed the warnings that were issued (Duong Lien Chau 2000). Indeed, it has been reported that the scale of the losses at sea was largely the result of fishing boats heading out towards the incoming cyclone to take advantage of good catches in the upwelling zone ahead of the storm. Even in cases where there is an appreciation of the risks associated with storm impacts, preventative action may not be taken for one reason or another. For example, raising the height of house roofs would provide better protection in the case of flooding but local populations are reluctant to accept the additional costs involved and no subsidies are available.

Having stressed the need for greater public involvement, it should be emphasised that there is an extremely high level of community participation in the system as far as the response when a cyclone strikes is concerned (contributing materials, labour, and so on) and in providing relief in the aftermath, resonant with the perspective that risk-bearing communities should be empowered through greater involvement in the process of risk reduction (cf. Pidgeon *et al.* 1993; Sime 1997: 171). It is, incidentally, interesting to contrast the modern view that the public should become more 'involved' in the process of risk reduction with the historical perspective that 'ownership' of civil defence measures lies with Vietnamese society as a whole.

Improved communication is a critical requirement with regard to many aspects of the system. For example, as a storm approaches the point of landfall, hourly telephone communication between the provinces and individual members of staff in the national forecast office represents the main, and often only, way to acquire the latest information. As noted, improving communications between the central and provincial authorities, making use of computer communications backed up by satellite links, is a priority of the Disaster Management Unit. Better communications at the local level are also needed to coordinate the response as a storm makes landfall. Conventional telephones are used for communication between the teams of workers 'on the ground' and the command centres. As overhead telephone cables are highly vulnerable to strong winds, this is not an effective means of communication; the use of mobile telephones was frequently mentioned as an effective option ruled out by cost. Any form of communication with small boats at sea was repeatedly cited as the single most effective way to provide protection to the fishing community. Without radio, fishers rely on their own forecasting ability and sporadic contact with larger radio-equipped ships. We have focused this discussion of communication difficulties on technological issues but, as discussed above, improved communication between organisations is often more a matter of sociological management and engineering than the injection of new technology.

Finally, a recurrent theme during the field interviews was the pressing need to reduce degradation of the natural environment as this can seriously exacerbate cyclone impacts. Deforestation, for example, has increased the risk of flash floods in coastal areas backed by mountainous country and the loss of the coastal mangrove has increased dike maintenance costs and the probability of saltwater flooding (Nguyen Hoang Tri *et al.* 1998).

### Threats

The storm response system faces new challenges as Vietnam goes through a period of rapid social and economic change. The process of economic renovation, *doi moi,* has had a profound effect on the capacity of the agrarian communities to respond to disaster, particularly with regard to the erosion of forms of collective action, the broader loss of social capital and the withdrawal of state support, as has been shown throughout this book.

The human resource on which storm protection has depended for centuries is being undermined. It is now possible to opt out, by paying a substitute tax, of the centuries-old obligation to provide labour as part of the communal system of dike maintenance and repair (Chapter 2). As the traditional method of earthen dike construction relies on low labour costs and cheap materials, the loss of this resource could represent a serious threat to the integrity of the protection system. It would require a considerable increase in financial resources to upgrade the dikes structurally (however desirable this may be) so that their reliability is not dependent on abundant labour. In Yen Hung District in Quang Ninh Province, the result has been that there has been insufficient labour available in recent years as the opt out is very attractive on economic grounds and the additional tax income has not covered the replacement labour costs; the opportunity to opt out of the labour requirement has, therefore, had to be restricted. In some cases, there has been a reduction in the resources available for dike maintenance at the local level as monetarisation of the system has permitted the diversion of finances into other areas (see Chapter 2). Informants at the district level expressed considerable concern regarding new systems for charging for forecasts, historically provided as service by the state at no charge. At a more fundamental level, Benson (1997: 92) argues that coping strategies that minimise risk, typical of present Vietnamese agriculture, may mitigate against income maximisation in the new economic circumstances, creating a poverty trap for the more vulnerable members of the community.

Nevertheless, the development, or re-establishment of, informal institutions has offset some of the negative consequences of market liberalisation and the reduction of the role of government by evolving collective security from below, for example, through risk spreading in credit unions, particularly in fishing communities (Adger 2000a). And, of course, improvements in the economy should tend to reduce vulnerability, at least at the community level. Adger and Kelly (1999) and Kelly and Adger (2000) discuss the effects of the process of *doi moi* on vulnerability to environmental stress in more detail.

In the longer-term, Vietnam faces the pervasive threat of long-term climate change and any related rise in sea level induced by the changing composition of the atmosphere (Granich *et al.* 1993; Asian Development Bank 1994; Tran Viet Lien and Nguyen Huu Ninh 1999). Tropical cyclone characteristics may well be affected as sea surface temperature plays an important role in determining whether or not tropical disturbances form and intensify. Any rise in sea surface temperature could induce an increase in storm frequency or strength. There are, however, many other factors which influence storm development and track, as we have seen in the definition of ENSO effects, and it is difficult to predict what the overall consequences might be as far as the behaviour of tropical cyclones is concerned (Walsh and Pittock 1998).

The Intergovernmental Panel on Climate Change (IPCC), the scientific and technical review body guiding the development and implementation of the United Nations Framework Convention on Climate Change, concluded that 'there is some evidence from model simulations and empirical considerations that the frequency per year, intensity and area of occurrence of tropical disturbances may increase . . .' but qualified this rather tentative statement with the rider that the evidence is 'not yet compelling' (Houghton *et al.* 1990). The later assessment by the IPCC was even more cautious: 'Knowledge is currently insufficient to say whether there will be any changes in the occurrence or geographical distribution of severe storms' (Houghton *et al.* 1996).

Of course, lack of knowledge does not mean that adverse changes will not take place and the costs of providing protection against the enhanced risks of a greenhouse world may prove substantial.[13] But, with storm frequencies varying from zero to twelve in a particular year in the present epoch, it could well be some time before any long-term trend in storm frequencies has a tangible impact. Coping with the pronounced natural variability of the present-day must remain the priority and represents a substantial precautionary first step in dealing with any change in the longer-term (Kelly 2000).

## Conclusions

We have seen how, even today, the storm protection system retains characteristics developed centuries ago – in the pragmatism with which the system evolves, in its multi-sectoral approach that reaches into all aspects of Vietnamese society, in the emphasis on anticipation and forward planning, in the sense of neighbourhood responsibility and moral justice – characteristics which, having stood the test of time and changing circumstances, can be considered fundamental strengths. What lessons can be drawn from this review of Vietnam's storm protection system regarding the issue of coping with the broader impact of the El Niño and La Niña? We select five points. The list is not exhaustive but illustrates a range of considerations.

- *Public awareness, involvement and commitment* has been highlighted as a critical issue. Indeed, the changing relationship between the disaster protection system and Vietnamese society, expressed in terms of moral duty and its later institutionalisation, provides a range of lessons regarding different modes of operation with regard to public participation and, arguably, degrees of effectiveness. At this time, we consider the issue of 'ownership' of the system to be particularly important in terms of both the attitude of the public to the system itself and the attitude of the Vietnamese institutions within the system to increasing foreign involvement.
- *Communication* is clearly an important aspect of any disaster management strategy, not only in terms of the technology involved but also with regard to effective communication between institutional players and with stake-holders, raising issues of mutual understanding, respect and trust. Effective communication is essential to ensure that perceptions of risk are not distorted and the societal response is consistent with the nature of the hazard. A sound institutional infrastructure provides the necessary foundation.

- *Anticipation* is a key component of Vietnam's storm protection system and this feature must be carried over to any effective process for coping with the broader impact of El Niño and La Niña. In fact, in the context of ENSO impacts, there are few nations with any level of forward planning comparable with Vietnam's disaster management system. It is necessary to promote pro-active, rather than re-active, attitudes.
- *Monitoring and evaluation* have provided the basis for continuous upgrading of the Vietnamese storm protection system since its inception. In view of the uncertain future, all means of facilitating the continual evolution of a disaster management system must be seen as highly desirable. For example, flexibility is essential to rapid modification of a system on the basis of experience.
- Finally, *duplication* or *redundancy* emerges as an important factor that underpins effective communication and disaster management, based on the realistic assumption that components of a system may fail under pressure, particularly when a system must be multi-sectoral in scope and extended in spatial scale.

The fact that Vietnamese society is prone to many different forms of natural hazard means that a rich variety of methods and structures for limiting the impact of environmental stress has emerged over time. To characterise Vietnam as particularly vulnerable because of its dependence on agriculture in the two low-lying deltas, as often occurs, is to neglect the resilience its society has developed over the centuries. Nevertheless, that resilience is itself under stress as social and economic developments combine with increasing environmental degradation to heighten social vulnerability, particularly amongst the poorer members of the community. The challenge for Vietnamese society is to retain what has proved effective from the past while taking advantage of the opportunities afforded by modern practice and continuing to adapt as circumstances alter.

## Acknowledgements

The authors would like to thank Nguyen Huu Ninh for his irreplaceable assistance in organising the fieldwork and Sarah Granich for her invaluable contribution in the field. Neil Adger and Cecilia Luttrell also made important contributions to the planning and implementation of this study. Finally, the authors gratefully acknowledge support from the Economic and Social Research Council, under the Global Environmental Change Programme (Award No. L320253240), and the British Academy Committee for South East Asian Studies and thank the many contributors to this research.

## Notes

1 Translated by Nguyen Ngoc Binh (Nguyen Ngoc Binh *et al.* 1975).
2 Vietnam Disaster Management Unit, http://www.undp.org.vn/dmu
3 The NOAA/US Navy database is the most comprehensive global collection of tropical cyclone observations available at this time. It contains a range of parameters for each

storm including, as available, location at six-hourly intervals, stage of development, wind, speed, central pressure, direction of travel, speed of travel, and so on.

4 Tropical storms form and develop in areas where there is strong upward motion and surface convergence. During El Niño events, the suppression of convection and vertical motion over the western Pacific is associated with an eastwards shift of the major area of upward motion and surface convergence that normally affects this region. Consequently, storms tend to form further east and the related disturbance in the flow over China and the western Pacific steers these cyclones away from Southeast Asia (Chan 1985). In contrast, during La Niña events, the major area of upward motion and convergence shifts westwards, storms form closer to Southeast Asia and the dominant westward flow here steers these cyclones towards Vietnam.

5 As an aside, lagged relationships between different local climate variables associated with the El Niño/La Niña signal may provide the basis for the traditional forecasting rules used by farmers in northern Vietnam, as reported by McArthur *et al.* (1993).

6 Vietnamese storm number 9726.

7 Vietnam Disaster Management Unit, http://www.undp.org.vn/dmu/events/1997-en.htm.

8 Timelines documenting seasonal activities and the passage of an individual storm provided the main focus for the interviews, alongside the identification of serious storm impacts in the past and key historical developments, strengths and weaknesses in the storm protection system.

9 The Central CFSC is chaired by the Ministry for Agriculture and Rural Development and this ministry also provides the Vice Chair on duty. Agencies represented on the committee include the Office of the Government of Vietnam (Vice Chair), the General Department of Hydrometeorology, the Department of Post and Telecommunication, the Voice of Vietnam, Vietnam TV and the Ministries of Commerce; Construction; Defence (Vice Chair); Finance; Fisheries; Health; Industry; Labour, War Invalids and Social Welfare; Planning and Investment; Public Security; Science, Technology and the Environment; and Transport.

10 The lower scores would fall mostly in the categories concerned with technological aspects of the system.

11 According to the World Bank (1995b), for example, the funding that is available in Thanh Hoa Province can only support a small fraction, less than 10 per cent, of the maintenance effort that is required along this province's 937 km of dikes.

12 Improved radar coverage along the coast has, in fact, occurred in recent years, spurred by the impact of Tropical Storm Linda in 1997.

13 One study has estimated that an annual investment of US$172 million would be required over the next 20 years in view of the projected sea-level rise alone, though this assumes a 'hard engineering' response based on improved physical protection (Asian Development Bank 1994).

# Part 3

# Development pathways

This section of the book examines key trends, opportunities and challenges facing modern Vietnam as the nation plots its development path for the early twenty-first century. One of the most significant driving forces of change for the foreseeable future will undoubtedly be the accelerating trend towards economic globalisation and a number of chapters consider the implications of Vietnam's increasing engagement with this process, focusing on the urban and industrial sectors. Relations between Vietnam and its neighbours over shared water resources are also considered in this section, with analysis of contemporary debates surrounding the management of the Mekong River, a significant resource for southern Vietnam and for the country as a whole. Drawing on the themes of vulnerability and resilience developed throughout the book, the section ends with consideration of prospects for sustainable development and the future well-being of the Vietnamese people.

# 11 Trade, investment and industrial pollution: lessons from Southeast Asia

*Rhys Jenkins*

## Introduction

The Southeast Asian economies have taken advantage of the globalising tendencies in the world economy over the past quarter century. Exports have grown rapidly, foreign capital has been attracted on a significant scale and industrial production has expanded at a phenomenal rate. Until the economic crisis beginning in 1997, Malaysia, Thailand and Indonesia were regarded as 'miracle' economies and held up as an example for other countries to emulate.

Many commentators regard these three countries as a second tier of newly industrialising economies following in the footsteps of the earlier industrialisers in the region, Hong Kong, Singapore, South Korea and Taiwan. Indeed, for the World Bank (1993a), the experience of these seven together with Japan constitute the 'East Asian miracle'. Although there is considerable debate over the key factors which led to the exceptional economic performance of this group of countries over the past two or three decades, there is a degree of consensus concerning some of the major features of the East Asian miracle. The defining characteristics of the miracle are high rates of economic growth combined with relatively equitable distribution of income compared to other developing countries, particularly Latin American countries (Page 1997). Additional features include macroeconomic stability, high rates of capital accumulation, significant growth of human capital, considerable success in penetrating world markets reflected in rapid export growth, and a substantial expansion of the industrial sector.

One of the main areas of controversy concerning the East Asian economies is the role of the state in promoting economic growth. The view that the state played a minimal role was subjected to intense criticism in the 1980s by authors such as Amsden (1989) and Wade (1990) who documented the active trade and industrial policies implemented in South Korea and Taiwan. The current debate is no longer about whether or not states intervened in the economy, but rather whether the interventionist policies that they did adopt were in fact an important contribution to economic success.

There is little doubt that in all the East Asian countries governments gave a high priority to economic growth and sought rapid industrialisation as a means of achieving this. In this context, the environmental consequences of industrial development received little attention. It is only relatively recently that the environmental consequences of unregulated industrial growth have become an issue and questions have begun to be raised about the 'sustainability' of the industrialisation process (O'Connor 1994). The collapse of the exchange rates, the reversal of capital inflows and the slowdown in economic growth in 1997 has led to a more critical view of these countries' economic performance. Moreover, at almost exactly the same time that the economies were experiencing an economic crisis, much of the Southeast Asian region was covered in haze and smog triggered by forest burning in Indonesia. This served to highlight some of the environmental consequences associated with economic growth.

Rapid, and often uncontrolled, growth in the region brought with it a major increase in environmental problems. In Japan, serious industrial pollution was already evident by the mid-1960s and led to more stringent environmental controls being introduced in the 1970s (O'Connor 1994: 21). Similar problems emerged in Taiwan and South Korea in the 1970s and 1980s (Bello and Rosenfeld 1992). Malaysia, Indonesia and Thailand have long depended on their natural resource base for their development and in recent years rapid urban and industrial growth has added pollution to their list of environmental problems.

The experience of the Southeast Asian economies over the past two decades is of particular relevance to Vietnam today. Vietnam has adopted an economic strategy involving export promotion, the attraction of foreign capital and the promotion of industrial growth that has much in common with the policies of the Southeast Asian economies. There are, moreover, many similarities between the level of development of Vietnam in the 1990s and Thailand in the 1970s or Taiwan in the 1960s (Riedel 1993). Given this context, it is of special interest to examine the trade, investment and pollution relationship for countries in Southeast Asia whose economic performance over the past two decades Vietnam hopes to emulate. The first aim of this chapter is, therefore, to examine the links between economic growth and increased participation of the Southeast Asian economies in international trade and investment, and the growth of industrial pollution in the region. The second objective is to bring out some of the implications of this experience for Vietnam in terms of economic growth, environmental quality and wider societal goals.

The chapter begins by looking at the relationship between income levels and environmental quality and asks whether rising per capita income in the region has led to increases in pollution levels. The second part of the chapter considers the environmental impact of international trade and, in particular, whether the countries of the region have a comparative advantage in more polluting activities. This is followed by a discussion of the impact of different industrial strategies on emissions, which seeks to test the view that more open strategies are less environmentally damaging than import substituting industrialisation. Given the significant role played by foreign investment in the region, evidence is then presented concerning the environmental consequences of the activities of multinational corporations in the region. Finally, some of the implications of the Southeast Asian experience for future developments in Vietnam are considered.

This chapter does not attempt to give an overview of all the environmental issues associated with trade and investment in the region. Its focus is specifically the question of industrial pollution and, as such, it does not touch on many other aspects of environmental degradation such as deforestation which are of major concern (see, for example, Hirsch and Warren 1998). It does, however, seek to focus attention on a relatively under-researched area which is certain to grow in importance in the future.

## Economic growth, environmental quality and industrial pollution

### The environmental Kuznets curve

A number of writers have suggested that there is an inverted-U relationship between environmental degradation and the level of per capita income. In other words, as countries' income levels increase, the quality of the environment first deteriorates but then at a certain point pollution levels begin to fall and environmental quality improves. This is often referred to in the literature as the 'environmental Kuznets curve' because of the parallel with the relationship between income distribution and per capita income which the economist Simon Kuznets in the 1950s predicted would first deteriorate (increasing inequality) and later improve during the process of economic development.

Several reasons have been advanced to support the thesis that environmental quality will improve at higher levels of income. First, changes in the composition of output which arise with economic development may lead to a lower level of pollution-intensity as heavy industry gives way to less polluting hi-tech industries, and as services increase in importance relative to manufacturing. Second, the process of technological change brings with it environmental benefits as more efficient production requires less raw materials and energy per unit of output and creates less waste. Third, it is argued that the demand for aspects of environmental quality, such as clean air, pure water and unspoilt habitats, are highly income elastic. At low levels of income the more immediate material needs are such that effective demand for environmental quality is limited but this rises rapidly with growing income. Finally, it is argued that higher levels of income are necessary in order to generate the resources which are required to invest in environmental protection.

The essence of the environmental Kuznets curve is that some combination of the above factors which tend to reduce pollution per unit of output at a certain level of per capita income start to outweigh the effects of increased output levels on the absolute volume of emissions and the resulting concentration of pollutants. In its strongest version this can be taken to imply that economic growth, far from being a cause of environmental degradation, will bring about environmental improvements.

During the 1990s, a number of empirical studies of the relationship between environmental degradation and income have been carried out. Most of these focus on indicators of pollution, although a few have also looked at other issues, most notably deforestation. Some of the studies of pollution have focused on concentrations of various pollutants in the atmosphere and in water (e.g. Shafik 1994; Grossman and Krueger 1992; Grossman 1994; Panayotou 1997) while others have been concerned with the levels of emissions (Panayotou 1993; Selden and Song 1994; Cole *et al.* 1997; Hettige *et al.* 1998). Although most of these studies claim to find evidence of an inverted-U relationship between both ambient quality indicators and pollution loads and per capita income,[1] several recent review articles have cast doubt on the validity of any such generalised relationship. Stern and colleagues (1996) identify a number of econometric problems such as simultaneity bias and heteroskedasticity in the main studies. Barbier (1997) argues that the studies offer little support for the conclusion that economic growth is the solution to environmental problems, and that the evidence for an inverted-U holds only for certain pollutants, mainly local air pollution. Ekins is even more critical stating that:

> The principal conclusion of this investigation into the environment–income relationship as revealed in the various econometric studies is that the hypothesis that there is an environmental Kuznets curve is not unequivocally supported for any environmental indicator and is rejected by the OECD and European studies of environmental quality as a whole.
>
> (Ekins 1997: 826)

A similar conclusion emerges from a special issue of *Ecological Economics* in 1998 where the editors summarise the thrust of the papers:

> They show that using different indicators, more explanatory variables than income alone and the estimation of different models, the environmental Kuznets curve results are not generally reproduced.
>
> (Rothman and de Bruyn 1998: 145)

One further weakness of the major studies of the environmental Kuznets curve is that they are mainly based on cross-section or panel data for a sample of developed and developing countries or based exclusively on developed country data. As far as the latter type of study is concerned, the range of per capita incomes from which the relationship is drawn is generally significantly higher than the income levels of most developing countries, making it difficult to use them to predict trends in developing countries. For the first type of study, the results may be misleading since they in effect involve pooling two samples with very different mean per capita income levels. In effect, the inverted-U may be a statistical artefact reflecting a positive relationship between pollution and income in developing countries and a negative one in developed countries (Vincent 1997). This suggests that it may be more useful to consider the historical experience of other developing countries in order to predict the likely trend in pollution. In the rest of this section, therefore, trends in environmental quality and emissions in Indonesia, Malaysia and Thailand in recent years will be reviewed with a view to identifying the implications of future growth in Vietnam for the environment.

## Environmental quality

Industrialisation, together with other aspects associated with urbanisation such as the growth of vehicle ownership, has led to deteriorating air quality in urban areas. In Kuala Lumpur, for instance, concentrations of carbon monoxide, nitrogen dioxide, particulates and ozone have increased significantly from the late 1980s. During the 1980s, air pollution increased rapidly in Jakarta, with concentrations of suspended particulates rising from 40 per cent above World Health Organisation recommended standards in 1978 to more than twice the standard a decade later, while sulphur dioxide concentrations doubled between 1981 and 1988 (Afsah and Vincent 1997). Air quality in Bangkok is also extremely poor with reports that suspended particulates in the atmosphere exceed recommended levels by 250 per cent (World Bank 1993b). Recent World Bank estimates put the health costs of air pollution in 1995 at 12 per cent of urban income in Jakarta, 7 per cent in Bangkok and 4 per cent in Kuala Lumpur, and on current trends these are likely to double over the next 25 years (World Bank 1997a: 17).

Similar deterioration has been observed in terms of water quality. Of rivers analysed in Malaysia between 1988 and 1994, 77 per cent had deteriorating water quality. In Indonesia, biochemical oxygen demand (BOD) measured at water-quality monitoring stations doubled in the 1980s (Afsah and Vincent 1997). Pollution of the Surabaya River was so extreme by the 1980s that it was virtually untreatable by conventional methods, and, although the situation improved in the 1990s, concentrations of BOD and COD remained above acceptable levels (Brandon and Ramankutty 1993). Water pollution is also a problem in Bangkok where accelerated industrialisation and urbanisation has led to a serious deterioration in the water quality in the canals and the lower length of the Chao Phraya River (Sachasinh *et al.* 1992: 23).

Industrial pollution makes a considerable contribution to these air quality problems. In Jakarta, for instance, the industrial sector accounts for 15 per cent of total suspended particulates (TSP), 16 per cent of nitrogen oxides ($NO_X$) and 63 per cent of sulphur oxides ($SO_X$) loadings, while in Indonesia's second city, Surabaya, the share of the industrial sector was 28 per cent, 43 per cent and 88 per cent respectively (World Bank 1994: 76). In Thailand, industry accounted for 56 per cent of emissions of suspended particulate matter, 22 per cent of sulphur dioxide, 23 per cent of carbon dioxide and 12 per cent of nitrogen oxides in the early 1990s (Sachasinh *et al.* 1992: 21). Similarly, in Malaysia the industrial sector made up 70 per cent of particulates, 39 per cent of sulphur oxides and 28 per cent of nitrogen oxides emissions nationally (Malaysia Department of Environment 1996). Moreover, these figures do not include the indirect contribution of industry to pollution through its purchase of generated power.

Although domestic wastewater is the major contributor to poor water quality, industrial effluent is also significant. On Java, between 25 per cent and 50 per cent of the total pollution load in a number of rivers was the result of industrial discharges (World Bank 1994), while in Bangkok 25 per cent of BOD load in the lower Chao Phraya River comes from industry (Sachasinh *et al.* 1992: 23). In Malaysia, the contribution nationally has been somewhat lower with only 10 per cent of BOD loading attributable to industry (Malaysia Department of Environment 1996), although undoubtedly this would be much higher in major industrial centres.

Similar problems have been found in coastal areas. Substantial proportions of fish and shellfish caught in Jakarta Bay exceeded World Health Organisation guidelines on concentrations of lead, mercury and cadmium (World Bank 1992: 46). In Malaysia, the Department of Environment has observed that 'the assessment of heavy metals throughout the country revealed that the overall quality of marine waters is still very much influenced by the degree of the discharge of partially treated or untreated industrial wastes and other land-based sources' (Malaysia Department of the Environment 1996: 18).

### Emissions of pollutants

The most readily available internationally comparable estimates of pollution are of carbon dioxide emissions. Figure 11.1a shows that these have increased more than seven-fold in Indonesia, six-fold in Malaysia and over nine-fold in Thailand over the past quarter century (data from Oak Ridge National Laboratory).[2] Figure 11.1b shows that despite substantial population growth in all three countries, per capita carbon emissions have also grown significantly. Indeed, the indications are that, over the past decade, the trend has been for an increasing rate of growth of per capita emissions. Thus, there is no evidence in Southeast Asia that higher levels of income are leading to reductions in carbon dioxide emissions.

Data on other pollutants are much more fragmentary but generally point in the same direction of substantial increases in emissions and effluents. In Thailand, during the 1980s, sulphur dioxide emissions and BOD releases from the industrial sector doubled, while nitrogen oxides almost trebled and suspended particulate matter increased almost fourfold (Sachasinh *et al.* 1992). In Malaysia, emissions of sulphur oxides and nitrogen oxides almost doubled between 1988 and 1995, while emissions of particulates increased more than three times (Chang Yii Tan and Leong Yueh Kwong 1990; Malaysia Department of Environment 1996). There is evidence of a significant increase in toxic emissions to water sources from industry in the 1990s in Malaysia (Malaysia Department of Environment 1996).

All three Southeast Asian countries face a growing problem of hazardous industrial waste. In Malaysia, in 1987, such waste accumulation was estimated at 100 000 tons (World Bank 1993a: 47) but by the mid-1990s it had risen to over 400 000 tons (Malaysia Department of Environment 1995: 29) and was expected to grow further as the structure and scale of industrialisation changes (Malaysia Ministry of International Trade and Industry 1995: 196). Thailand is also facing

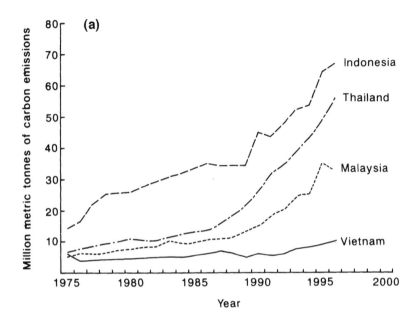

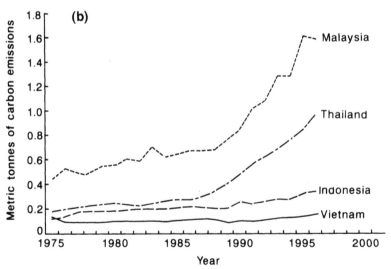

*Figure 11.1* Carbon dioxide emissions in Indonesia, Malaysia, Thailand and Vietnam, 1975–1997 by: (a) total emissions; and (b) per capita emissions

increasing problems with hazardous waste as a result of changes in industrial structure. The total volume of industrial hazardous waste increased from just over half a million tons in 1986 to more than 1.6 million tons in the early 1990s and is expected to grow to almost three million tons by the year 2001 (Parasnis, 1999, Table 1). The most important industries in terms of volumes discharged in the early 1990s were basic metals, fabricated metal products, transport equipment, electrical machinery and chemical products (O'Connor 1994: 31). In Indonesia, quantities of toxic and hazardous waste are deposited in uncontrolled landfills and dumped into rivers with other industrial wastes. An estimate in 1993 put the total volume of hazardous wastes generated in West Java and Jakarta at about 2.2 million tons a year (World Bank 1994: 77).

Given the limited availability of in-country data, and in some cases the problematic nature of the estimates which are available, another indicator of the growth of pollution which it is useful to look at comes from estimates obtained using the World Bank's Industrial Pollution Projection System (IPPS).[3] This involves applying coefficients derived from US data for 1987 on emissions of various pollutants per thousand dollars of output to industrial production data in Southeast Asia. This tends to underestimate the level of pollution in the Southeast Asian countries which are likely to have higher levels of emissions per unit of output than in the United States because of less stringent regulation. However, given the lack of direct data, it provides some indication of the extent and growth of pollution problems in the region.[4]

The data shown in Figure 11.2 confirm that the past two decades have seen a massive increase in emissions of major pollutants from the manufacturing sector in all three countries,[5] as a result of both increases in scale but also the changing industrial structure.

### The income–pollution relationship in Southeast Asia

It is interesting to note the relationship between pollution and income per capita in the three Southeast Asian economies during this past decade in the light of the earlier discussion of the environmental Kuznets curve. It is clear that economic growth in this period has been accompanied by increased pollution in all three countries. This does not of course necessarily contradict the environmental Kuznets curve hypothesis. It may rather simply reflect the fact that the Southeast Asian economies are on the rising portion of the curve.

The empirical studies cited earlier in this section estimated the turning points at which concentrations or emissions of pollutants begin to fall as income rises. In a number of these, the turning point for several local air pollutants have been estimated as occurring at levels of income in the region or US$3000 to US$6000 while some water pollutants have a turning point at between US$7000 and US$8000 (Barbier 1997; Ekins 1997). In the mid-1980s, per capita income (in purchasing power parity terms) was about US$4700 in Malaysia, US$2900 in Thailand and US$1700 in Indonesia. By the mid-1990s these had risen to US$8900 in Malaysia, US$7100 in Thailand and US$3700 in Indonesia. On

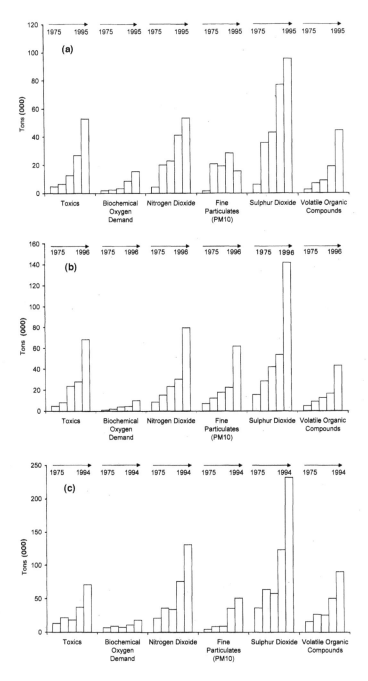

*Figure 11.2* Estimates of emissions of selected pollutants in: (a) Indonesia (1975, 1981, 1985, 1990, 1995); (b) Malaysia (1975, 1981, 1985, 1990, 1996); and, (c) Thailand (1975, 1981, 1985, 1990, 1994), based on **IPPS** coefficients

this basis, both Malaysia and Thailand might be expected to have experienced reductions in air pollution, and Malaysia might also have expected reductions in water pollution during this period. In the case of Indonesia, however, given its lower level of per capita income, environmental deterioration is consistent with the environmental Kuznets curve.

## Trade and pollution

Vietnam is seeking not only to accelerate its economic growth but also to significantly change its insertion in the international economy. The collapse of the Soviet Union and the changes in Eastern Europe, and the ending of the United States embargo on Vietnam, have changed substantially the options which confront the economy. From a relatively closed economy, with minimal links with the world capitalist economy, Vietnam is increasingly becoming more export-oriented. It is, therefore, important to consider the links between trade and environment in Southeast Asia in order to see what light this might throw on the environmental impact of greater openness on Vietnam.

The link between trade and pollution is a matter of some debate, both theoretically and empirically (Nordström and Vaughan 1999). Within orthodox trade theory, pollution can be thought of as an input into the production process. Although often regarded as an output of production, emissions reflect the fact that environmental resources have been used up in production. 'Each unit of emission indicates that a unit of environmental resources has been depleted in the process of production' (Rauscher 1997: 30). Thus, it is possible to think of traded goods as embodying a certain quantity of environmental resources. A similar approach is adopted by Lee and Roland-Holst (1993), who used the concept of Embodied Effluent Trade. The implication is that trade flows involve international transfers of environmental resources.[6]

A prediction of this approach is that countries that are well endowed in terms of environmental resources will tend to export goods that use such resources intensively; in other words, they will export goods with a high level of embodied effluent. In a North–South context, this suggests that low income countries which have less stringent environmental regulation than high income countries (which is assumed to reflect a relative abundance of environmental resources) will tend to specialise in more polluting industries. Indeed, a number of theoretical models have been developed based on such assumptions (Copeland and Taylor 1994; Copeland 1994; Rauscher 1997).

An alternative view plays down the impact of environmental regulation on the pattern of international trade. Under this perspective, the costs associated with pollution prevention or abatement are a very small part of total costs and, therefore, the key determinants of the location of production and comparative advantage lie elsewhere (Leonard 1988; Dean 1992). In particular, it is argued that differences in labour costs internationally play a far more important role than differences in environmental regulation, so that the comparative advantage of low income countries lies in labour-intensive production. Since it is also argued that

*Table 11.1* Ratio of embodied effluent in Indonesian exports and imports, 1990

| Pollutants | Japan | Rest of world |
| --- | --- | --- |
| Sulphur dioxide (SO₂) | 9.27 | 2.34 |
| Nitrogen dioxide (NO₂) | 3.45 | 1.61 |
| Carbon monoxide (CO) | 2.24 | 1.37 |
| Volatile Organic Compounds | 0.91 | 0.80 |
| PM10 (fine particulates) | 0.61 | 0.95 |
| Total suspended particulates | 3.71 | 1.91 |
| Biochemical Oxygen Demand | 4.01 | 0.91 |
| Total suspended solids | 1.22 | 0.60 |
| Lead | 10.09 | 2.51 |
| Linear Acute Human Toxicity Index | 6.20 | 2.17 |

Source: Lee and Roland-Holst (1993).

Notes
Column 2 is ratio of effluent embodied in Indonesian exports to Japan to effluent embodied in Indonesian imports from Japan.
Column 3 is ratio of effluent embodied in Indonesian exports to the rest of the world to effluent embodied in Indonesian imports from the rest of the world.

there is a positive correlation between capital-intensity and pollution-intensity, countries with a comparative advantage in labour-intensive industries will utilise fewer environmental resources by specialising according to their comparative advantage (Birdsall and Wheeler 1992).

The empirical evidence concerning the link between international trade flows and environmental regulation is very mixed. One approach models international trade flows within a conventional trade model but with the addition of a variable designed to capture differences in the strictness of environmental regulation between countries. A study of this kind by Tobey (1990) found no relationship between trade flows and regulation whereas a more recent study by van Beers and van den Bergh (1997) did find a link between regulation and trade in non-resource based, pollution-intensive industries. Another approach is to calculate the pollution content of a particular country's trade to see whether it is a net importer or exporter of environmental resources. A number of studies of United States trade have found that exports are less pollution-intensive than imports so that the United States appears to be a net importer of environmental resources (Kalt 1988; Robinson 1988). It is of interest, therefore, to consider the trade patterns of some of the Southeast Asian economies to see whether they have specialised in exports of relatively polluting goods or not.

Here, again, the evidence is rather mixed. An Organisation for Economic Cooperation and Development (OECD) study of Indonesia shows that the country's exports embody a substantially higher level of effluent than do its imports (see Table 11.1). This is particularly marked in the case of trade with Japan, its major developed country trading partner. The case of Malaysia is in contrast to that of Indonesia in that evidence on trade in manufactures indicates that exports embody less effluent than imports. When IPPS coefficients were

*Table 11.2* Net effect of imports and exports of manufactures on pollution in
Malaysia, 1994

|  | Imports (million lbs) | Exports (million lbs) | Balance (million lbs) |
|---|---|---|---|
| Sulphur dioxide ($SO_2$) | 196.9 | 105.1 | −91.8 |
| Nitrogen dioxide ($NO_2$) | 99.5 | 50.5 | −48.9 |
| Carbon monoxide (CO) | 166.3 | 71.2 | −95.1 |
| Volatile Organic Compounds | 72.1 | 48.3 | −23.9 |
| PM10 (fine particulates) | 27.8 | 30.7 | 2.8 |
| Total suspended particulates | 41.4 | 64.0 | 22.6 |
| Biochemical Oxygen Demand | 25.3 | 10.4 | −14.9 |
| Total suspended solids | 605.7 | 135.7 | −470.0 |
| Toxic to air | 54.8 | 31.2 | −23.6 |
| Toxic to land | 89.1 | 47.3 | −41.8 |
| Toxic to water | 7.7 | 3.4 | −4.3 |
| Total toxic | 151.5 | 81.9 | −69.6 |
| Linear Acute Human Toxicity Index | 0.4 | 0.2 | −0.1 |

Source: Author's own elaboration from UNIDO trade data and IPPS coefficients.

applied to Malaysian trade, it was found that there was a net import of environmental resources.

Table 11.2 shows that for most pollutants, the manufactured goods which Malaysia imports create more pollution overseas than those that it exports generate in Malaysia. Only in the case of particulates (both fine and total suspended) do exports generate more pollution than imports. Although Malaysia has a trade deficit in manufactures, this does not account for the difference in pollution associated with imports and exports. In other words, even if trade were balanced, given its present composition, Malaysia would be a net importer of environmental resources except in the case of particulates.

Unfortunately, there are no similar studies for Thailand, and it is not possible to say whether this nation is a net importer or exporter of environmental resources. The contrasting cases of Indonesia and Malaysia, however, indicate that there is no general pattern of comparative advantage in terms of pollution-intensity in Southeast Asia. This can be interpreted as evidence that environmental factors are not the critical determinant of the pattern of comparative advantage in the region and that, in some cases, comparative advantage favours relatively low-polluting industries, while in others it favours more pollution-intensive industries.[7]

## Industrial strategy and pollution

### *The debate*

Closely related to the question of the links between trade flows and pollution is the debate over whether different types of industrial strategy have different implications in terms of industrial pollution. It has been argued by the World

Bank and others that an export-oriented industrialisation strategy is less environmentally damaging than an import-substituting strategy. Since a shift to export-orientation is also regarded as economically advantageous in these quarters, changing industrial strategy is believed to be a 'win–win' situation.

The impact of industrialisation on pollution is a combination of three factors (Grossman and Krueger 1992; Birdsall and Wheeler 1992). First, it depends on the total volume of industrial production, in other words a *scale* factor. Second, it depends on the weight of different industries within the industrial sector. Clearly, where more polluting industries such as petrochemicals or cement make up a large share of production, the overall level of pollution will be higher than where clothing and footwear account for the bulk of industrial output. This is referred to as the *composition* effect. Finally, pollution levels will depend on the extent to which pollution abatement equipment is used, better environmental management is practised and cleaner technologies are employed. These will tend to affect the level of pollution per unit of output within a firm or industry. This is usually referred to as the *process* or *technology* effect.

There are a number of reasons why it is argued that greater openness will tend to result in lower levels of pollution. First, as indicated in the previous section, it is possible that the comparative advantage of less developed countries lies in labour-intensive sectors which tend on the whole to be less polluting than more capital-intensive industries. Thus the *composition* effect will tend to lead to a reduction in the pollution-intensity of production.

It is also claimed that the *technology* effect will tend to lead to less polluting production in more open economies. Two major reasons are thought to contribute to this. First, production for export may lead to the adoption of clean technologies because of the requirements of international markets. Cases in point arise where dioxin needs to be eliminated in the pulp and paper industry, and chromium in tanning. Anecdotal evidence is reported from Chile of firms reducing emissions in order to meet foreign product standards, e.g. in the pulp and paper industry (Birdsall and Wheeler 1992). This may be reinforced in the not too distant future through the introduction of the ISO 14000 series of environmental standards, if they become a requirement for exporters. Second, more open economies have greater access to the latest foreign technology and so will be in a better position to incorporate the latest waste or emissions minimising technologies, which will be diffused much more rapidly than in closed economies. Evidence of this comes in a study of pulp and paper which found that the cleaner thermo-mechanical pulping process was adopted more quickly in more open economies (Wheeler and Martin 1992). It is not clear how far this finding can be generalised, however, since the technology concerned was both cheaper and more environmentally friendly.

Against this, it is argued that more open economies find it more difficult to implement high environmental standards because of the need to minimise costs in order to be internationally competitive. Where producers are heavily reliant on exports, they are able to argue strongly against tighter environmental regulation on the grounds that these might jeopardise exports. If this is true of all countries, there is pressure to reduce standards to the least common denominator. Although the World Bank has produced evidence to support the claim that more open economies tend to have a lower level of pollution-intensity (Birdsall and Wheeler 1992; Lucas *et al.* 1992; Wheeler and Martin 1992), their findings have been seriously questioned (Rock 1996). Thus, the relative merits of both sides of the debate have still to be determined empirically, and further evidence needs to be sought.

### Industrial strategy in Southeast Asia

Southeast Asia is a good area in which to seek evidence of a link between openness in the economy and pollution-intensity. Indonesia, Malaysia and Thailand have all experienced rapid industrial growth during which the state has played an active role in promoting the industrial sector without much regard for the environmental impacts. They have also gone through contrasting phases of industrial policy with greater or lesser emphasis on import substitution and export promotion.

Indonesia in the period since 1970 has gone through three major periods of economic policy. Up until the early 1980s, growth was propelled by the oil and commodity boom and public investment, with relatively high levels of protection giving an inward-oriented strategy. In the first half of the 1980s, the government began some policy reforms in response to falling oil revenues but the strategy remained inward oriented until the mid-1980s. After 1985, however, there was a significant shift towards export-orientation and deregulation (Hill 1996: 14–17; Wheeler and Martin 1993; World Bank 1993b: 136–9).

Malaysia, which had pursued import substitution in the 1960s, switched to a more export-oriented strategy in the 1970s. In the first half of the 1980s, government policy once more emphasised import substitution with the development of heavy industries under the Fourth Malaysian Plan. From the mid-1980s, however, there was a shift again to a second phase of export-oriented industrialisation and a move away from the protectionist policies of the early 1980s (Alavi 1996).

During the 1970s, Thailand emphasised import substitution. Attempts were made to promote exports during this period but protection of the manufacturing sector remained high (Suphachalasai 1995). The World Bank sees the major shift in policy towards export promotion in Thailand as having occurred in 1981, although protection remained significant up to the mid-1980s (World Bank 1993b: 140). A more broadly based trade liberalisation took place in the 1990s. Other authors have commented that it was not until after a major recession in 1984–5 that the government took a decisive step towards a manufactured-export led strategy (Phongpaichit and Baker 1996: 27).

### Pollution-intensity and industrial strategy

Given the changes in industrial strategy in the three countries during the period, it is relevant to ask whether this has been accompanied by changes in the pollution-intensity of manufacturing production. In order to illustrate the compo-

*Table 11.3* Change in estimated pollution-intensity by selected pollutants in Malaysia

| | 1974–81 (%) | 1981–85 (%) | 1985–96 (%) |
|---|---|---|---|
| Total toxics | −2.6 | 165.1 | −38.5 |
| Total metals | 10.7 | 27.6 | −24.3 |
| Sulphur dioxide (SO$_2$) | 10.6 | 31.0 | −28.1 |
| Nitrogen dioxide (NO$_2$) | 6.5 | 38.2 | −27.7 |
| Carbon monoxide (CO) | 6.9 | 20.0 | −24.6 |
| Volatile Organic Compounds | 12.4 | 25.6 | −23.2 |
| PM10 (fine particulates) | −3.4 | 30.4 | −29.6 |
| Total suspended particulates | −0.9 | 10.6 | −33.0 |
| Biochemical Oxygen Demand | 19.0 | 89.1 | −45.2 |
| Total suspended solids | −21.7 | 39.8 | −36.2 |

Source: Author's own elaboration from UNIDO manufacturing value added data and IPPS coefficients.

sition effect, estimates were made of the pollution-intensity of production using IPPS coefficients for each of the three countries.

The most detailed calculations were carried out for Malaysia where data was available at the 4-digit level of the International Standard Industrial Classification (ISIC). The data generated above for Malaysia enables us to consider this issue in relation to changes in trade orientation within a single country. As was indicated above, Malaysia has gone through several stages in terms of its industrialisation strategy. The early period of import substitution from Independence to 1970 is not considered here because sufficiently detailed data on the composition of manufacturing output are not available. There have, however, been contrasting periods since then with the 1970s being a period of export-oriented industrialisation, the first half of the 1980s seeing further import substitution, and the period from 1985 onwards seeing a return to export-oriented industrialisation once more.

Table 11.3 shows the changes that have taken place in the pollution-intensity of manufacturing industry in Malaysia during three distinct periods of time. The first period from 1974 to 1981 coincides largely with the first stage of export-oriented industrialisation.[8] The period 1981–5 represents a period of more inward-oriented industrialisation while the years 1985 to 1994 (the latest year for which data was available) is again a period of export-orientation. Table 11.3 shows that in the 1970s, there was no uniform trend in the indicators of pollution-intensity, with some showing an increasing trend and others falling. Thus, although this was a period in which Malaysia was emphasising export promotion, this was not associated with a general reduction in pollution-intensity as has been predicted in some quarters.

The data for the 1980s shows a clear distinction between the two periods. All the pollutants covered showed an increase in pollution-intensity during the first half of the decade. From 1985 onwards, however, there was a tendency for all indicators to fall. The evidence from the 1980s lends *prima facie* support to the

*Table 11.4* Change in estimated pollution-intensity by selected
pollutants in Indonesia

|  | 1975–85 (%) | 1985–95 (%) |
|---|---|---|
| Total toxics | 11.2 | 16.2 |
| Total metals | 155.3 | 9.1 |
| Biochemical Oxygen Demand | −19.4 | 27.8 |
| Total suspended solids | 343.4 | 7.9 |
| Nitrogen dioxide (NO$_2$) | 7.0 | −34.6 |
| PM10 (fine particulates) | 10.9 | −77.1 |
| Sulphur dioxide (SO$_2$) | 16.0 | −37.9 |
| Carbon monoxide (CO) | 32.8 | −9.2 |
| PT | 13.7 | −53.2 |
| Volatile Organic Compounds | 8.8 | 37.4 |

Source: Author's own elaboration from UNIDO manufacturing value added data and IPPS coefficients.

World Bank's view that more open trade policies will lead to a less polluting composition of output. In contrast, the situation in the 1970s indicates that, contrary to this view, an export-oriented strategy can be associated with increasing pollution-intensity based on some indicators.

Indonesia is cited by the World Bank as an example of a country where a shift towards a more open economy has had beneficial effects in terms of reducing pollution-intensity (Munasinghe and Cruz 1995). Two groups of industries were identified, assembly industries and processing industries, with the latter having much higher levels of pollution per unit of output. Liberalisation in the 1980s was then shown to have led to a shift in the composition of output towards the less polluting assembly industries, although of course the overall growth of industry has meant significant increases in the absolute level of pollution. This finding is not quite as clear-cut as might appear from the summary contained in Munasinghe and Cruz (1995). The case study on which the summary is based is rather more cautious. While the share of assembly industries did indeed increase with liberalisation, this was to be expected in terms of a general international relation between the share of assembly industries in manufacturing and per capita income levels. Wheeler and Martin, therefore, concluded that there is 'little evidence of a policy impact on the share of assembly in total production' (Wheeler and Martin 1993).

Further questions are raised concerning the impact of liberalisation on pollution in Indonesia from a simulation exercise carried out in an OECD study. This looked at the effects that the removal of tariffs would have on the sectoral composition of output in Indonesia and hence on the pollution-intensity of production. It was found that for ten of the twelve pollutants examined, liberalisation would lead to increased pollution-intensity, the only exceptions being biochemical oxygen demand and fine particulates (Lee and Roland-Holst 1993).[9]

In view of this mixed evidence, it was decided to carry out our own estimates of the pollution-intensity of Indonesian manufacturing using the IPPS coefficients.

*Table 11.5* Change in estimated pollution-intensity by selected pollutants in Thailand

| | 1975–85 (%) | 1985–94 (%) |
|---|---|---|
| Total toxics | −5.0 | −0.8 |
| Total metals | −21.2 | 33.8 |
| Biochemical Oxygen Demand | −22.9 | −38.7 |
| Total suspended sediment | −14.5 | 39.4 |
| Nitrogen dioxide ($NO_2$) | 12.4 | −1.7 |
| PM10 (fine particulates) | 34.6 | 47.6 |
| Sulphur dioxide ($SO_2$) | 11.3 | 1.4 |
| Carbon monoxide | −2.3 | 3.0 |
| Total suspended particulates | 9.7 | 18.1 |
| Volatile Organic Compounds | 12.7 | −9.8 |

Source: Author's own elaboration from UNIDO manufacturing value added data and IPPS coefficients.

Given that the two studies previously cited covered the period up to 1989 and 1990, respectively, it is also desirable to extend the study. This was done using United Nations Industrial Development Organisation (UNIDO) data for 1975, 1985 and 1995.[10]

Table 11.4 shows the increase in the level of pollution per dollar of value added for ten pollutants in Indonesia in two periods. The first is the period up to the mid-1980s when an inward oriented strategy was dominant. Only BOD pollution showed a declining trend during this period. In contrast, in the later period, when economic policy was liberalised, a number of indicators of air pollution fell. However, water pollution, toxics and metals continued to rise. Thus, by looking at rather a longer period of time after liberalisation than the earlier studies, there is evidence that liberalisation has been associated with a less pollution-intensive industrial structure, at least as far as air pollution is concerned.

A similar exercise was carried out for Thailand. This was more difficult than in the Indonesian case. First of all, problems arose because the change in economic policy was less clear cut and there is, therefore, less agreement on the identification of different phases. For the purposes of the exercise, however, it was decided once more to compare the period from the mid-1970s to the mid-1980s which was regarded as more inward-oriented with that from the mid-1980s to the mid-1990s which was more outward-oriented. A second difficulty arises from some inconsistencies in the data for value added.[11]

Table 11.5 shows a much less clear picture than that for Indonesia. In the period 1975–85, the indicators of toxic, metal and water pollution all declined, while five of the six criteria air pollutants were increasing relative to value added. The pattern in the second period is even more mixed, but overall six of the ten measures showed increasing pollution-intensity. This was particularly marked for heavy metals, total suspended solids and particulates (both fine and total). Thus, in contrast to Indonesia, the later, more outward-oriented, period in Thailand has been associated with greater rather than lower pollution-intensity.

*Table 11.6* Pollution-intensity of Malaysian industrial sector by level of Effective Rate of Protection, 1987

| ERP | >50% (lbs per $m of value added) | 0–50% (lbs per $m of value added) | <0% (lbs per $m of value added) | All industries (lbs per $m of value added) |
|---|---|---|---|---|
| Toxics | 3 200 | 25 500 | 2 200 | 9 900 |
| Metals | 1 000 | 900 | 100 | 700 |
| SO$_2$ | 22 900 | 16 300 | 7 000 | 15 600 |
| NO$_2$ | 13 600 | 11 000 | 2 600 | 9 200 |
| CO | 13 100 | 8 800 | 600 | 7 600 |
| VOC | 3 800 | 9 200 | 2 400 | 5 000 |
| PM10 | 14 800 | 600 | 4 200 | 6 900 |
| TSP | 12 100 | 2 700 | 6 800 | 7 400 |
| BOD | 400 | 4 600 | 200 | 1 600 |
| TSS | 44 200 | 6 400 | 500 | 18 000 |

Source: Author's own elaboration from IPPS data and Alavi (1996).

### Protection and pollution

The argument that a more export-oriented industrialisation strategy would lead to a change in the sectoral composition of production towards less pollution-intensive industries implies that during the period of import-substituting industrialisation, the structure of protection has a 'brown bias'. In other words, it is suggested that highly polluting industries tend to receive higher protection than less pollution-intensive industries. Evidence of such a bias has, indeed, been found in the case of Turkey (Kosmo 1989, quoted in Birdsall and Wheeler 1992).

Does such a relationship hold in the Southeast Asian countries? Wheeler and Martin (1993) tested this for Indonesia. They estimated rank correlations between three measures of pollution-intensity (BOD, total suspended particulates, and toxics) and effective rates of protection in three years (1971, 1980 and 1989).[12] They found no evidence that pollution-intensive industries were more heavily protected and, indeed, some indication that the reverse was the case. This is consistent with the results of the previously cited OECD study of Indonesia which concluded that 'the historical asymmetry in the effluent content of trade was not a result of the existing pattern of Indonesian protection, (Lee and Roland-Holst 1993: 29). It may help explain why the shift to a more outward-oriented strategy from the mid-1980s in Indonesia did not lead to a clear-cut shift to a less polluting industrial structure.

Looking at the Malaysian case, however, where the shift towards less polluting industries from the mid-1980s is much clearer, there is some evidence of a link between protection levels and pollution-intensity. Table 11.6 presents the average level of pollution-intensity (in terms of pounds of emissions per million dollars of value added) of Malaysian industries according to the level of effective protection which they received in the mid-1980s.[13] Three groups of industries were identified: those that are highly protected (Effective Rate of Protection (ERP) >50 per cent); those with intermediate levels of protection (0–50 per cent); and those receiving negative effective protection. They had approximately equivalent shares of manufacturing value added in 1987: 36 per cent highly protected; 31 per cent medium; and 33 per cent negative.

The classification by protection level shows that for seven of the ten pollutants, the highly protected group of industries are the most polluting. The exceptions are for toxics, volatile organic compounds and BOD where the intermediate group has the highest pollution-intensity. For all the pollutants listed, the industries for which effective protection was negative were less polluting than the average. It can be concluded, therefore, that the protection structure which existed in Malaysia in the mid-1980s tended to encourage the more pollution-intensive industrial activities and that therefore the shift away from import substitution was a factor in reducing the overall pollution-intensity of manufacturing.

## *Industrial strategy and technology effects on pollution*

So far we have concentrated exclusively on the *composition* effects of industrial strategy on pollution. However, as was indicated above, there are also arguments that suggest that a more open economy will tend to encourage less polluting production within each sector. The first of these arguments suggests that export-oriented firms will, as a result of the demands of the markets to which they sell, adopt cleaner technologies than firms that produce for the domestic market. They are also more likely to introduce environmental management systems than non-exporting firms. Despite these *a priori* arguments, there are no empirical studies that have tested this hypothesis in any of the Southeast Asian countries. This in part reflects the difficulties of research on environmental issues at the firm level and a resulting paucity of any empirical studies. My current research on Malaysia aims to provide some evidence on this issue.

It is not clear that producing for export necessarily results in better environmental performance. The pressures from advanced industrial markets are still relatively restricted to particular industries where eco-labelling is being introduced, and is not a major factor across the board. ISO 14000 is only just getting off the ground, and although it may have an important impact in developing environmental management systems among exporting firms in the future, its impact so far has obviously been limited. On the other hand, firms producing for export will be more subject to competitive pressures which make them reluctant to increase costs through spending on pollution abatement. It has been reported that efforts to clean up the textile industry in Thailand have been hampered by a perception that the industry is losing competitive advantage to countries such as Vietnam and China (United Nations Conference on Trade and Development and United Nations Development Programme 1994: 146). This is clearly an area where further empirical research is required.

The second argument that is advanced in support of the view that a more open economy leads to lower pollution is in terms of the greater ease of access to the latest, and hence least polluting, technologies, from the advanced industrial economies. However, the industrial strategies of all three countries considered here have, even when emphasising import substitution, been relatively open to imports of machinery and equipment in order to encourage the industrialisation process. Preliminary evidence from interviews with firms in Malaysia does not suggest that access to technology is a major obstacle to improving environmental performance.

## Foreign direct investment and pollution

Much the same kind of debate exists concerning the impact of direct foreign investment on industrial pollution as was described for the case of trade in the previous section. First, there is the question of the sectoral concentration of foreign capital. In other words, does foreign investment tend to be represented disproportionately within the most pollution-intensive sectors?

There are also issues related to the *technology* effects of foreign investment. First, it is often claimed, that multinational companies adhere to their own corporate environmental standards which are higher than those of the developing countries in which they operate (Gladwin 1987). Thus, increased inflows of foreign capital tend to bring with them higher environmental standards. The extent to which multinationals do in practice require their subsidiaries to observe higher environmental practices is unclear and two recent surveys came to quite different conclusions.[14] Second, even when multinational companies do not have explicit corporate environmental policies, their tendency to use parent company technology, which has been developed to meet the stricter regulatory requirements of their home countries, will lead to them having less polluting production than local firms in developing countries (Ferruntino 1995). Again the limited empirical evidence on this issue is mixed with a UN survey supporting the hypothesis (United Nations Centre on Transnational Corporations 1992), while a World Bank study found no difference between the performance of private local and foreign firms in terms of their pollution abatement efforts (Hettige *et al.* 1995).

Against these benign effects, it has been claimed that developing countries maintain lax environmental regulation in order to attract investment – the 'pollution haven' hypothesis. A weaker variant of this argument suggests that, although governments may not deliberately adopt lax environmental policies to attract investment, they will be constrained in their ability to introduce stricter environmental controls by their desire to attract foreign capital (Mabey and McNally 1999).

### Concentration of foreign investment

Previous studies of foreign investment and the environment in Southeast Asia have argued that transnational corporations (TNCs) are concentrated in the most polluting industries. The United Nations, for instance, claimed that in the 1980s in Malaysia and Thailand, foreign subsidiaries were predominantly engaged in the

*Table 11.7* Share of foreign ownership in manufacturing in Indonesia and Malaysia

| *ISIC 3 | Industry | Indonesia 1988 (%) | Malaysia 1993 (%) |
|---|---|---|---|
| 311 | Food products | 9.4 | 30.7 |
| 312 | Food products | 27.6 | |
| 313 | Beverages | 26.6 | 63.6 |
| 314 | Tobacco | 3.3 | |
| 321 | Textiles | 24.8 | 43.9 |
| 322 | Wearing apparel, except footwear | 1.8 | |
| 323 | Leather products | 0.0 | 47.8 |
| 324 | Footwear, except rubber or plastic | 12.9 | |
| 331 | Wood products, except furniture | 13.0 | 27.0 |
| 332 | Furniture, except metal | 6.4 | 34.6 |
| 341 | Paper and products | 39.7 | 11.4 |
| 342 | Printing and publishing | 11.3 | |
| 351 | Industrial chemicals | 12.8 | 37.9 |
| 352 | Other chemicals | 38.6 | |
| 353 | Petroleum refineries | 0.0 | 54.1 |
| 354 | Misc. Petroleum and coal products | 0.0 | |
| 355 | Rubber products | 17.6 | 38.7 |
| 356 | Plastic products | 8.5 | 49.4 |
| 361 | Pottery, china, earthenware | 23.2 | |
| 362 | Glass and products | 7.9 | |
| 363 | Cement | 13.7 | 31.0 |
| 364 | Structural clay products | 7.1 | |
| 369 | Other non-metallic mineral production | 0.0 | |
| 370 | Basic metals | 4.9 | 29.3 |
| 381 | Fabricated metal products | 26.8 | 38.1 |
| 382 | Machinery, except electrical | 37.1 | 51.8 |
| 383 | Machinery electric | 26.3 | 86.7 |
| 384 | Transport equipment | 27.2 | 30.3 |
| 385 | Professional and scientific equipment | 23.0 | 94.8 |
| 390 | Other manufactured products | 11.5 | 49.0 |
| 3 | Total manufacturing | 12.4 | 44.7 |

Source: Hill (1996) and Malaysian Industrial Development Authority.

Notes
Data for Indonesia refers to the share of foreign firms in value added.
Data for Malaysia refers to foreign share of paid-up capital.
*ISIC is the International Standard Industrial Classification.

more pollution or hazard-prone industries (Economic and Social Commission for Asia and the Pacific and United Nations Centre on Transnational Corporations 1988: 99, 291).

Table 11.7 indicates the share of foreign firms in the manufacturing sector in Indonesia and Malaysia in the late 1980s and early 1990s, respectively. This shows a different pattern from the one commonly assumed. There is no relationship between the pollution-intensity of an industry and the share of foreign capital.[15] For example, although in Indonesia the two industries with the highest level of foreign ownership (pulp and paper and other chemicals) are both relatively polluting industries, those that follow are not so highly polluting. In the case of Malaysia, the top two sectors for foreign ownership are scientific equipment and electrical and electronic products, neither of which ranks very high in pollution terms. Thus, there is no evidence that foreign firms have tended to be concentrated in the most polluting industries.[16]

### Technology effects of foreign investment

There is some fragmentary evidence concerning the first issue that was raised above, the environmental standards which multinational corporations set for their subsidiaries in Southeast Asia. A study of two environmentally proactive United States TNCs – Du Pont and Occidental Chemical – in Thailand found that in both cases the firms behaved in an environmentally responsible way, both because of notions of product liability and firm image and for reasons of quality control and cost-effectiveness. In the case of Du Pont, whose technological processes replicated similar ones in the United States, any change in the basic design and management system was viewed as a potential threat to product quality and a potential source of increased cost (Brown *et al.* 1993: 146).

A limitation of this study is that the corporations analysed were self-selecting, and they both involve toxic processes in the chemical industry where the pressures towards corporate responsibility have been particularly acute. Somewhat broader surveys in Thailand and Malaysia carried out for the United Nations, present a less clear-cut picture. A study of 33 TNC subsidiaries in Thailand, while recognising that TNCs have made some contribution to environmental management, concluded that: 'Overall, it seems unlikely that unilateral action from TNCs has contributed to environmental management in the country to any significant extent' (Economic and Social Commission for Asia and the Pacific and United Nations Centre on Transnational Corporations 1988: 325). Similarly, the study of Malaysia concluded that:

> the situation in Malaysia reflects that generally observed in other international studies of a dominant pattern of 'local accommodation', by TNCs with subsidiary companies left very much to adapt to local host country regulation. Guidance, control and intervention by the parent companies tend to be not very powerful, purposive or cohesive in the field of environmental management.
>
> (Economic and Social Commission for Asia and the Pacific and United
> Nations Centre on Transnational Corporations 1988: 118–19)

A recent survey of 59 TNC subsidiaries in Malaysia presents a slightly more positive picture. This found that just over a quarter of these had environmental standards set by the parent company and a similar proportion had environmental targets set by headquarters (Hansen 1999, p. 25 and Table 13). My own current research on Malaysia has found that in the textile and iron and steel industries only one of the seventeen foreign subsidiaries interviewed followed an environmental policy set out by its parent. In the electronics industry, by contrast, where it would be expected that a high proportion of firms would be environmentally conscious, the corresponding figure was only four out of fourteen. This evidence from Malaysia supports the view that the impact of corporate policy in raising environmental standards generally is quite limited. It also suggests that there may well be an industry specific dimension to the setting of corporate environmental standards. This is also supported by Hansen's study (which also covers TNC subsidiaries in India and China as well as Malaysia). He finds that parent company environmental standard and target setting vary by industry and are above average in the chemical industry (Hansen, 1999, Tables 12 and 15).

Although the evidence indicates that the diffusion of high environmental standards through the activities of TNCs may not be as extensive as the advocates of corporate responsibility suggest, it is nevertheless possible that foreign firms out-perform locally owned companies in their environmental management and performance. The findings of Brown *et al.* (1993) suggest that even in the absence of explicit corporate environmental policies, foreign firms may perform better because of considerations of costs and product quality.

There are a number of methodological issues which arise in any comparative study between foreign and local firms (Jenkins 1989). Particularly important in the context of environmental management issues is the fact that foreign subsidiaries are often larger than their local counterparts. Since both for organisational reasons and because of economies of scale in pollution abatement equipment, large firms are likely to have more sophisticated environmental management systems, and they are also more likely to invest in environmental protection, comparisons of large foreign firms with small local firms is likely to favour the foreign subsidiary. Similarly, the fact that there may well be considerable inter-industry differences in the extent of environmental management and protection, also needs to be taken into account in making any comparisons.

Unfortunately, there are very few studies of any kind which have compared the environmental performance of foreign subsidiaries and local firms in Southeast Asia. The Economic and Social Commission for Asia and the Pacific and United Nations Centre on Transnational Corporations (1988) study of Malaysia compared twenty TNCs with six local firms. Although all the local firms were of comparable size to the TNCs, the number surveyed was very small. The study concluded that:

> In part because all the local firms covered in the survey were relatively large and in part because they tended to be in different groups of industries, there were no glaring differences in their environmental management policies and practices as compared to those of the TNCs. Where there were differences, these appeared to be a matter of approach rather than substantive environmental protection differentials.
>
> (Economic and Social Commission for Asia and the Pacific and United
> Nations Centre on Transnational Corporations 1988: 130)

A more comprehensive study of BOD pollution by 250 factories in Indonesia found that foreign ownership had no effect on pollution-intensity at the plant level. This study controlled for the effects of size, efficiency and plant vintage and concluded, therefore, that:

> anecdotes about 'clean' multinationals may well have confused 'outsider' status with the effects of size and efficiency. Our results suggest that big, efficient domestic firms are not significantly more pollution intensive than their multinational counterparts.
>
> (Pargal and Wheeler 1995: 16)

Similarly, a study of twenty-six pulp and paper plants in Indonesia, Thailand, Bangladesh and India, found that foreign ownership was not a significant determinant of pollution abatement efforts among firms (Hettige *et al.* 1995). Although the study included South Asian as well as Southeast Asian countries, in view of the paucity of relevant studies, it provides valuable additional evidence.

Finally, preliminary results of research on the textile, iron and steel and electronics industries in Malaysia indicate that when other factors such as the size of firm and its export-orientation are taken into account, foreign ownership is not associated with significantly better environmental performance.

### Environmental regulation and foreign investment

The view that host governments deliberately utilise weak environmental regulation as a means of attracting foreign capital has been extensively debated but there is little empirical evidence to support the strong version of this thesis. On the other hand, it is certainly the case that regulation in the three Southeast Asian countries considered here is relatively weak, particularly in terms of enforcement, compared to that of the countries of origin of the bulk of the foreign investors which they seek to attract.[17] Moreover, when new foreign investment is being attracted, relatively little consideration is given to environmental factors.

The extent to which new foreign investment is subject to environmental review varies from country to country and industry to industry. In all three countries, certain investments require an Environmental Impact Assessment (EIA) to be carried out. However, it has been observed that these have not been a particularly effective measure in Southeast Asia generally (O'Connor 1994). Moreover, only a minority of investments are subject to EIAs. In Malaysia, in 1994, for instance, out of a total of 973 applications for manufacturing licences, only fifteen were subjected to EIA for approval (Rahim and Nesadurai 1996: 5–7).

In the majority of investments, which do not require an EIA, environmental factors do not play a major part in investment approval. In Thailand, for instance, it was found that environmental factors were not an issue in negotiations between TNCs investing in the country and the government (Brown *et al.* 1993: 147). The Industrial Estate Authority of Thailand (IEAT) is the sole authority for monitoring compliance with environmental regulations on industrial estates, and clearly, since its primary role is to attract foreign investment, it is unlikely to prioritise environmental factors over economic concerns. Indeed, the IEAT has recently been criticised for its failure to control pollution on the industrial estates for which it is responsible (Bangkok Post 1997b).

In Malaysia, the role of the Malaysian Industrial Development Authority (MIDA) as a one-stop investment approval agency, means that economic aspects are likely to outweigh environmental factors in its decisions. It has been reported that: 'In previous years when Malaysia was "hungry" for foreign investment, there were problems in trying to integrate environmental considerations into the approval process' (Rahim and Nesadurai 1996: 5–6). Although the situation has improved recently, and the Malaysian Department of the Environment now has an officer within MIDA who can provide investors with advice concerning their environmental responsibilities, this role is largely advisory.

Although there is little evidence that foreign investors find existing environmental regulations burdensome, some do indicate that tighter regulation would hamper their operations (Economic and Social Commission for Asia and the Pacific and United Nations Centre on Transnational Corporations 1988: 121). A perception that foreign investment would be discouraged may, therefore, constrain governments in the region from embarking on more stringent regulation or stricter enforcement.

## Conclusion: some lessons for Vietnam

This review of the evidence concerning the consequences of economic growth and increased integration into the global economy for industrial pollution in Southeast Asia provides a rather mixed picture.

- First, there is no evidence to support the view that economic growth in the region has led to a reduction in environmental damage. On the contrary, in all three countries, rising incomes have been accompanied by increased emissions and concentrations of most pollutants.
- Second, although there is evidence that in some countries, at some periods, the Southeast Asian economies have had a comparative advantage in less polluting industries, and that the adoption of more outward-oriented strategies have led to a less pollution-intensive production structure, there is no generalisable relationship. Thus, those who claim that increased openness will contribute to the amelioration of pollution problems are over-optimistic.
- Finally, although some transnational corporations are environmentally proactive and introduce advanced environmental management systems and higher standards through their subsidiaries, this is by no means the case for all foreign companies.

What lessons then can be drawn for Vietnam from the experience of the Southeast Asian economies?

First, the experience of the Southeast Asian economies shows that industrial pollution in Vietnam is likely to increase significantly over the next two decades. It should be noted that Vietnam already suffers from significant pollution problems. In Hanoi, in the early 1990s, carbon monoxide levels were almost twice, nitrogen dioxide three times and suspended particulates up to ten times the World Health Organisation's permissible levels. It was also reported that 27 per cent of patients in three provinces suffered acute respiratory ailments (Reed 1996: 277). Degradation of surface and sub-surface water supplies through industrial waste is also an increasing hazard (Forsyth 1997). Even if it were the case that an environmental Kuznets curve exists, at least for some pollutants, the income level at which pollution levels begin to fall are well above the levels which Vietnam will achieve in the foreseeable future. In the mid-1990s Vietnam's per capita income in purchasing power parity terms was around US$1200. Even if per capita income were to grow at an average annual rate of 7 per cent, it would take over fifteen years to reach the current income level of Indonesia and over twenty-five years to get to that of Thailand. Since neither of these countries has begun to experience falling levels of pollution it must be assumed that economic growth in Vietnam will intensify and not reduce its environmental problems. In fact, the recent Asian crisis makes projections of 7 per cent growth in per capita GDP highly optimistic at present. A lower growth rate would, of course, reduce the impact of the scale factor on Vietnam's physical and biological environment. A lower growth rate may also slow down the rate of investment, however, potentially leading to a slower reduction in pollution-intensity, and lower growth means that Vietnam will be even further from the point at which pollution might start to fall.

A second issue is whether or not the change in economic strategy towards greater outward orientation will lead to reduced pollution-intensity as a result of changes in the composition of output and access to cleaner technologies. The discussion of Southeast Asia indicates that there can be no general presumption that a shift to an export-oriented strategy will lead to a less polluting industrialisation trajectory. The outcome depends very much on which sectors develop under a more open economic strategy. It has been suggested elsewhere that trade liberalisation in Vietnam would lead to a small increase in income and employment but that the environmental benefits from such a change would be small. It has also been suggested that greater emphasis on investment in light industry and tourism would lead both to lower levels of pollution and greater employment creation (Reed 1996: 286–94). It is clear, from both the experience of Southeast Asia and Reed's (1996) simulation exercises, that any environmental benefits from an outward-oriented strategy are small compared to the increased pollution associated with the growth of output. Further, other policies are required to avoid substantial environmental damage from such strategies.

Not only has Vietnam adopted a more export-oriented strategy but it is also seeking to attract substantial inflows of direct foreign investment. In so far as these firms bring in technologies that have been developed in countries where environmental regulation is stricter than in Vietnam, there will be a shift towards lower pollution-intensity. This is more likely to be the case where investment comes from

Japan, the United States or Europe. However, Taiwan, Hong Kong and South Korea were three of the four leading investing countries in Vietnam in the mid-1990s. These countries have less stringent pollution controls than the more advanced industrial countries and environmental concerns are relatively recent, so that it is less likely that foreign investment will bring with it high environmental standards. In practice, the picture in Vietnam is mixed with a number of foreign firms failing to comply with environmental laws and regulations (Sund 1998).

The lessons from this review of the experience of Southeast Asian industrialisation are stark. Sustainable industrial development in Vietnam will not come about as a result of the new economic strategy that is being followed. If other measures are not adopted, pollution will increase substantially over the next two decades. Given the already significant levels of air and water pollution in major urban and industrial centres, the impact of further deterioration on human health will be significant. Environmental improvements in urban areas require both stricter enforcement of pollution controls and investment in treatment facilities. In the absence of such measures, increased levels of income will be accompanied by deterioration in other components of the quality of life.

## Acknowledgement

This paper is based on research funded by the ESRC Global Environmental Change Programme under the project 'Pollution, Trade and Investment: Case studies of Mexico and Malaysia' (Award No. L320253248).

## Notes

1 The study by Hettige *et al.* (1998) is an exception. It considers discharges of organic pollutants to water finding that pollution levels stabilise but do not fall at higher levels of income.
2 Available at http://cdiac.esd.ornl.gov/ftp/trends
3 Available at http://www.worldbank.org/nipr/ippsdata.htm
4 It tends to overestimate the growth of pollution since it does not take into account any reductions in emissions per unit of output which may be achieved through technological improvements over time.
5 For similar estimates for Thailand, Philippines and Indonesia for the period 1975–88, see Brandon and Ramankutty (1993).
6 This could be applied more widely than just effluents, but since the specific focus of this chapter is industrial pollution, the point is not developed here.
7 It should be noted that the Malaysian estimates refer to trade in manufactures, while the Indonesian case included all trade. Since almost two-thirds of Indonesian exports were accounted for by petroleum, this had an important impact on the overall outcome (Lee and Roland-Holst 1993).
8 Strictly speaking, it would have been more appropriate to take the period 1970–80. However, industrial production data was not available at a sufficiently disaggregated level before 1974, and there was no Manufacturing Survey carried out in Malaysia in 1980 which made 1974–81 the nearest approximation.
9 A more recent study of the likely impact of Uruguay Round and APEC trade liberalisation on pollution in Indonesia however concluded that the effect would be to reduce pollution (Strutt and Anderson 1999).

10 The UNIDO data are only available at the four-digit level of the International Standard Industrial Classification from 1981 onwards. For the purpose of comparability, the growth of pollution-intensity from 1975–85 is calculated using three-digit data, and is, therefore, less detailed than that used in the Malaysian case. For the 1985 to 1995 period, four-digit data are used.

11 Value added data at the three-digit level of the ISIC was used, which makes the estimates less reliable than those for Malaysia or Indonesia (after 1981).

12 It is not clear from the report how many sectors were involved in the correlation exercise nor the level of disaggregation at which it was carried out.

13 The estimates of the ERP are from Alavi (1996) and refer to 1987. Alavi gives estimates for 62 five-digit industries. Where more than one five-digit industry is given in the same four-digit group, these have been combined to give a weighted estimate of protection at the four-digit level. Where only one five-digit industry is reported within a four-digit group, it is taken to be representative of protection in that group. As a result, estimates of protection are obtained for forty-five four-digit industries that accounted for 94 per cent of total manufacturing value added.

14 The majority of TNCs surveyed by the United Nations Centre on Transnational Corporations Benchmark Corporate Environmental Survey reported having corporate environmental policies which went beyond those required by national legislation of the host country (United Nations Centre on Transnational Corporations 1992: 234). However, a Malaysia Ministry of International Trade and Industry survey of Japanese TNCs found quite the opposite with the majority of firms only taking the measures required to meet local environmental standards (World Bank 1993b).

15 In the case of Indonesia this was tested statistically. The Spearman's Rank Correlation Coefficient between the level of foreign ownership and pollution-intensity for a number of pollutants is not significant at the 5 per cent level in any of the cases tested.

16 For a similar conclusion for Malaysia, see Rahim and Nesadurai (1996).

17 Rasiah (1999: 21) claims that although environmental reasons were secondary in the relocation of export-oriented subsidiaries to Malaysia, electronics and textile firms reported transferring machinery because of lax environmental conditions there.

# 12 Sustainable urbanisation and environmental issues in Vietnam

*David Drakakis-Smith and Andrea Kilgour*

## Sustainability and urbanisation: the concepts

This chapter discusses the recent rise to prominence of the concept of sustainable urbanisation and examines the relationships between its three main components – development, sustainability and urbanisation. We attempt to define sustainable urbanisation and stresses its dynamic nature. We then examine environmental issues in Hanoi, in the context of other important and related aspects of sustainable urbanisation, particularly poverty, and the nature of overall patterns of urbanisation in Vietnam. The final section of the chapter revisits the concept of sustainable urbanisation, assesses the responses to environmental problems and suggests some future directions for management debate.

Urbanisation and sustainability have rarely been discussed together other than in the context of the contribution of rapid urbanisation to the world's environmental problems (Mitlin and Satterthwaite 1996), at least not until the 1990s. The ecological footprint of cities has, indeed, made a considerable impression on their surrounding regions, but, despite the centrality of cities to the development process, the debates on urbanisation and sustainability were largely constructed as separate entities. There was, on the other hand, a central role envisaged for cities in sustained economic growth. As we will see, however, sustained urban growth is not the same as sustainable urbanisation. There is, of course, an extensive literature on what might constitute the basic components of sustainable urbanisation – on poverty, basic needs provision, environmental and infrastructural issues, urban management and, inevitably, economic growth – but seldom were these components pulled together into any conceptual discussion centring upon urban sustainability. By the mid-1990s, this had changed substantially with several major texts being given over to urban sustainability (e.g. Stren *et al.* 1992; Houghton and Hunter 1994; Girardet 1996; Burgess *et al.* 1997) and the concept being adopted as a major theme during the United Nations Conference, Habitat II. So how should sustainable urbanisation as a process be defined?

One approach could be to adopt the accepted principles of sustainable development *per se* and use them as a framework for studying urban development, in relation to meeting the needs of the present without impoverishing future generations, and to identify areas of concern. There is, of course, a considerable debate about the nature and discourse of environmental sustainability. Shiva (1991) and Goldman (1998), for example, represent a growing critique of how sustainable development is interpreted by many of the global institutions and advanced economic powers – as global resource management oriented towards the concerns and priorities of a northern agenda. Satterthwaite (1997) has often drawn attention to this contradiction in its urban setting. Another important issue in this context, is that there is as strong a case for reducing contemporary inequality as there is for worrying about the status of future generations. As outlined in Chapter 2, equity is important when it comes to the use of current resources – it should not always be 'jam tomorrow' for developing countries. Indeed, 'meeting the needs of the present' in an urban context must imply considerable change at the local level to ensure that economic, social and political equity, together with stability and harmony, make a future that is worth sustaining.

There are two levels at which we can address the question of the definition of sustainable urbanisation. The first represents a broad approach in which the principles of sustainability in an urban context are established, regarding our objectives in pursuing sustainable urban development. It is at this point that we must recognise cities not so much as assemblies of functions and buildings but as places where people live and to which they increasingly migrate in search of a better life. Sustainability in this context becomes humanised and implies the desire and determination to achieve a set of goals which benefit individuals and their households as much as enterprises or governments. At this generalised level, therefore, there are certain prerequisites which the pursuit and management of sustainability in an urban context must satisfy. These can be defined as equity in the distribution of the benefits of economic growth, access to adequate basic human needs, social justice and human rights, environmental awareness and integrity, and an awareness of linkages and representations of change over space and time

Although the ability to achieve these goals will be related to the economic well-being of the state, policies encouraging sustainable urban development can and should be introduced to any management strategy, whatever the economic situation. Above all, however, it is important to recognise that these goals are closely related and need to be addressed as far as possible in a comprehensive manner. Isolating one area for attention will not necessarily result in lasting improvements. This has been clearly illustrated in the environmental domain, where legislation and encouragement to households to behave more responsibly will have little impact if the underlying poverty that affects attitudes towards the environment by many urban residents is not addressed. As Forsyth (1997) has observed, the ecological modernisation that is beginning to appear in some developing countries, i.e. the raising of some environmental standards, often at the insistence of, or to impress, external investors, not surprisingly reflects the agendas of local elites, foreign enterprises and international agencies. Low income groups often have quite different priorities, related more to their immediate living or working environments than to global warming or to falling water tables.

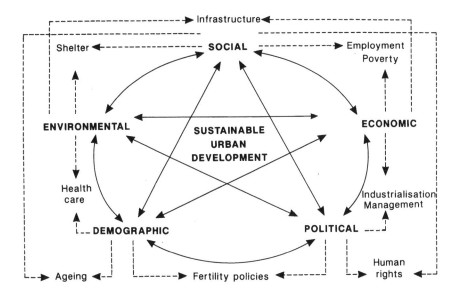

*Figure 12.1* The main components of sustainable urbanisation

The micro dimension of the definition of sustainability leads on from the principles outlined above to identify particular issues or components around which research, debate and policy can be structured. However, through the framework provided by the macro guidelines, it takes on board the need to be aware of the complex integrative nature of effective urban sustainability. Figure 12.1 illustrates just a few of the micro-level issues that can be raised within the overall conceptual framework of urban sustainability; crucially important to the investigation of such issues is the interlinked nature of the framework itself. Thus, in the economic sphere, it is essential that not only are the regional, national and global roles of urban economic activities considered, but also what this means in terms of returns to labour. The role of the informal sector in this context is seen to play a crucial role in creating income for groups in or vulnerable to poverty, as well as providing flexibility to the manufacturing process and service sector. A series of similar and parallel issues can be raised for each of the basic components of the sustainability concept – all of which have multiple linkages. Figure 12.1 is, of course, meant to be illustrative not comprehensive or rigid, and many other components and linkages could be added to it (for further discussion see Drakakis-Smith 1995; 1996; 1997).

Mitlin and Satterthwaite (1996) suggest that the issues and concerns described above might be taken to constitute the 'development' aspect of sustainable development in the context of human settlements. The 'sustainable' component, they argue, 'requires action to prevent depletion or degradation of environmental assets so that the resource base for human activities may be sustained indefinitely'

(United Nations Centre for Human Settlements 1996: 422). Specifically, they identify minimising use or waste of non-renewable resources, including cultural and historic assets alongside energy and material consumed in industry, commerce and domestic use. Sustainable use of renewable resources, such as water, crops or other biomass products, and the importance of keeping wastes within the absorptive capacity of local and global 'sinks', such as rivers, the sea or the atmosphere are also recognised.

These are eminently sensible comments but they can be, and have been, interpreted by many as equating sustainable development only with the brown agenda. What is clearly needed as well is a programme for sustaining and developing 'human resources' through education or health care. Part of the problem in these interpretations is the nature of the discussion of sustainable urbanisation itself. In order to give substance to the nature of sustainability in the urban milieu, sustainability needs to be discussed in the context of the urbanisation process as a whole. Hence, 'sustainable urbanisation' is a more useful conceptual tool than either that of 'sustainable development and cities' or 'sustainable cities'. In particular, the debate needs to be cognisant of the vigorous and varied dynamics of both urbanisation and development.

This chapter will first address the relationship between urbanisation and development in general terms, particularly with the growth of ideas related to mega-cities. This phenomenon has been especially addressed by scholars writing about Asian cities, and has been identified as a process currently occurring in Vietnam by McGee (1995a). The subsequent sections of the chapter will, therefore, examine urbanisation in Vietnam in this context, with particular attention being paid to Hanoi where rapid urban growth is taking place in the context of an already heavily populated delta region. The social and environmental consequences of this surge in population form the substance of the remainder of the chapter.

## The changing nature of urbanisation and development

The nature of the interrelationship between urbanisation and development is often more assumed than real, with the simple positive correlation between levels of urbanisation and development being sufficient for most development strategists and national planners to look towards urban-based economic growth as the path to development. In the early 1990s, the United Nations Development Programme (United Nations Development Programme 1991) and the World Bank (1991) both issued major policy statements the fundamental basis of which is the direct relationship between urban and national economies. However, notwithstanding this link, there is conflicting evidence to indicate the way the relationship works. Indeed, if rates rather than levels of economic growth and urbanisation are compared, a very different picture emerges in which some of the fastest urban growth rates are being experienced by some of the poorest countries. Clearly, therefore, there is more to urbanisation than economic growth and *vice versa*, and this has enormous and important impacts on the nature of sustainable urbanis-

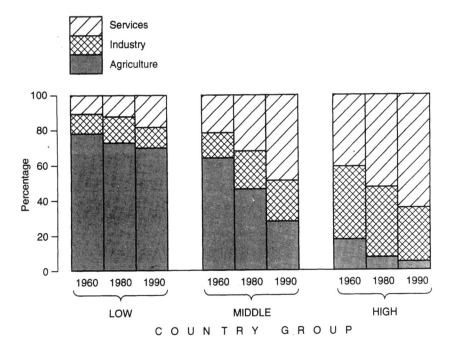

*Figure 12.2* Major changes in the composition of Gross Domestic Product in developing
countries over time

ation. It is particularly important in this context to recognise the fact that not only
do economic processes evolve and change over time, but that urbanisation itself is
also changing its nature in many developing countries, particularly with the rise
to prominence of mega-cities. McGee (1995b) has identified three broad forces
operating in concert on the urbanisation process which has encouraged their
emergence – structural change, globalisation and the transactional revolution, and
each of these affects the nature of sustainability in the urbanisation process.

Of all these changes, those involving structural transformation of the economy
are the most well-known and almost without exception there has been a substantial shift away from primary production to secondary and tertiary activities. As
Figure 12.2 indicates, however, the rates of change have not been the same across
the developing world. They have been fastest in the middle-income economies as
manufacturing growth begins to accelerate, consolidating in the high-income
developing countries through a further transformation to service activities. Both of
these economic changes are caused by and contribute to urbanisation, with the
growth in tertiary activities in particular underpinning both the growth and
concentration of mega-cities. We must also note, however, that the much slower
transformation of low-income economies does not preclude urban growth.

Indeed, it is urban growth without parallel economic development within these cities that poses some of the most difficult questions with regard to sustainable urbanisation.

Whilst the emergence of mega-cities is usually discussed in the context of economic structural transformation they are but one phenomenon in a series of changes which have been taking place in the spatial order of development. These include the polarisation of national development to core regions, transborder regional developments and the emergence of international growth corridors. The processes creating mega-urban regions within this new spatial order are clearly linked to major changes at the global level, although the nature of each mega-city must also be rooted in its local history, culture and politics.

Globalisation has been closely interwoven with changing regional and international divisions of labour, being characterised by the emergence of world cities (Sassen 1994; Yeung and Lo 1998) and by the growth of intense global and regional competition. These processes are, of course, very closely interconnected and mega-cities compete with one another in the quality and efficiency of their built and transactional environments in order to attract investment. The more they offer, the more concentrated becomes this investment and the more the biggest cities acquire their world trade centres, telecommunication complexes, international schools, prestige retail stores and the like. Of course, globalisation is not the deterministic process that some commentators have suggested. Whilst some cities are favoured by world status, all cities are linked to the global economy to some extent and, moreover, may find reasonable and continued prosperity through successful management of sub-global processes. Furthermore, even within the so-called global cities not all elements of society have equally strong bonds to the most dynamic elements of the global economy, some being linked weakly via low-paid work or through the petty commodity system. In such circumstances sustainable urbanisation not only means different things for different cities but also to different people in those cities.

This more complex interpretation of globalisation and urbanisation is reinforced by the third of McGee's influences on mega-city formation, the transformation of space-time through new means of communications, allegedly bringing about a shrinking of the world. However, not all such changes are occurring to the same extent. Much information, capital and decision-making may be transmitted electronically, but much of the interchange of increasingly complex information must be undertaken face-to-face, whilst people and commodities still have to be physically transported (Rimmer 1995). The impact of these transactional changes can, therefore, be contradictory. Thus, the development of cheap reliable minibus transportation has exacerbated greater long-distanced migration or commuting, accentuating this concentration across a range of cities, but the cost of new technologies often means that electronic nodes are concentrated into those cities that can afford them. On the other hand, cheap, personalised developments, such as mobile telephones, fax machines and personal computers, mean that within cities, individuals are becoming less place dependent. The consequence is what may be best described as unfocussed concentration into mega-urban regions that have multiple nodes rather than a single centre. Hanoi is rapidly developing such a pattern.

There is an irony here. Whilst the various transactional revolutions have been fundamental to the creation of mega-urban regions in terms of both concentration and polynucleation, they have been so unevenly layered and poorly managed as to be responsible for some of the most pressing urban environmental problems, such as congestion and pollution. Perhaps even more important, however, is the fact that comparatively little research and investment has been put into essential parallel flows and movements. It is not just money, ideas, people and commodities that need to be shifted, but also energy, water, sewage and solid waste. The brown agenda still awaits its transportation revolution.

## Urbanisation before *doi moi*

Although pre-colonial Vietnam had notable cities from which the country and its regions were ruled, by and large it was a rural-based economy and society. As in other parts of Southeast Asia, urbanisation essentially began during the colonial period when an export economy was established and the structures of collection and marketing were introduced. Apart from plantation owners in the south, the French colonials were essentially an urban administrative group and began to construct new spacious quarters adjacent to the old imperial urban centres. Around these grew the Vietnamese and Chinese urban populations engaged in commerce, retailing and light industries. In many ways, this petty bourgeoisie, prospering within the colonial system, inhibited independence movements but it also produced aspiring nationalists who chafed at the sight of French and Chinese domination of Vietnam's economy (Thrift and Forbes 1986).

Socialism, when it arrived in the 1940s, affected urbanisation and development in a variety of ways, primarily through its association with the division of the country for some two decades. Between 1954 and 1975, two quite different political and economic systems existed in the North and South. Substantial chronological and spatial differences in the urbanisation process have resulted from this situation. It has been suggested, in this context, that many non-European socialist governments had ambivalent attitudes towards urbanisation. On the one hand, it was recognised as an essential pre-requisite for the promotion of industrialisation; on the other hand, many socialist states outside Europe were more reliant on agricultural production. The result, it has been suggested (Murray and Szelenyi 1984), was a much slower rate of urban growth and a more even spread of this growth down the urban hierarchy through the development of small regional centres (Figure 12.3). More rural-based movements, such as those found in China, Tanzania or Vietnam, will, of course, have had an even more limited impact on urban growth. Although, as the situations in Tanzania and Vietnam reveal, it may well have accentuated the process of urban primacy. However, social systems alone do not shape the urbanisation process and in Vietnam this has certainly been the case.

Despite its longer history of urbanisation, North Vietnam as a colonial and

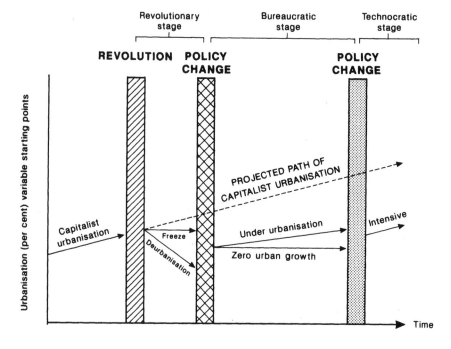

*Figure 12.3* A model of urban growth in a socialist economy
Source: Murray and Szelenyi (1984).

socialist state was characterised by a slow rate of urban growth (Figure 12.4). The lack of parallel infrastructural support for urban industries meant that many workers did not live in the city but commuted on a daily basis. The intensive bombing campaigns between 1968 and 1973 also meant that many urban economic activities were shifted out of cities such as Hanoi and Haiphong to surrounding villages. As a consequence of this combination of conflict and socialist urban management, and in spite of its history as a capital city, Hanoi had less than half a million people in the mid-1980s. Its administration constituted a network of units from precinct or district level, through wards and sectors, to neighbourhood committees. Effectively, the ward committees were the focal point in this system. At this time, the cities were not considered a locus of economic growth in the north of the country, with migration away from urban centres and the densely populated Red River delta common.

In South Vietnam, urbanisation had developed under different circumstances. Saigon (now Ho Chi Minh City), in particular, had grown very rapidly prior to reunification as the northern areas became less secure (Figure 12.4). Effectively, by the mid-1970s, the city housed some two-thirds of the South's population, plus large numbers of Americans, and had effectively become a heavily subsidised service economy, with an extensive informal sector. After reunification, the state

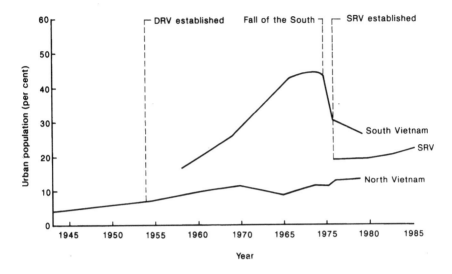

*Figure 12.4* Urbanisation trends in Vietnam, 1945–1985

attempted to disperse this 'surplus' urban population to the New Economic Zones. Temporarily, the growth of Ho Chi Minh City was slowed down, but at least one-third of these people eventually returned to the city.

These circumstances in both North and South Vietnam meant that, in essence, the two largest cities in the country failed to contribute substantially to the post-colonial regeneration of the economy. Prior to *doi moi*, they were demographically but not economically prominent in the development of Vietnam. Indeed, during the period between the mid-1950s to the mid-1980s substantial policy emphasis was placed on small towns. In the South, prior to reunification, this was for security reasons, but in the 1970s small towns were also envisaged as playing an important role in linking urban and rural development. Paralleling this conceptualisation of small town development were the on-going attempts to socialise further the rural production system throughout the country. There is, therefore, little evidence prior to *doi moi* of a comprehensive and positive state policy towards urbanisation and its role in development, nor were the planning and management of large cities given much prominence.

## Current patterns of urbanisation in Vietnam

According to official statistics, in 1995 some 13 million people, or 19.8 per cent of the population, were classified as urban; of these 8.5 per cent (1.18 million) were resident in Hanoi, and 24.9 per cent (3.2 million) in Ho Chi Minh City (Figure 12.5). The two-city primacy index of 0.33 is high by international standards. Vietnam, thus, has one of the lowest levels of urbanisation in the

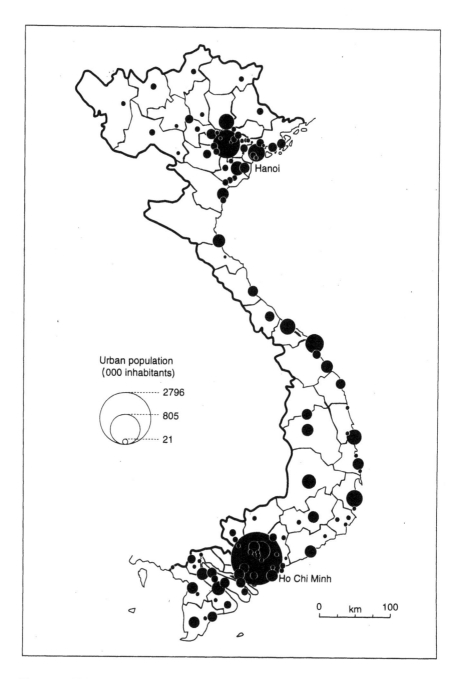

Urban population
(000 inhabitants)

2796

805

21

*Figure 12.5* Urban centres in Vietnam, 1995

world and yet one of the more highly concentrated (Drakakis-Smith and Dixon 1997).

While the official figures outlined above appear to be accepted by the international agencies, there are reasons to believe that the level of the urban population is underestimated, even for Hanoi and Ho Chi Minh City. The Vietnamese definition of urbanisation excludes those resident in the urban areas who are engaged in rural activities. Perhaps more significantly, the rapid expansion of the urban areas since the late 1980s has overtaken the administrative definitions, thus excluding large numbers of people who could legitimately be regarded as urban. In addition, there is evidence to suggest that the urban population registration system has broken-down, thus large numbers of people have moved into the urban areas but are not included in the official figures. This situation would appear to suggest that gross underestimates affect urban data, and this is implicit in the official suggestion that the level of urbanisation has declined by 0.4 per cent to 19.5 per cent. While the revised definition of urban places adopted in 1990 undoubtedly contributed to this situation, the implied decline in the level of urbanisation runs counter to the informed opinion of Vietnamese planners, officials, academics, locally-based agency staff, consultants and foreign researchers. This is particularly the case with respect to the major centres where the rapid development of parts of the inner city areas, the peripheral growth, especially along main route ways, the expansion of squatter settlements, particularly in Ho Chi Minh City, signs of people living on the streets, and the level of traffic and movement of people and goods, all give the impression of dynamism and very rapid growth. This results from in-migration, high birth rates and the incorporation of the already densely populated rural areas.

Despite the limitations of the data on growth in the urban centres, there is little dissent from the view that growth is heavily concentrated in Hanoi, Ho Chi Minh City and their immediate environs. There is, however, not surprisingly, considerable uncertainty over the size and rates of growth of these centres. In consequence, a variety of wider definitions of the urban areas have been suggested. For Ho Chi Minh City, the addition of peripheral areas results in the population in 1994 being estimated at around five million and around eight million in 1999 (Vietnam Ministry of Construction 1995; Yates 1999). A quarter of a million under-registered people in the core area could be added, even at a conservative estimate (GKW-Safege 1995). Similarly, a broader definition of Hanoi would raise the population to around three million, with no estimate available for the unregistered population; the incidence is, however, probably lower than for Ho Chi Minh City. These estimates include only the officially recognised urban population within the peripheral areas. Not only are these administratively based definitions becoming rapidly outdistanced by the spread of the urban area, but many of those resident in the adjacent rural areas could well be regarded as urban. Perhaps, in total, a realistic estimate for the cores and peripheries of the two major centres at the end of 1994 would be nearer to nine million rather than the four or five million indicated in most official sources.

It may well be that even the use of the broader definitions and allowances for under-registration still seriously underestimate the area and population of the two major centres, as urban land uses and economic activities have moved extensively into the surrounding areas of the Red River basin and Mekong Delta. Industrial, commercial and residential developments have spread along route ways, engulfing already densely populated rural areas, connecting and expanding large numbers of smaller centres. These observations tend to confirm the view put forward by McGee (1995a) that the growth of Hanoi and Ho Chi Minh City is giving rise to rapidly emergent Extended Metropolitan Regions similar to those observable elsewhere in Pacific Asia. While the process is much more advanced in the case of Ho Chi Minh City, the familiar spread of linear development connecting emergent sub-centres in areas of already comparatively dense population, with resultant extremely complex patterns of land use, is also clearly present in the Hanoi area. In addition, the population density of the Red River delta is so high – exceeding 1000 persons per square kilometre in some districts – that the speed of growth is likely to be very rapid. The emergent extended metropolitan regions as delimited by McGee (1995a) are each estimated to contain over 20 million people.

Within these emergent metropolitan regions serious problems of congestion, pollution, inadequate drainage, water supply, power and communications are emerging. Urban expansion and redevelopment are largely uncontrolled, resulting in chaotic patterns of land use. In these respects, the growth of Hanoi and Ho Chi Minh City is beginning to mirror that found in the Bangkok Metropolitan Region (Wongsuphastawat 1997). Indeed, as in Bangkok, the infrastructure was already inadequate before the period of rapid growth was initiated. In both Hanoi and Ho Chi Minh City, the infrastructure is deteriorating rapidly in the wake of the demands placed upon it and limited upgrading and maintenance. In such circumstances, there is an urgent need to reappraise the urbanisation process in Vietnam within the context of its sustainability. If this is not done the development process itself will be placed under threat.

In this context we must be very wary about confusing 'sustained growth' with 'sustainable development'. Confusing these terms leads to a contradictory set of urban management policies. There is, on the one hand, a paternalistic concern for the plight of the poor and the quality of life of all urban residents, and on the other hand, an underlying determination that cities should continue to play a leading role in sustaining economic growth. The development strategies of the major international agencies reflect this in their emphasis on enhancing urban productivity as the principal mechanism for tackling poverty (Department for International Development 1998) and see high rates of national economic growth as the best policy solution to assist in managing cities. Such philosophies are clearly evident in the current development strategies being recommended and pursued in Vietnam and will be discussed below. There is as yet both insufficient, and insufficiently reliable, information for us to comprehend fully the complex inter-linkages of the urbanisation process in the context of sustainability. However, we can throw some light on this and offer comment on current management practice by examining the impact that the headlong rush for market-led growth is having on related aspects of the urbanisation process, jeopardising the sustainability of such growth. We will focus here on environmental issues, given the theme of the volume. We will, however, then set this within the broader context of sustainability in an urban context.

## Environment and urbanisation in Hanoi

Research on environmental matters in Hanoi, as in other urban areas, has been very patchy. A few comprehensive household surveys of environmental conditions and household perceptions were conducted, however, by the Ministry of Construction's National Institute for Urban and Rural Planning in 1994 (Vietnam Ministry of Construction 1996; Pham Tan 1997). This provided useful information against which current and future management responses may be assessed. The backdrop to the environmental situation is a rapidly growing urban population in which migration has played an increasingly important role in the 1990s. Even in the 1980s squatters were beginning to appear, despite the regulations governing migration and by the mid-1990s these controls had broken down almost completely. New migrants to Hanoi now rely more on their families and friends or on their own initiative, rather than the state to provide assistance in finding work and shelter. There is also an increase in the 'floating population' who have no permanent residence in the capital.

### *Housing issues*

Official urban housing in Vietnam until the mid-1990s mostly comprised old one- to three-story buildings, indicating the slowness of development after the introduction of reform. Maintenance and improvements were also lagging. As a result, by 1995, almost two-thirds of Vietnam's urban areas comprised slums and dilapidated houses in need of repair and improvement (Nguyen Quang Vinh and Leaf 1996). In Hanoi, the city on which this overview will focus, almost one third of the population lived in accommodation with less than 3 m² per capita (300 000 have less than 2 m² per capita) (Drakakis-Smith and Dixon 1997). The mean area per capita has been calculated at under 4 m² (Hiebert 1992; Trin Duy Luan 1997). Alarming as these figures seem, they are probably underestimates of the real housing situation. The rapidly growing urban population has increasingly affected the position of planners and residents alike in response to the increasing housing and environmental problems witnessed in Hanoi during the early 1990s. However, the growing urban population is only one factor affecting housing development. Administrative mechanisms for housing discourage state employees from developing their own housing or adding new buildings to the housing stock, land-use administration has followed a top-down approach and policies have become accountable only since the revision of the Hanoi Master Plan after 1986 (Nguyen Manh Kiem 1996; Pham Quynh Huong 1997).

Again, the formal legislation which should have preceded the Ordinance failed to materialise at the same pace as construction. Many could not wait to receive official permission as bureaucratic procedures could take months or years to gain acceptance and, as a result, houses were constructed illegally on state-owned land and in green spaces, contravening official guidelines on size, height and facilities. It has been estimated that at least 80 per cent of recent private residential construction, regardless of less rigid guidelines, still received no official permit (Trin Duy Luan 1995).

By the end of 1992, the last government subsidy for housing was abolished with the inclusion of housing rent subsidy in state salary. At the same time, the rent of state housing increased 45 times (United Nations Development Programme 1995). Although not covering all costs, this increase in state housing rents should have created a potential source of revenue for the housing sector for further operation and maintenance. This appears to have failed to find its way into official budgets, with informal reports suggesting that most has been used on salary increases. In fact, the state contribution towards new housing construction has officially fallen, down from 25 per cent to 16 per cent between 1991 and 1995, with informal estimates suggesting that the proportion was as little as 10 per cent in 1997 (Vietnam Ministry of Construction, personal communication, 1998). But any construction actually occurring has also failed to tackle the problems of home-lessness apparent in Vietnam's major cities.

The policy of total renovation rather than slum upgrading in the inner city and the construction of new blocks on the outskirts of the city has incurred difficulties as most poor residents wish to live and work in the city centre. Moreover, many of the developments are far too expensive for those most in need of shelter. As a result, more and more migrants and locals either return to undesirable areas in the centre or look beyond this and build houses in peripheral zones, again of a temporary and illegal nature. Furthermore, as the cost of materials soared and resource supply was limited, poor households were forced to seek innovative solutions. In Hanoi, for example, the city's main flood control dike was removed and re-used for house construction.

In response to the flooding risks, the Ministry of Construction launched plans for a National Shelter Strategy in 1995 in an attempt to identify issues and provide guidance to improve Vietnam's shelter situation in areas of land, housing, infrastructure, building materials and mobilisation of financial resources. In Hanoi, today, it would appear that the original programme has done little to achieve its objectives. Those remaining in accommodation constructed by various government ministries have taken the opportunity to use the extra space in corridors and on balconies to increase the size of their rooms. Typically, as it can be afforded, a small cubicle is built outside the home of the family including either a kitchen stove and sink or a toilet. This removes the need to share the communal facilities at the end of each floor but does not intrude on the apartment's already limited space. Furthermore, external balconies are being added in an extended 'birdcage' style to gain room at the rear. In some instances, the bedroom has been divided at the height of approximately two metres and an extra wooden shelf added to act as another room. All of these additions contravene official regulations, but with market prices increasing in the city, people are prepared to improve the size or conditions of their existing home rather than endure the hardships of moving and constructing (probably illegally) in another area (Kilgour 2000).

Notwithstanding such processes, the survival of the city's 'tube houses' and pre-colonial shop-houses has been steadily eroded by commercial redevelopments, designed in particular to capitalise on the capital's chronic hotel shortage and the popularity of its historic area. The 36 streets of the ancient quarter, dating originally from 1010 AD, currently house some 90 000 residents at densities of 1900 per hectare. The living conditions are poor for many but the commercial opportunities are extensive. As a result the construction of mini-hotels and flat roofed shop-houses with grate-metal doors, popular in other Asian cities (Hiebert 1992), has been adopted, removing traditional Vietnamese styles. Moreover, these developments virtually ignore the height and sanitation controls that now exist in this area. For the government, concerns regarding preservation of this cultural area have been slow to materialise. However, recent foreign pressure and offers of investment, particularly from Sweden, mean that there is a chance that some form of conservation may take place. Any form of change is met with much scepticism particularly by the older residents in the community who fear an increase in rents in return for improved living standards (Pham Tan 1997).

### Residential environmental issues

The broader impact of this population and residential growth alone has been enormous (see Trin Duy Luan and Nguyen Quang Vinh 1997 for a national overview of housing issues). Most green spaces have been built over and in some districts of Hanoi construction densities reach as high as 85 per cent of land surface. One third of the households contain three generations and the same proportion comprise six persons or more. Moreover, most of the existing housing stock is in poor condition. That built by the state has not been maintained because revenues from rents were too low, whilst tenants were not allowed to make their own improvements. In all, some 55 per cent of Hanoi's housing is classified as 'needing repair' or 'not safe' (see Pham Quynh Huong 1997 for further discussion).

Little coordination has occurred between housing construction and infra-structural provision so that only one half of the households have access to a private toilet, about one quarter use shared toilets and the same proportion have to use public latrines (usually located *inside* residential blocks). Lack of coordination with infrastructure also means that water supply is poor (Nguyen Duc Khien 1995). Less than 70 per cent of the population of Hanoi have direct access to house taps. Leakages from this system have been estimated at between 50 and 70 per cent. Many of the pipes were laid during the colonial period. As a result, many households suffer from inadequate supplies, despite the recent completion of an extensive improvement under a Finnish aid scheme. Meanwhile, pressure on water resources has meant that aquifer reserves have fallen, in addition to becoming more polluted, with growing consequences on consumer health.

The other side of the problem of water supply is the discharge of waste water. In Hanoi, waste water is combined with sewerage and industrial liquid waste. In the inner city, the discharge of this waste is into a system built in the 1930s, which currently serves only 60 per cent of the city, and much of which is in poor condition. All told, about one quarter of the waste water is discharged directly onto the ground or into water bodies, add to this the leakages from pipes and considerable pollution of aquifers results. In the sub-urban areas of the city the situation is even worse, with some 65 per cent of waste water going directly into the ground. Poor drainage also results in frequent flooding given Hanoi's flat topography, with about one-third of households suffering frequent floods in the rainy season, worsening the health environment.

Other infrastructural service problems which affect the urban environment relate to refuse collection. Half of Hanoi's refuse is organic which means that fermentation and decay occur rapidly and daily collection is necessary. Whilst some improvements have occurred in recent years, more than one third is not collected daily and the process is still manual rather than mechanical. In the suburban areas, the situation is worse and some 44 per cent of households dispose their waste casually onto the ground or into a water body. Most collected waste is simply dumped, untreated, at the three land fill sites, one of which supports the scavenging activities of some 50 people (Nguyen Duc Khien 1996a; 1996b; Vietnam Ministry of Construction 1996). All of these infrastructural problems have severe environmental impacts on the city. The four main rivers that drain Hanoi are all seriously polluted, as are most of its lakes. This is caused largely by the discharge of domestic and industrial waste water, together with solid waste. Not surprisingly water-borne diseases, such as diarrhoea, are widespread, affecting labour force capacity and attendance.

Air pollution appears to have become more significant. Although industrialisation has contributed to situations where two-thirds of the households surveyed by the Ministry of Construction complained about noxious smells, most of their complaints relate to their immediate environment of decaying wastes and inadequate sewerage. Another air pollution problem affecting Hanoi is from dust, with levels ranging from two to seven times permitted levels (Nguyen Duc Khien 1996a). Much of this relates to increased transport, particularly low-quality, polluting motorcycles, and by the continued use of coal and charcoal for cooking and heating. Industrial sources also contribute to the situation since technology is old and controls are virtually non-existent. Altogether, toxic gases are two to three times the recommended levels. Clearly, the growth of industry has been an important factor in increased pollution levels (Nguyen Duc Khien 1996b) but so has the sheer increase of population and the inadequate infrastructural framework existing in the city. Indeed, domestic pollution far outweighs that of industry at present (Mol and Frijns 1997; 1998). Perhaps even more importantly, the Ministry of Construction (Vietnam Ministry of Construction 1996) survey reveals that the principal concerns of the residents themselves are those that relate more to their immediate living environment, such as rubbish dumping and inadequate water supplies.

Notwithstanding these personalised concerns, the environmental policy responses that have been introduced in the 1990s have largely been related to industrialisation. Forsyth (1997) suggests that this is largely the result of economic liberalisation and the transfer of global environmental concerns to Vietnam via the state and an emerging elite – a process that has already occurred in neighbouring states in Southeast Asia, such as Singapore and Malaysia. The National Law on Environmental Protection, passed in 1994, has thus encouraged the introduction of environmental impact assessments for new and old industries but has had relatively little to recommend on more domestically-oriented environmental problems. Nevertheless, there is a growing monitoring system building up at ministry level for domestic enterprises (Pham Quynh Nhu, personal communication, 1999).

As Satterthwaite (1997) notes, this is not an unusual situation in the context of developing countries, where elite and foreign concerns, such as global warming, have received much greater priority than the day-to-day problems of the urban poor. In Vietnam, Forsyth (1997) suggests that local environmental movements have had little success in seeking policy responses to their immediate needs. We must note, however, the exception to this provided by the Vietnamese Women's Union, a national organisation which has links to the state and the communist party and therefore commands some respectability with both government and the poor. The Union is used by some non-governmental organisations (NGOs) as a successful route to the grass roots because of the enhanced awareness its education programme has brought to women in low-income areas. In part, the lack of overall success is allegedly due to the persistence of public obedience to the state; in part too, however, it is the result of the fact that most of the limited number of NGOs support elitist environmental goals. Local agendas will remain unaddressed until supported by more powerful groups in society. Forsyth suggests that this may come from the local municipal level. This, however, implies more decentralisation of democracy than the present government may be willing to permit. Pollution abatement measures in Vietnam are currently based on the 'command and control' approach of the traditional socialist system, and yet enforcement is limited and weakened by the rise of the independent market economy, particularly at the small-scale level. Other problems restricting enforcement relate to the poor data quality, lack of funds and shortage of a trained labour force.

Despite the growing environmental problems posed by rapid industrial growth, domestic issues resultant from rapid population growth in Hanoi are still the basis for most environment pollution and are the prime concern of the residents of the city. This implies that domestic environmental problems, such as improved water supply and sanitation, ought to receive greater priority than industrial pollution controls, or at least equal priority. There are few signs that this is occurring since both national and municipal concerns seem to relate more to industry. The need for greater attention to domestic environmental matters confirms that effective responses to urban environmental issues would be more appropriately coordinated by municipal authorities (United Nations Centre for Human Settlements 1996), but for this to happen priorities need to be changed, funds need to be made available and the people need to be involved.

Reports on the involvement of citizens in environmental planning are contradictory. On the one hand, citizens are said to be 'eager to participate'; on the other hand, they allegedly 'lack the habit of keeping their city clean' (United Nations Development Programme 1994: 65–6). Meanwhile, some campaigns to raise awareness of urban environmental problems have met with 'relatively disappointing' responses (United Nations Development Programme 1996), whilst others, particularly those using the Vietnam Womens' Union, have had more success. Mol and Frijns (1997) claim that 'the missing link' in Vietnam is the presence of NGOs to activate local populations, an absence that Eccleston and Potter (1996) attribute to the lack of space within the monolithic socialist system for such organisations to emerge. Individuals were expected to make their complaints to local representatives of the state. Effectively, the restricted lobbying capacity that currently exists comes largely from academics and has limited and largely indirect grass roots involvement (see Coit 1998).

Recent research has indicated that although efforts have been made at the national and city level to improve environmental conditions, at the household level this is confined to areas of the city where Vietnamese *nouveau riche*, expatriates or foreign visitors can be found. Here, improved sanitation and refuse collection is apparent and street clearance is visible. It would appear that insufficient resources are channelled in this direction and as the agencies become separately managed no extra money is supplied by the bureaucracy to improve such services. Investment remains within the industrial sector, allowing pressing urban issues to be overlooked often in the interests of the state. However, other domestic issues remain unassociated by the poor with environmental or health deterioration. Kitchen conditions, in particular, require further consideration. In traditional urban houses, kitchens tend to be situated in the middle of a small courtyard and shared by all households living in the 'tube house'. Few have any form of chimney and conditions are generally dark and wet. The kitchens have tended to be located beside toilets and have small areas, if any, for any form of food preparation. Maintenance and hygiene appear to be overlooked or assumed to be the responsibility of others as the facilities are shared.

Until recently, conditions have been very similar in state apartment blocks. After 1986, conditions improved as private facilities in hallways were erected. These divide cooking and washing areas but are generally cupboard-like and about 1 m² in area. Ventilation, natural lighting and smoke evacuation remain poor (Nguyen Huu Dung 1993). Even in newly constructed private houses, kitchen standards depend entirely on the culture and educational standards of the owner. It has been estimated that over 60 per cent of such house owners, if building their kitchen inside, still continue to combine it with other utilities including the toilet (Nguyen Huu Dung 1993). If a water tap is present in or near the home, there remains the risk of clean and used water becoming mixed within the system from illegal tapping of water pipes, creating waterborne diseases.

Household problems primarily affect women and young children and are related to injuries and illnesses created by kitchen fumes, mould, cramped and unhygienic conditions. In particular, respiratory problems prevail, although burns and scalds in the home are also common. Children play around these cooking facilities as there are few other places to go to. In the overcrowded city, green spaces have been built upon and in the streets even if pavements exist, they are often used as parking spaces for motorbikes and as overspill areas for crowded houses, private enterprises and street selling. As much a cause for concern, however, as cooking and washing facilities, is the lack of insulation or protection from natural hazards. Lack of insulation and sleeping in cold, damp rooms leads to cold, influenza and pneumonia. A similar number of people, some 350 000, die each year in North Vietnam alone from temperature-related problems as from waterborne diseases (Vietnam General Statistical Office 1997). Poor health, related to inadequate housing standards and insufficient infrastructure prevents dwellers from earning sufficient wages to improve their housing situation and encourages savings or domestic expenditure to be spent on health care rather than family necessities.

## Sustainable urbanisation revisited

Reorienting the growing policy responses more towards urban domestic environmental problems is only a part of what is needed. If the basic principles of sustainable urbanisation are to apply we must look towards other linked dimensions to determine policy responses. In this context, many observers have claimed, not incorrectly, that the urban poor are the source of much of the domestic pollution that is found within rapidly growing cities, such as Hanoi – building on marginal land, discarding wastes anywhere, and using inappropriate sanitary facilities. No one is more aware than the poor themselves of the deficient environment in which they live, it is just that they have alternative priorities, and, moreover, can do little to improve the structural facilities related to their living environment. And yet, with the retreat of the state, increasing pressures are being put on the poor to 'help themselves' with the state acting as a facilitator. It is at this point that we must return to the role of cities in sustained growth. Cities are seen as the main engine of economic growth in most developing countries, with substantial benefits for the state. On the other hand, the benefits accruing to the labour force from this growth are often minimal with wages being kept low in order to attract investment. Working conditions too are often very poor, with many environmental health hazards being present. In these circumstances, fairer distribution of the profits from industrialisation would considerably help the poor improve the quality of their domestic environment; whilst improved working conditions would reduce vulnerability to poverty by reducing health risks.

The situation in Vietnam is little different from the generalised situation described above, despite a widespread reluctance to admit to the existence of extensive urban poverty. In part, this is a function of varying data bases. The government's *Statistical Yearbook* (Vietnam General Statistical Office 1997) classifies less than 20 per cent of the population as poor, whilst the World Bank (1995a),

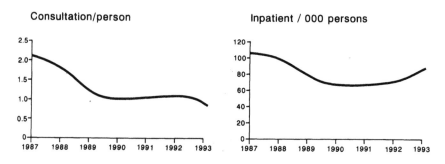

*Figure 12.6* The use of and access to health services in Vietnam, 1987–1993

on the basis of the *Vietnam Living Standards Survey*, puts the figure at 51 per cent. What most sources emphasise is that poverty has been falling since *doi moi* but social commentary within Vietnamese analyses of the development process is relatively rare. Certainly, it is clear that the benefits of economic growth have been spread unevenly, both spatially and socially (Hiebert 1994). In the latter context, Tranh Khan (1994) claims that the disparity between rich and poor was three or four times in the 1970s, six to eight times in the 1980s and Pandolfi (1997) suggests in the late 1990s this has grown to 20 times in rural areas and 40 times in urban areas. The United Nations Development Programme claims that *doi moi* has significantly improved the situation with regard to poverty in Vietnam but other studies, such as Tranh Khan (1994) and Tuong Lai (1995), refute this.

Hanoi replicates this conflicting situation. On the one hand, average incomes are well above the national average. Most of the recent increases, however, are accruing to a small groups of *nouveau riche* (Fforde 1998), with most of the capital's growing population, particularly the poorer households, being left with expenditure levels which are very similar to those in rural areas. It is against such a background that we should begin to evaluate the response to environmental problems within Hanoi. Clearly, low-income groups are those most affected by deteriorating urban environment conditions. Poor health, consequential on the inadequate water and sewerage systems, means that cost recovery for suggested improvements is unlikely to be realised. Nor can the poor help themselves to improve their housing situation. The retreat of the state and rise of speculative housing construction for foreign and local elites means that overcrowding is increasing and average living space per capita in Hanoi dropped from 6 m² in the 1950s to 4 m² in the 1990s (Trin Duy Luan 1995). Immigration exacerbates this situation, with densities worsening and squatters becoming more evident. The integrative nature of sustainability suggest that there will be further consequences from the persistence of poverty – malnutrition, for example, is on the increase because of the inability of many families to feed themselves within the cash economy. The introduction of medical and health fees, too, has meant that the use of these services has declined (Figure 12.6), with consequent short-term and long-term impacts on vulnerability.

At present, urban environmental issues are usually seen as a distinct and separate matter from urban poverty. In both cases, there are those who can make optimistic noises about both the present and the future. New environmental laws, for example, will reduce pollution from recently set up factories; whilst urban poverty is seen as 'a necessary transition of the urban scene' (United Nations Development Programme 1995: viii) and further reduction of poverty will come from the deepening of *doi moi*. Both of these are wildly optimistic statements. Furthermore, real improvements in the quality of the urban environment will not come until the priorities of the poor themselves are recognised.

There must be two sets of responses to this situation: first, an increased role for the state (national and local) in improving its investment in the larger-scale infrastructural networks which are beyond the capacity of the poor to change; and, second, the enabling of the poor to make positive improvements to their own situation. The most direct means by which this can be done is to ensure a more even distribution of the economic benefits of *doi moi*. This presents a challenge. Raising minimum wage levels will be difficult in a climate of state reductions in domestic expenditure, and a strong belief in the fact that low wages are Vietnam's principal competitive advantage. Furthermore, many people are not employed in the formal organisations that might be affected by such legislation. Indeed, most people work in an informal sector in which exploitation is often more acute. Others have no employment at all. Nevertheless, more employment opportunities and fairer wages, in order to reduce poverty and vulnerability is one way to improve the local environment. As Hewitt (1997) has pointed out, those most vulnerable both to natural hazards, environmental degradation and to cuts in welfare services are those living in the worst housing, in marginal locations and with little economic or social capital to absorb adversely changing circumstances. Community groups can contribute to small-scale projects, pooling limited funds and using their own labour to construct drains, repair sewerage or install latrines. However, there is little experience of such movements within a system of governance that has been dominated until recently by a highly centralised state. In this context, there is an urgent need for an increased presence of non-governmental or community-based organisations to lead by example.

Such measures will only be able to have an impact if they are, once again, linked to the other elements of sustainable urbanisation such as improved housing, health and education, sensible demographic policies and a better informed urban management. The environmental problems of Hanoi are not self-contained, they have a complex cultural and socio-economic context. Responses to these problems should be seen in this light, rather than through a direct 'sticking-plaster' reaction. Most environment concerns in Hanoi are domestic, rather than industrial, in nature and origin. The present response should reflect this situation and look towards helping households and communities to help themselves, thereby building resilience and adaptability, and in the process, improve the overall environment of the city.

# 13 Politics, ecology and water: the Mekong Delta and development of the Lower Mekong Basin

*Chris Sneddon and Nguyen Thanh Binh*

## Introduction: the Mekong Basin as transboundary resource

This chapter examines the potential downstream impacts of cooperative, transnational development of the Lower Mekong Basin on the Mekong Delta region of Vietnam. While much of the discussion focuses on ecological issues, it is argued that the dialectic character of social and ecological transformations is central to an understanding of the evolution of political and environmental outcomes. This must not be lost in the litany of potential ecological concerns over human alteration of the basin (Harvey 1996). In other words, ecological transformations are crucially dependent on the political and economic changes occurring within and external to the riparian states of the Mekong. How these states conceive and implement river basin development in turn hinges on the complex biophysical processes at work in the basin. In the past, states have tended to 'see' environmental systems with a simplifying and utilitarian gaze (Scott 1998), one that masks their inherent dynamism and ignores the uncertainty of knowledge claims about their future patterns and processes. Predictive models of Mekong futures which overlook this complexity and uncertainty jeopardise both the integrity of the basin's ecological systems and the resilience and security of the livelihoods of the basin's residents.

The Mekong Basin extends through or abuts the national territories of six countries (China, Burma, Laos, Thailand, Cambodia and Vietnam), encompassing a 795 000 km$^2$ area and approximately 60 million inhabitants (see Figure 8.1 in Chapter 8). The Mekong River, the hydrological backbone of the basin, travels 4200 km from its headwaters in the Tibetan plateau to its delta in Vietnam. Despite nearly five decades as the focus of ambitious transnational development plans touting the river's hydroelectric, irrigation and navigational potential, the Mekong remains the least-altered large river system in the world. This status may, however, change in the not-too-distant future due to a convergence of political and economic trends within the basin states and societies.

With finalisation of the Agreement on the Cooperation for the Sustainable Development of the Mekong River Basin in April 1995, the riparian states of the lower basin (Laos, Thailand, Cambodia and Vietnam) reaffirmed their:

determination to continue to co-operate and promote in a constructive and mutually beneficial manner in the sustainable development, utilisation, conservation and management of the Mekong River Basin water and related resources for navigational and non-navigational purposes, for social and economic development and the well-being of all riparian States, consistent with the needs to protect, preserve, enhance and manage the environmental and aquatic conditions and maintenance of the ecological balance exceptional to this river basin.

(Mekong River Commission 1995a: 2)

Shortly after the agreement, the newly reformulated Mekong River Commission (MRC) (formerly the Mekong Committee) heralded this 'new era of Mekong cooperation'. It noted that 'much of the potential from the Mekong's water and related resources is still untapped and needs to be managed for the welfare of the people and sustainable development of the riparian countries' (Mekong River Commission 1995b).

Yet what exactly does 'sustainable development' of the basin entail? Prior to the 1995 agreement, the Mekong Secretariat issued a feasibility study carried out by two international consulting firms which identified up to eleven sites on the mainstream of the river suitable for large-scale hydroelectric projects (Mekong Secretariat 1994). Critics of the proposed development programme argue that the report, like the myriad of previous Mekong development schemes that prioritise large-scale hydroelectric development, exhibits biased methodologies, incomplete economic analyses and neglect of key environmental and social costs (Roberts 1995; Williams 1996). Researchers sympathetic to the development goals of the riparian states and the MRC respond that large impoundments are the most realistic approach for dealing with the pressing social and economic needs of the 60 million basin residents (Jacobs 1996). Non-governmental organisations (NGOs) in the region counter that it is precisely basin residents who have been systematically excluded from the decision-making bodies where transformation of the river basin is planned and promulgated (Towards Ecological Recovery and Regional Alliance 1995; Mitchell 1998).

These conflicting views on the efficacy of mainstream development highlight perhaps the central conundrum facing multiple interests in the basin: the likely impacts of large-scale development projects on the ecology, and hence the livelihoods, of the basin (Öjendahl 1995: 175–6). Just what are the 'needs' of the river basin in terms of ecological sustainability and economic development, and who has historically been the steward of these needs? Given that the Mekong is an international river, how have interactions among the basin states influenced the institutional context for cooperative development and environmental management? How and to what extent will massive alterations of the upstream portions of the Mekong River influence downstream regions such as Vietnam's delta region?

Vietnam is situated as the ultimate downstream territory of a river that begins in China, briefly forms the border between Laos and Burma, widens and slows as the border of Thailand and Laos (where it becomes the 'Lower' Mekong), and continues into Cambodia and Vietnam's delta region before emptying into the Bien Dong Sea (South China Sea). By virtue of its geographical position, the Mekong Delta stands out as a socio-ecological component of the Mekong Basin especially dependent on hydrological interactions.

> While the cycle differs from place to place reflecting various local conditions, the cycle is everywhere characterised by definable periods of water deficit and surplus along with expected but less predictable episodes of extreme deficits and surpluses. In the same way, there are also definable variations in water quality, reflecting complex relations among water, land, climate, and human activities. In the context of the hydrologic cycle, how water resources are defined and managed is therefore an especially crucial issue for understanding natural resource management and agroecosystem relationships.
>
> (Koppel 1990: 3)

At the time these words were composed, cooperation among the Mekong's riparian states that might lead to implementation of large-scale water projects appeared rather dim. Current prospects for large-scale development of the Mekong's middle and upper reaches seem much more immediate. There are a number of proposed large-scale water resource development efforts by upstream states such as hydroelectric dams and interbasin transfer projects for irrigation development. They represent a pressing set of hydrological and ecological concerns for the environmental managers of the delta.[1] Those environmental problems already at work in the delta, such as salinity intrusion, acidic soils, fisheries decline, and flooding and scarcity in some areas, will certainly be influenced, if not exacerbated, by alteration of upstream hydrology.

In signing the 1995 agreement on use of the basin's resources, the Vietnamese government engaged in speculation with the delta's socio-ecological system in two ways. First, they wagered that state representatives would be able to influence the agreement's mandated water utilisation rules and development planning frame-work in ways that will minimise risk to the delta's complex hydrologic system. Second, they assumed, even in case of an unsatisfactory outcome to the first 'risk', that the hydroelectric and navigational benefits of mainstream development will outweigh any negative social and ecological impacts rendered on the delta. Vietnam's sanctioning of the agreement thus raises important questions: What are the potential impacts of future alteration of the Mekong system on delta ecosystems and livelihoods? What types of institutional mechanisms are necessary to ensure that development activities both in the delta and in the entire Mekong Basin do not do irreparable harm to ecological integrity and livelihood capabilities?

This chapter proceeds with a brief description of the biophysical character of the Lower Mekong and analysis of the evolution of the institutional mechanisms for managing the basin's resources. We give particular attention to the crucial roles of development discourses and hydropolitics as reflected in the changing circumstances and institutional evolution of the Mekong Committee, the inter-governmental agency created to oversee cooperative development initiatives and coordinate scientific research in the basin. Next, we turn to discussion of the Mekong Delta, outlining existing ecological concerns and the potential impacts of upstream alterations of the Mekong River. In addition, we bring attention to salient issues concerning environmental management institutions in the delta. We conclude by stressing two critical aspects of water resource development, ecological complexity and participatory decision-making, that are under-represented in discussions of both the basin and the delta.

## The Mekong as large-scale socio-ecological system

The Mekong system in its entirety is best conceived as a complex mosaic of land-forms and ecosystems linked through the timing, duration and intensity of water movement throughout the drainage basin. Indeed, from a hydrological standpoint, the defining characteristic of the Mekong basin (as in many tropical rivers) is its pronounced seasonality. With the onset of the southwest monsoon, the rainy season from mid-May to early October ushers in a period of high flows and flooding. By contrast, the dry season period from November to April witnesses a tremendous decrease in discharge rates (Pantulu 1986a: 702–3). To illustrate, the flow of water into the Mekong Delta in September is on average thirty times greater than during low flows in March and April (Lohmann 1991: 44). Average flow rates range from as little as 1764 cubic metres per second in the dry season to 52 000 cubic metres per second in the wet season (Binson 1981). This seasonality of water flows implies that ecological processes and structures throughout the basin have co-evolved with the basin's hydrological dynamics. Changes in these dynamics will inevitably result in alterations in the relationships among living organisms and the physical environment. By perceiving the value of water resources primarily in terms of hydroelectric generation and irrigation develop-ment, past and present development schemes within the Mekong Basin have neglected water's critical ecological functions.

### *'Developing' the Mekong: the Mekong Committee, 1950–1998*

The Mekong River emerged as the focus of intense interest to the riparian states and international development institutions beginning in the 1950s. Following several reconnaissance studies and investigations centred on the potential for hydroelectric development and flood control in the basin, the United Nations' Economic Committee for Asia and the Far East sponsored formation of the Committee for the Coordination of Investigations in the Lower Mekong Basin, or the Mekong Committee.[2] From its inception, the Mekong Committee exhibited a high-modernist ideology in its proposed development schemes and institutional activities. The guide-posts of this 'high modernism' are an almost evangelical faith in science and technical progress, a belief in unlimited productive potential through the human control of nature, and a commitment to the 'rational design

**Discourse**

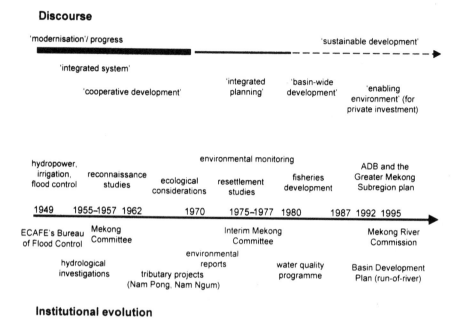

*Figure 13.1* Institutional evolution and discourse in the 'Mekong Project', 1949–1995

of social order commensurate with the scientific understanding of natural laws' (Scott 1998: 4).[3]

The rationale for creation of the Mekong Committee was not the need to resolve an existing conflict, but rather a shared desire to harness the river's tremendous development potential in order to spur economic growth, making it something of a rarity among international river management institutions (Bernauer 1997). Figure 13.1 represents the Mekong Committee's institutional evolution in terms of concepts employed in institutional discourses and key activities that reflect adoption of new concerns.[4] Throughout its history, the Mekong Committee has sought to integrate a multitude of environmental and social aims into its organisational and discursive structures, culminating in the current 'sustainable development' phase of the MRC. Despite the multiplicity of discourses surrounding different aims, the overarching goal has been and remains the transformation of the Mekong into an impounded river capable of hydro-electric generation, irrigation development and, in general, water control.

Early feasibility studies, technical reports and economic analyses generated by the UN-sponsored Mekong Committee emphasised the need for basic hydrological information regarding the river's flow regime and general biophysical character.[5] The early years of the 'Mekong Project' (the late 1950s through the mid-1960s), as the collection of schemes for large-scale dam construction came to be known, were dominated by hydrological investigations and reconnaissance studies to identify likely tributary and mainstream dam projects. Planned development projects for the mainstream of the river, representing a cascade of massive dams, were put on hold until such a time as the information base was sufficient, international funding for the projects could be secured, and all of the riparian states were agreeable to the idea of comprehensive development of the basin as a whole. As a result, the riparian states concentrated water resource development efforts primarily on tributary projects such as the Nam Pong in Northeast Thailand and Nam Ngum in Laos both built in the mid- to late-1960s.

Armed conflict in the region significantly scaled back the river basin development goals and activities of the Mekong Committee in the late 1960s and early 1970s. In addition, increasing concern over the social and ecological impacts of water resource development – particularly the construction of large dams – prompted the Mekong Committee to incorporate broader, non-economic development aims into its institutional structure.[6] The Interim Mekong Committee (IMC), consisting of Vietnam, Thailand and Laos, replaced the Mekong Committee following Cambodia's withdrawal in 1975 and ushered in a period of relative inactivity during the 1980s. The cessation of armed civil conflict heralded the return of more stable governance structures in Cambodia in the early 1990s. Combined with a state desperate for economic development and amenable to resource development as a 'quick fix', this new government revived discussions of regional river basin projects and the need for a reformulated coordinating structure that included all riparian states (Öjendahl 1995: 161).

Throughout the period that culminated in the 1995 Mekong treaty on 'sustainable development' and continues today, much of the momentum for cooperative development derives from several complementary political–economic processes. At a national level, one process relates to ongoing transformations in the formerly socialist societies of Laos, Cambodia and Vietnam as their governments move toward the establishment of more market-driven economies (Brown 1993). In response, international financial capital has swarmed to the Mekong region in the late 1990s to take advantage of these emerging Southeast Asian (and Chinese) markets and the already investment-friendly Thai economy. Since the turn of the decade, a variety of transnational and national actors have initiated four development forums and six capital investment funds designed to attract private investment to the Mekong region (Stensholt 1996: 201).[7] To cite just one example of the sums involved, investments in priority projects (dams, telecommunications infrastructure, transport systems) were expected to exceed US$1 billion annually by the year 2000 (Gill 1996: 5).

Thus, the hydroelectric and irrigation development enabled by the construction of large-scale, mainstream impoundments on the Mekong have emerged as crucial components in the broader economic development of the Mekong 'subregion' envisioned by international financial institutions such as the Asian Development Bank (ADB). Hampered by deficient domestic resources and limitations on aid packages, development planners have turned to private capital. Mitsuo Sato, President of ADB, succinctly explained the logic of the argument:

The financial resources required for subregional projects far exceed the capacity of governments and donor agencies to provide. While it is true that donors have very limited funds, there is a huge international pool of private capital looking for investment opportunities.

(quoted in Bangkok Post 1997a).

It is within this context that the MRC requested US$90 million to meet the objectives in its 1996 work programme (Towards Ecological Recovery and Regional Alliance 1996a: 22). This trend towards seemingly 'neutral' private capital to fund Mekong development is likely to spur a concomitant tendency towards less transparent and less accountable forms of basin management and governance institutions (Bakker 1999).

### Mekong hydropolitics

One of the more remarkable observations in the case of the Lower Mekong is the degree to which the basin states have held on to a vision of cooperative development despite a history of violent conflict and seemingly intractable differences in political philosophies and societal organisation. However, the common aim of *developing* the Mekong to appropriate its potential energy for use in economic growth appears to have over-ridden any qualms related to cooperation among ideologically opposed states. For example, electricity purchased by Thailand continued to flow uninterrupted from Laos' Nam Ngum dam to its destinations throughout periods of intense regional conflict in the 1970s (Sharp 1982: 34). In this case and others, the discourse of modernisation and progress through resource development passed seamlessly between the global capital-friendly Thai state and the socialist regimes in Laos, Cambodia and Vietnam.

In fact, throughout the early years of the Mekong Committee in the late 1950s and early 1960s, the cooperative development of the basin was invoked as a means of *bringing* peaceful relations to the region (Brady 1993: 90–2). To cite one more detailed example, the United States government's promotion of the Mekong Project was a key part of its more general strategy to funnel aid dollars towards economic development projects in the region as ideological and material counterpoints to a socialist path. At a speech given in 1965 designed to explain continuing United States involvement in Southeast Asia, then-President Lyndon Johnson opined that the 'vast Mekong River can provide food and water and power on a scale to dwarf even our own TVA [Tennessee Valley Authority]' and asked Congress for a billion-dollar investment in the Mekong Project (quoted in Price 1966). Although Johnson's billion-dollar proposal was never realised,[8] this example points out how deeply the prospects for development, and hence for material transformation of the basin's ecology and hydrology, are intertwined with political–economic concerns. As mentioned above, the Mekong Committee itself was highly influenced by United States resource development agencies and to some degree internalised the theme of 'peace through development' encapsulated in the Johnson speech (Nguyen Thi Dieu 1999).

While extra-basin political processes have been influential in the evolution of the Mekong Committee, state–state political and economic interactions within the basin comprise the most important frame through which to understand institutional dynamics. The changing status of Cambodia vis-à-vis the Mekong Committee from the 1970s to the early 1990s was mentioned above as one pertinent example. Yet perhaps the most contentious issue regarding past and present sharing of the basin's water resources hinges on the downstream–upstream relationship between Vietnam and Thailand and the key importance of the Mekong to each nation's economic development strategy. Put bluntly, disputes have historically focused on this relationship. Vietnam desires to ensure the agro-ecological integrity of the Mekong Delta due to its critical importance as a food-producing region. Thailand aims to use the water of the Mekong to generate hydroelectricity and to augment dry season water shortages in the Khorat Plateau (the portion of the Mekong Basin that falls within Thailand's national boundaries).

The negotiations over the 1995 agreement on shared use of the Mekong's water hinged on the ability of negotiators to achieve a framework for cooperation that recognised both positions to the satisfaction of the Vietnamese and Thai states.[9] A statement by Dr Prathes Sutabutr, Thai representative to the MRC, exemplifies Thailand's persistent attitude towards cooperative river basin development:

> Unlike the Indochinese countries, Thailand is completely prepared to develop the water resources of the Mekong. We are far advanced in technology and have faced water shortages in the Northeast. So if the Mekong Committee would jeopardise our projects, it would have been better not to have the Committee.
>
> (quoted in Traisawasdichai 1995a)

Since the onset of the Mekong Committee and particularly during the last three decades of rapid economic growth, Thai development planners and politicians have focused on the basin's capacity for hydroelectric generation to meet burgeoning domestic energy demand and irrigation development to increase agricultural production in the country's water-deficit Northeast region. Only mainstream projects (hydroelectric dams and water withdrawal mechanisms) would provide energy and irrigation water in sufficient quantities to meet these goals. For their part, Vietnamese officials engaged in Mekong cooperation have regarded the Thai agenda with mistrust due to historical differences related to the politics of the region (Brady 1993; Öjendahl 1995) and the potential impacts of upstream projects on the socio-ecological systems of the Mekong Delta. This latter concern forms the major theme of the second half of this chapter.

While analysis of disagreements between Thailand and Vietnam over water usage is crucial for understanding the hydropolitics of the Mekong region, another centres on the potential for financing the incredibly expensive development schemes in the basin proposed under the umbrella of the Mekong River Commission. This need for investment has long been recognised within the

Mekong Secretariat, the main administrative organ of the Mekong Committee. A former Senior Planning Engineer of the Mekong Committee headed a list of the 'most serious . . . obstacles to development' with the utter 'scarcity of investment funds both in foreign and domestic currencies' (Hori 1993: 113). Arguably, much of the rationale for the 1995 agreement on 'sustainable development' came from the donor community who pushed for more regional, mainstream projects (as opposed to tributary projects) that demonstrated the full potential of the Mekong Committee as a forum for cooperative river basin development. The new institutional framework of the Mekong River Commission also reflects the importance of donor support for any development schemes in the basin. The Donor Consultative Group is a prominent feature of the Commission's organisational structure and is in a structurally powerful position to influence its direction (see Mekong River Commission 1995b).

### Mekong ecology: a state of uncertainty

The type of large-scale river basin development envisioned in the *Mekong Mainstream Run-of-River Hydropower Report* – involving eleven large mainstream dams (Mekong Secretariat 1994) – will drastically alter the river system's hydrologic regime and ecological functioning. Given the 1995 agreement's call to maintain the 'ecological balance' of the Mekong River within this developmental context, it is imperative to raise questions regarding the information base on which ecologically 'sustainable' transformations of the basin are founded. Despite more than three decades of scientific investigation, the availability of reliable and comprehensive information regarding the basin's hydrology and ecology remains in doubt. For example, one researcher described the basin's water monitoring programme as suffering from the 'data-rich but information-poor' syndrome. Numerous stations have been established throughout the basin devoted to statistical sampling, but the validity, reliability and systematic interpretation of these data is all highly uncertain (Eriksson 1992).

The paucity of environmental information cannot always be explained as simple deficiency in the organisational resources of the Mekong's development institutions. It must be grounded in an understanding of the ways in which environmental information is deployed within the context of specific political and economic aims. For example, flood control has been a frequently cited benefit of large-scale impoundments since formation of the Mekong Committee in the 1950s and remains so today. However, this appears to contradict the conclusions of a team of engineers from the United States assigned to carry out a reconnaissance field study of the basin in 1955. Their report states that flood control is 'not a subject of major interest in the Mekong River Basin'. After talking with local officials, particularly in the delta region, the team concluded that 'the people have learned to live with the wet season floods by elevating their homes and they have developed a "floating" rice which accommodates its growth to the floods' (US Bureau of Reclamation 1956: 26). In addition, *any* predictions regarding the hydrological impacts of dams, positive or negative, are hindered by the limited amount of long-term rainfall and runoff data, and the questionable reliability of data collected in the early years of the Mekong Committee (Institute of Hydrology 1982: 92). The case of flood control points out the inherent unpredictability of large-scale river systems and the need for caution in the face of limited data.

The case of fisheries research in the basin further demonstrates the links between scientific knowledge, politics and development. Throughout its institutional history, the Mekong Committee has never considered fisheries a priority area in the overall development of the basin. This neglect is linked to the historically dominant position of engineers and economists within the Mekong Secretariat (the Committee's technical and administrative body), and the prioritisation of economic growth through hydroelectric and irrigation development by the riparian governments. The head of Thailand's National Mekong Commission has expressed a view common among those institutional actors that have sought rapid large-scale development of the mainstream: 'If we worry too much about fish, we will have no development. Building dams will only hurt a few fish. Nevertheless, with technology, we can produce [fish] artificially in the hatchery' (quoted in Traisawasdichai 1995b).

The ecological and social value of fish within livelihood systems in the basin belies this attitude. As noted earlier, the Mekong is one of the most diverse freshwater systems in the world, with an estimated 800 to 1000 species of fish. Many of these species undertake spectacular migrations, the routes and timing of which are almost completely conjectural at this point. Furthermore, the fish species of the Mekong have co-evolved with the pronounced seasonal changes in river flows. The majority of species are dependent on the river's annual flooding cycle for crucial productive inputs and, more critically, for reproductive success. An estimated 90 per cent of the Mekong's fish species spawn in the floodplains of the mainstream and tributaries (Lohmann 1991: 44). In the context of this impressive diversity, the Mekong is experiencing 'a long-term, progressive decline of fish and fisheries . . . due to a combination of human impacts including but not limited to deforestation, dams, and pollution' (Roberts 1995: 9). If plans to construct even some of the large-scale dams on the mainstream are implemented, this decline will be compounded and accelerated largely due to the additional obstacles to fish migration routes.[10]

Aside from the significant reduction in biological diversity that construction of large-scale impoundments on the Mekong would most likely entail, any reduction in the fish populations of the river system will have a direct and immediate impact on the millions of small-scale farmers and fisherfolk whose livelihoods are either wholly or partly dependent on fisheries. Table 13.1 demonstrates the importance of fish on a country-by-country basis and in particular the tremendous significance of fisheries in Vietnam. Studies estimate that capture fisheries production in the Mekong amounted to more than 887 000 tons including 248 000 tons from coastal capture fisheries in the Vietnam delta region (Interim Mekong Committee 1992: 10), although these figures may 'seriously underestimate' actual subsistence production (Interim Mekong Committee 1992: 5). An assessment by the Mekong

*Table 13.1* Estimated annual fish production and consumption in countries of the Mekong Basin, 1988–1990

| Country | Population within the delta (million) | Fish production (thousands tons per year) | Mean fish consumption (kg per capita per year) |
|---|---|---|---|
| Cambodia | 7.5 | 100 | 13.3 |
| Lao PDR | 4.2 | 27 | 6.5 |
| Thailand* | 21.1 | 322 | 5.3 |
| Vietnam* | 14.6 | 438[†] | 30.0 |
| **Mekong Basin** | **47.4** | **887** | **18.7** |

Source: Adapted from Interim Mekong Committee (1992: 10).

Notes

*Figures for Thailand and Vietnam refer only to those portions of the country within the Mekong Basin.

[†]Of this figure, 190 000 tons are from inland capture fisheries, 218 000 tons are from coastal capture fisheries and 30 000 tons are from coastal culture fisheries.

Secretariat's Executive Director in the mid-1990s placed total basin freshwater fisheries production near 1 million tons annually worth US$700–800 million per year (Matoba quoted in Mekong Secretariat 1997: 13). Fish and fish products comprise the major source of animal protein in the diets of basin residents, accounting for anywhere between 40 to 80 per cent of total animal protein intake (Interim Mekong Committee 1992: 9).

Recognising the potentially harmful effects of mainstream development on basin fisheries, the Mekong Secretariat commissioned a report in the early 1990s examining the current status of fisheries ecology in the basin and some possible impacts from run-of-river hydroelectric development (Hill and Hill 1994). Despite this acknowledgement, the political clout necessary to place fisheries on a par with hydroelectricity or irrigation as a basin-wide development concern seems to be lacking within the institutional mechanisms for managing the Mekong. For instance, the 'most serious negative impacts and reservations' expressed in the fisheries study were excluded in the Executive Summary and Main Report sections of the final study on the proposed run-of-river cascade of dams (Roberts 1995: 12). These are the portions most likely to be read by riparian officials and international donor representatives. By underestimating the crucial role that fish assume in the lives of the vast majority of Mekong residents and downplaying the ways this valuable resource will be affected by mainstream development, the Mekong states risk alienating those small-scale environmental managers their cooperative plans are purportedly designed to benefit. This neglect of fisheries reflects the manner in which the riparian states have deemed hydroelectric and irrigation projects as superior routes to rapid economic growth and industrialisation, apparently with little regard for equitable distribution of benefits and environmental impacts.

## The Mekong Delta

The delta region, or Cuu Long ('nine dragons'), of the Mekong is distinguished by extraordinarily imprecise boundaries between land and water. The hydrological and ecological characteristics of the region reflect this fuzziness, from the small inland watercourses called *rach* that serve as an impermanent, natural drainage system to the mangrove forests of the coasts that bind the silt deposited by the river over centuries to the terrestrial environment (Brocheaux 1995: 4). The delta is a profoundly rhythmic socio-ecological system, one where human inhabitants have historically adapted to the daily and seasonal fluctuations of water. Furthermore, the delta is a geomorphologically and ecologically diverse region with several distinctive landforms and a myriad of ecological systems (see Figure 8.1 in Chapter 8).

Some fundamental parameters help explain the acute importance of the Mekong Delta in terms of human inhabitancy and food production. It encompasses an area of 39 000 km$^2$ and a population of around 15 million, a population density of 400 people per square kilometre (McAdam and Binh 1996). As of 1991, rice paddy production in the delta amounted to more than 10 million tons and the region was averaging a 9.2 per cent annual increase in paddy production (Nedeco 1993). Agricultural production in the delta represents 27 per cent of Vietnam's total annual GDP and 40 per cent of the country's total national agricultural output. Rice production in 1995 amounted to 13 million tons (Nguyen Nhan Quang 1996). More recent figures reflect even greater increases in the delta region's production (16.8 million tons in 1999) and overall contribution to the country's food output (51 per cent). As of 1999, the delta contributed 4 million out of the 4.5 million tons of exported rice (Vietnam News Agency 2000). Despite the region's already impressive contribution to the national economy, the Mekong Delta 'is still regarded as the agricultural region in Vietnam with the biggest untapped potential' (Öjendahl 1995: 167). It is expected to be a fundamental driver of the 'country's ambitions for a sustained economic growth of 8 per cent per year' (Nedeco 1993: 7), as evidenced by recent government plans to zone off approximately 1 million ha exclusively for the growth of high-quality rice for export (Vietnam News Agency 2000). Thus, the concerns of Vietnamese officials and delta residents regarding proposed upstream development and the possible effects of this altered water regimes on the delta's agro-ecosystems are immediately apparent.

Somewhat lost amidst the above figures is recognition that the delta region is an ecological system that has already undergone tremendous socio-ecological upheavals over the course of the last two centuries. From 1890 to 1938, forested area in the western portion of the delta decreased from an estimated 1.9 million to 170 000 ha, reflecting the successive waves of human settlement and expansion of rice lands in the region (Brocheaux 1995: 7–13). By the 1990s, only vestiges of the original forests of the delta remained (Le Cong Kiet 1994). Similarly, the hydrology of the delta has been permanently altered by the construction of a complex series of canals, much of which was carried out under the auspices of the

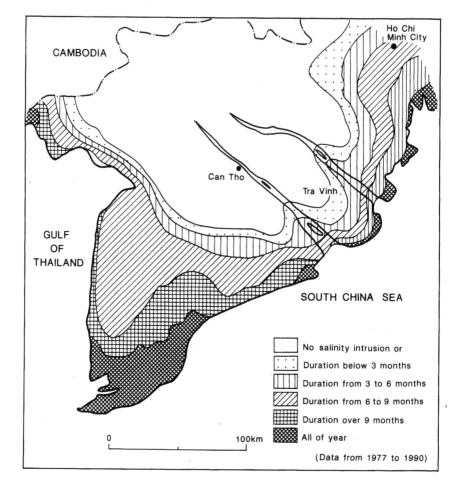

*Figure 13.2* Salinity intrusion in the Mekong Delta. Note salinity contour at >4g/l

French colonial state from 1870 to 1930 resulting in the drainage of 1.425 million ha of land (Brocheaux 1995: 17–22).

Changing land uses in the delta reflect national-level policy initiatives to stimulate agricultural development. Beginning in the early 1970s and continuing through the late 1980s, traditional rice culture changed quite rapidly from a double transplanting, single cropping system to one utilising direct seeding and year-round cultivation (Nguyen Huu Chiem 1994). This agricultural intensification was encouraged by the state and facilitated by the increasing availability of high-yielding varieties of rice, farm inputs, machinery and water diversion technologies. Within the *doi moi* period, the delta's agro-ecosystems have undergone further transformation primarily through the diversification of farming practices. Since about 1988, farmers in some regions of the delta – particularly the tide-affected floodplain – have increasingly adopted practices such as multiple cropping of rice and upland crops, integrated fish-rice farming systems, and vegetable and fruit production (Tanaka 1995: 371–8). These changes have not come without social consequences. Agricultural production changes have been accompanied by rapidly changing land ownership policies under which 'processes of social stratification are reappearing (assuming they ever disappeared) at an equally rapid rate' (Koppel 1990: 3).

These observations imply biophysical changes brought about by changes in upstream flow regimes are likely to influence resource-based livelihoods in a differential fashion. Those rural residents living in agriculturally marginal areas or those without the economic resources to withstand fluctuations in food production will be most vulnerable to changes in water availability. When thinking about the ecological threats facing the delta due to upstream alterations, it is imperative to call attention to the direct and indirect links among resources, production systems, policy and social vulnerability. Ecological concerns cannot be divorced from socio-economic or political concerns, as will become apparent in the discussion that follows.

## Ecological concerns in the delta: threats from Mekong development

The influence of upstream water development on the ecological and social systems of the delta are not likely to present new dilemmas, but rather exacerbate already existing ecological concerns. The most pressing concerns for the delta include: salinity intrusion; acidic soils; flooding; and inland and coastal fisheries. This section looks at the extent of each concern and how upstream alterations of the Mekong's flow might influence the problem under consideration. We also direct attention to the manner in which emerging problems such as climate change and water pollution might complicate management of Delta ecosystems.

### Salinity intrusion

The extent of salinity intrusion from the Bien Dong Sea into the Mekong Delta can be seen in Figure 13.2. During the dry season, saline water (more than 4 parts per trillion) extends up to 70 km inland. If preventative measures are not taken, the area affected by salinity may increase from the current 1.7 million ha to 2.2 million ha in the near future (56 per cent of the entire delta) (World Bank 1996a). The duration of saline waters in the delta depends on the proximity of the area to the coast. Some areas are affected all year while others do not experience salinity intrusion (Nedeco 1993). Thus, the effects of salinity intrusion are distributed unevenly according to the geography of the delta.

The contributing factors to salinity intrusion in the delta are complex. They include: the periodic flooding of the coastal area due to regular tides; upstream freshwater intakes both in the upper part of the delta and in the upper reaches of the Mekong Basin; and the destruction of mangrove forests through shrimp cultivation and other processes leading to increased inflows of sea water.[11] Rice is a salt-sensitive crop, and while some species of rice tolerate a certain level of salinity, higher levels can make rice-growing very difficult or impossible. Surveys of one district in Tra Vinh province in 1995 found that rice yields had declined by 50 to 90 per cent over the past three decades primarily due to salt intrusion (Hopkins 1995). Salinity intrusion also exacerbates drinking water shortages for domestic use and spurs several ecological effects, such as increased sedimentation, in those river reaches newly exposed to fresh-water–sea-water mixing. Furthermore, salt water intrusion alters the ecological character of affected areas in unpredictable ways by transforming habitats leading to concomitant changes in flora and fauna.

Accordingly, many of the actors involved in the negotiations and discussions surrounding the 1995 Mekong accord were optimistic about the possibility that construction of large impoundments on the mainstream would allow for greater water level control and thus enable amelioration of salinity intrusion in the delta. This was the view held by Hoang Tran Quang, head of the Vietnamese National Mekong Committee:

> We want more fresh water to push away the sea water. So we support dams on the Mekong mainstream. We think dams will contribute to better control of sea water intrusion. Basically, we want to increase the flow in the dry season and reduce flooding in the wet season. If the projects [upstream dams] can fulfil these two functions, we will strongly support them.
>
> (quoted in Traisawasdichai 1995b)

However, a system of dams operating in the fashion described by Dr Quang demands a significant degree of transnational coordination entailing sophisticated communication and detailed knowledge of the hydrology and ecology of the river basin. As Dr Quang recognises, 'We still lack basic data about the flow regime in the delta and concrete knowledge how upstream dams and diversion schemes can regulate waterflow for the downstream countries' (quoted in Traisawasdichai 1995b). Coordination appears even more intractable when the Mekong is looked at as a single interrelated system. Seasonal flows are not only dependent on the river's mainstream, but on multiple tributary systems. The time of year when water is needed in the delta to flush out the unwanted salinity, the dry season, is precisely the time of year when water demands throughout the basin are highest. Interbasin transfer projects that seek to augment dry-season flows for irrigated agriculture via withdrawals from the Mekong (such as Thailand's Khong-Chi-Mun and Kok-Ing-Nan interbasin transfer schemes) must be particularly well-managed if addressing the salinity problem in the delta is a central goal of comprehensive basin development.

In a further complication, the Chinese government has already completed one large hydroelectric project (the Manwan Dam on the Lancang Jiang, or Upper Mekong), has one under construction (the Dachaoshan Dam) and has plans for five additional large-scale impoundments. Some researchers are optimistic that these dams will enable additional low-season flows through careful regulation (Chapman and Daming 1996). There has been little discussion, however, of what the coordinating procedures might be if conflicts arise between energy demands in Yunnan Province (where the dams are and will be operating) and downstream demands for water to mitigate salinity intrusion.[12] While the portion of the Mekong within Yunnan amounts to only 16 per cent of the river's mean annual discharge, the holding back of water for hydroelectricity generation in a low-flow year potentially involve serious ramifications. Currently, the Chinese government retains observer status in the cooperative framework of Mekong Basin development and does not seem particularly eager to sign on to the 1995 agreement. The downstream effects of the Chinese dams are relevant not only to concerns over salinity intrusion, but to all other ecological considerations in the delta as well.

*Acid soils*

Up to 40 per cent of the land area of the Mekong Delta exhibits soil characteristics described as acid sulphate (see Chapter 8 for a detailed discussion of acid sulphate soils in the delta). Many areas of the delta that maintained mangrove forests in the past were underlain by acidic soils. Deforestation and wetland drainage exposed many of these soils to higher temperatures and aeration which induced the chemical processes that result in acid sulphate soils. Farmers have adopted several strategies to deal with this undesirable soil condition reflecting a blending of approaches based on traditional and scientific knowledge. Researcher Le Quang Minh describes how some farmers near the coast allow salt water to flood their rice lands in the dry season in order to prevent the drying out of the soil (precursor to acidification). The first monsoon rains flushes out the saline water and allows a regular growing cycle (cited in Sluiter 1992: 143–4).

As in the case of salinity intrusion, the impacts of an altered water regime upstream are difficult to predict. Delta farmers recognise that soils in the region are highly variable, even over short distances (Sluiter 1992: 144). Farming practices in the delta have co-evolved with seasonal water fluctuations and farmers employ a variety of techniques specific to local conditions. What appears to be most crucial in the case of acid sulphate soils is the ability of individual environmental managers in the delta to use water flexibly in the treatment of affected areas. Insofar as upstream development hampers this flexibility, acidic soils will be more than less of a problem.

*Flooding*

The delta's hydrologic regime is profoundly seasonal. The July–November wet season brings excessive amounts of water, the amount of which determines the extent of flooding. An estimated 1.83 million ha of the Mekong Delta (47 per cent) are prone to flooding with major floods occurring every four to five years (Mekong River Commission 1997). As noted previously, initial hydrologic studies of the

Mekong system perceived flood control as a relatively minor issue (US Bureau of Reclamation 1956). An even earlier investigation carried out in the late 1940s noted the ways in which delta residents benefited from flood cycles through silt deposition and natural fish entrapments (Bureau of Flood Control 1950: 20). However, little research has been conducted on the extent to which the past century of intensive human alteration of the delta has influenced the flood regime. In many of the flood-prone areas, farmers have adapted to the cyclical flooding in ingenious ways. The widespread use of 'floating rice' and the seasonal patterning of different rice varieties provide salient examples (Nguyen Huu Chiem 1994). In other areas, especially those where people have been forced to live on marginal lands by economic circumstances, flooding causes severe problems through destruction of housing and rice fields, and contributes to several water-related health problems. A flood in 1994 caused over 300 deaths, affected over 300 000 people and inundated over one million ha of land (World Bank 1996a and see also Chapter 10).

Again, the major questions concerning the impacts of upstream development works on flooding in the delta are open to debate. There is little consensus regarding the benefits versus costs of dams as means to control flood waters. Advocates of large dams on the mainstream argue that greater control over water flows throughout the basin will afford greater ability to manage water levels in the delta and reduce the harmful effects of flooding (Mekong Secretariat 1988: 8). To reiterate, this entails a tremendous degree of coordination among water management institutions on a transnational basis, a coordination that has yet to be demonstrated in the Mekong region.

*Fisheries*

The potential impact of upstream water resource development on delta fisheries presents one of the most acute and unpredictable questions confronting the region's inhabitants. As highlighted in Table 13.1, annual fish consumption in the delta averaged 438 000 tons and measured 30 kilograms per capita from 1988 to 1990 (Interim Mekong Committee 1992: 10). The figures in both categories represent totals significantly higher than that of the other countries. Out of total fish production in the delta, 43 per cent are from inland capture fisheries, 50 per cent are from coastal capture fisheries and 7 per cent are coastal culture fisheries. All are dependent on the ecological and hydrological functioning of the river system. The delta's fisheries represent 44 per cent of Vietnam's total annual fishery production and account for 75 per cent of animal protein in the diets of delta residents (Interim Mekong Committee 1992: 11–18).

These numbers underscore the importance of fish to the livelihoods of people living in the delta. Currently, the major problem confronting inland fisheries is an apparent decline in fish abundance, both in terms of total biomass caught and species diversity (Petr 1994: 210). The lack of long-term baseline data on fish behaviour and aquatic ecology compounds the difficulty of identifying contributing factors to fish declines. Most estimates of fish populations in the delta place the total number of species at 200 of which 30 are economically important and frequently caught (Interim Mekong Committee 1992b). Some of the trends propelling recession of capture fisheries likely include: the increasing use of pesticides and inorganic fertilisers by farmers in the delta (Trinh Le 1992), the increasing number of fishing families and more widespread use of modern fishing technology, and a national policy that encourages increased fish production (Interim Mekong Committee 1992).

The rapidly changing socio-economic and institutional contexts for fishing practices make assessment of trends and clear identification of problems in delta fisheries even more difficult. Historically, people throughout Vietnam perceived capture fisheries as open access resources. During the Socialist Republic of Vietnam era, cooperatives were assigned specific fishing areas by district authorities under the auspices of the Ministry of Fisheries and Aquaculture Products (MFAP). Since reform of the collectives in the mid-1980s, MFAP fishery directives are rarely enforced and the frequency and form of household-level fishing activities have greatly increased, particularly in the Mekong Delta. This pertains primarily to inland fisheries. Traditional management practices are gaining in importance in coastal fisheries (see Ruddle 1998). The phenomenon of leasing reaches of the river as fishing grounds is presently practised in the Mekong Delta. District authorities are dividing small streams and canals into 500-metre sections and renting the fishing rights within each stretch to individual households in exchange for an agreed upon tax (field interviews 1997).

Upstream diversions and impoundments are likely to have a tremendous impact on the fish species of the delta, introducing further uncertainties to the rapidly fluctuating pattern of fishing practices and rights. Migratory species and species that depend on a particular flooding regime for part or all of their life histories face substantial reductions in numbers if not outright extirpation. Despite claims that the proposed cascade of dams promoted by the MRC will have lesser impact on basin fish populations due to their run-of-river character, a large run-of-river dam in the Thai portion of the basin, the Pak Mun project, has already exhibited deleterious impacts on fish (Roberts 1995). In addition, any changes in upstream flows, particularly those generated by impoundments, will undoubtedly result in changes in the estuarine and coastal fisheries of the delta. The ecologically 'vibrant state' of brackish and estuarine zones in the delta is 'sustained by organic matter from terrestrial and aquatic plant material inundated by the flood' and subsequently transferred to the river's mouth (Pantulu 1986b: 724).[13]

*Biological diversity*

The indigenous freshwater vegetation of the Mekong Delta exhibits a high degree of diversity. Specific plant communities include: riverbank vegetation; lowland forests on marshes; and aquatic vegetation associated with rivers and waterways. These native plant communities are disappearing rapidly due to the accelerating conversion of delta lands to agricultural land uses (Le Cong Kiet 1994). The delta's mangrove regions still support 46 species of trees despite several decades of increasing pressure from agricultural expansion and aquaculture development (Khoa and Roth-Nelson 1994: 739). In addition, much of the delta's remaining wetland ecosystems support several species of rare avifauna (Beilfuss and Barzen 1994: 454) including the endangered Eastern sarus crane (Choowaew 1993: 7). While it is unclear how alteration of the upstream hydrologic regime might affect these already threatened ecosystems, any perturbations are likely to place further stress on the native plant communities of the delta and the fauna that inhabit them unless carried out in an extremely sensitive way.

*Future threats: pollution and global climate change*

Water pollution concerns in Vietnam are concentrated in rapidly urbanising areas and industrial zones, such as the Thao, Cao and Thuong river basins (Pham Ngoc Dang and Tran Hieu Nhue 1995: 174–86 and Chapters 11 and 12). But recent trends indicate the need to closely monitor the waters of the Mekong delta for signs of contaminants. The increasing use of agricultural inputs such as fertilisers and pesticides are likely to have negative impacts on several water quality parameters such as dissolved oxygen and nitrate concentrations. In addition, there is increasing evidence of toxic compounds in upstream reaches of both the tributaries and the main Mekong River. Researchers working with the Mekong River Development Network found traces of dioxin and insecticides in fish caught in the mainstream on the Thai–Lao border. The researchers concluded the toxins originated from effluent discharges in a heavily industrialised area of Yunnan province in China (Tangwisutijit 1995).

The possible effects of global climate change on low-lying regions like the Mekong Delta (Chapter 3) add an additional element of unpredictability to upstream–downstream dynamics in the Mekong Basin. Working under an assumption that planned dams and reservoirs for water management on the mainstream of the Mekong will exist in the future, a modelling exercise has shown significant alteration in the hydrologic regime of the basin due to climate changes associated with a more potent greenhouse effect. Hydrological changes imply a much higher degree of uncertainty concerning appropriate water control practices (Riebsame *et al.* 1995: 66–73). The Mekong Delta, more than other parts of the basin, is physically vulnerable to climate change due to potential significant impacts on salinity intrusion and flooding. Jacobs (1996) argues the 'institutional resiliency' of the MRC and its antecedent the Mekong Committee, coupled with the generally low level of development in the basin, indicate an institutional ability to adapt constructively to future climate changes. However, this prognosis underplays the potential for much higher levels of development in the basin prior to the effects of climate change and, given this, the ways in which the politics of water use might disrupt basin-wide coordination of water management. Within Vietnam's rapidly changing political–economic context, the reduced role of state institutions in protecting communities from environmental events may serve to increase the vulnerability of delta communities to the combined effects of upstream development and climatic change (Adger, 1999b; and Chapter 2).

### Environmental policy and change in the delta: the institutional context

The Deputy Chairman of the State Planning Committee of Vietnam has noted the 'delta has more than 500 000 ha of unused lands' which accounts for about 12.6 per cent of the region's total area. He also asserts that these lands must be 'reasonably exploited' through appropriate crop and livestock farming, and current state plans envision the conversion of 300 000 ha to rice cultivation (Tran Khai 1994: 9–10) in the hope of increasing food production to 15–16 million tons per year (Nguyen Nhan Quang 1996: 124). This illustrates the primary challenge confronting Vietnam's government and people regarding the Mekong Delta: how to reconcile the proposed acceleration of agricultural development and economic growth with the long-term sustainability of the delta's ecological systems and the livelihoods of people dependent on those systems. Proposed upstream changes in the Mekong River are crucial precisely because both productive activities and biophysical processes in the delta are dependent on the seasonal flows from the Mekong. Thus, the institutions charged with managing the delta's environment – whether state or non-state – assume an essential role in preparing for and adapting to any upstream alterations of the Mekong's hydrology. This section describes some of the environmental management institutions within Vietnam and assesses their capacity for dealing with the delta's complex socio-ecological settings given upstream alterations of the Mekong. It is based on research carried out by one us (Nguyen Thanh Binh) in December 1997 that involved interviews with representatives from 18 domestic and international non-governmental organisations (NGOs).

From the perspective of NGOs working in Vietnam, the formulation of environmental policy continues to be plagued by several shortcomings that inhibit effective practices related to ensuring ecological sustainability. First, environmental policies and programmes tend to overemphasise technology-based approaches that are rooted firmly in modernisation theory and reductive science. Such approaches frequently fail to account for local ecological and social variability. Second, natural resource policies and programmes tend to be formulated in a highly centralised, top-down fashion that promotes rapid economic development as the highest priority while discounting the potential contributions of local people. Experience with resource development projects in Vietnam's Central Highlands shows that this hierarchical model of decision-making fails to generate the tools, organisational mechanisms and social platforms crucial to broad participation by communities most directly affected by land and water resource development projects (Vu Van Me and Desloges 1997: 7).

In a more positive vein, the need to balance economic development with preservation of the natural resource base is clearly represented in policy discourse at the national level. Vietnam's National Plan for Environment and Sustainable Development[14] describes a rather sophisticated interpretation of sustainability, advocating the necessity of satisfying 'the basic material, spiritual and cultural needs of all the people of Vietnam, both present and future generations, through the wise management of natural resources' (Vietnam State Committee for Sciences *et al.* 1991: 6). According to the Plan, 'wise management' places priority on the maintenance of 'essential ecological processes and life-support systems' and 'Vietnam's wealth of genetic diversity' as crucial national objectives, ones reflective of the conservation-oriented agendas, such as the *World Conservation Strategy*, that are dominant in international environmental debates. With regard to water resources, the Plan emphasises several areas for special concern. These include prioritisation of integrated watershed development; prevention of water pollution; further refinement of water quality standards to reflect multiple uses; and better communication between water planners and other resource planners and managers (Vietnam State Committee for Science *et al.* 1991: 45). The government-sponsored Mekong Delta Master Plan also stipulates the need for development planning to be 'environmentally sound and sustainable' (Nedeco 1993: 3).

   The most difficult task is how to translate laudable policy goals into specific delta projects that promote ecological sustainability and reduce social vulnerability. There remains a wide gap between the goals of environmental and resource management policies and the way in which government agencies actually implement specific programmes. The factors contributing to this gap include the ambiguous and contradictory nature of many policies leaving the possibility of multiple interpretations. In addition, the lack of managerial skills, planning capabilities and human resources within implementing agencies and an over-arching emphasis on achieving quantitative targets in programmes to the neglect of other intractable social and political concerns (Pham Ngoc Dang and Tran Hieu Nhue 1995) contribute to policy ambiguity.

### Two dilemmas: environmental information and coordination

Two additional components of this challenge are the availability of environmental information and difficulties of coordination at several institutional levels. A paucity of environmental data and lack of a comprehensive environmental monitoring system in Vietnam introduce further uncertainties into the calculus of sustainable environmental management. This is recognised in the National Plan which reports that:

> reliable and timely information . . . is crucial for establishing policies, decision-making and planning of the development and management of natural resources, and there is a great dearth of information in Vietnam which could be used for this purpose.
>
> (Vietnam State Committee for Sciences *et al.* 1991: 37)

   This limited knowledge base of biological and physical processes extends to the delta, suggesting that predictions about the effects of upstream development are even more tenuous.

   A further dilemma facing the institutions responsible for environmental management in the delta is an apparent deficiency of coordination within and between agencies. As is common in governments throughout the world, Vietnam's environmental management agencies have adopted a notion of 'resources' as single-use entities rather than components of complex ecological systems with important life-supporting functions. This has contributed to fragmentation in the way that government departments approach interrelated biophysical processes. For example, the Ministry of Irrigation plays a central role in the management of surface water while the General Department of Mine Resources is responsible for ground water resources. In addition, while several ministries at local and national levels are charged with determining water quality standards and enforcing pollution regulations, such activities are not systematic and there is widespread confusion over which regulations take precedence (Pham Ngoc Dang and Tran Hieu Nhue 1995).

The disjunctures among environmental management agencies are especially apparent when considering the need for coordination of activities in the delta vis-à-vis upstream development. The principal state agency with responsibility for environmental management is the Ministry of Science, Technology and Environment (MOSTE), formerly the State Committee for Sciences (Nguyen Dac Hy 1995: 228). The creation of MOSTE represents a positive step since it will increase the ability of the state to harmonise and coordinate environmental policy that was previously scattered throughout several different agencies. Yet the Vietnamese team charged with negotiating the 1995 agreement over water use in the Lower Mekong Basin was dominated by representatives from the Ministry of Water Resources,[15] particularly its International Relations Division, and the State Planning Committee (Browder 1998: 123–4). The relationships among these different agencies is unclear, but coordinating Vietnam's national-level environmental policy directives focused on the delta with transnational initiatives centred on the basin remains a major undertaking.

While much of this section has assumed a critical stance regarding the institutional context of environmental management in the delta, we also acknowledge that the Vietnamese government is attempting to negotiate a very difficult economic transition in what has proven to be a frequently hostile global economic context. Furthermore, the importance of ecological sustainability and the vital need to collect and analyse environmental information has been recognised by state planners as witnessed by the initiation of the National Research Program for Environmental Protection in 1991 (Le Thac Can and Vo Quy 1994). However, shortcomings in policy and implementation practices designed to encourage ecological sustainability are likely to be magnified as transformation of the delta accelerates. The previously mentioned Mekong Delta Master Plan and a Decision of the Prime Minister launching a five-year plan (1996–2000) both stress the overriding need to exploit the resources of the delta as rapidly as possible (Nguyen Nhan Quang 1996). In addition, there are numerous signs in the current political economy of development in Vietnam that the privileging of rapid economic growth above all other social goals is contributing to greater inequalities among different sectors of society (Kolko 1997). The reconciliation of economic priorities with their social and ecological costs in a rapidly changing context remains the fundamental problematic of the delta.

## Complexity and participation as emerging issues

We conclude with analysis of two crucial issues: the problems associated with the management of large-scale, complex ecosystems; and the prospects for participatory environmental decision-making. These dimensions have received scant attention in the literatures on Mekong development and the delta, yet the questions they engender speak to the heart of current debates over sustainability and economic development. We raise them to point out important areas of future inquiry.

### *'Managing' complexity in the Lower Mekong Basin*

An anthropologist examining the politics of cooperative development in the Mekong arena in the 1990s posed the question whether or not effective management was possible given the varied political interests at multiple scales operating within the basin (Hinton 1996). It is worth asking the same question from an ecological perspective. As understanding of large-scale ecological systems grows, more and more researchers stress the essential complexity and unpredictability of such ecosystems. The management implications of this 'new ecological paradigm' (Fiedler *et al.* 1992; Worster 1994) are immense. As one ecologist notes:

> We should expect the rate of change in ecosystems to accelerate or decrease very dramatically with little or no warning. Hence we should expect to be surprised. Better historical information about an ecosystem can help us to better design our monitoring techniques so as to reduce some surprises. However, the only way to deal with surprise is to have human systems that are adaptive and prepared to respond appropriately to surprises.
>
> (Kay 1991: 489)

In fact, the complexity and likelihood of surprise in any given ecosystem is compounded in the case of large-scale systems which are actually made up of a mosaic of interacting ecological systems subject to varying rates of evolution and external and internal dynamics. Unfortunately, there appears to be little evidence of such preparedness on the part of the institutional mechanisms designed to monitor the integrity of the Mekong as an ecological system.

It is also quite clear, from references to the Mekong's 'ecological balance' in the 1995 agreement, that the Mekong River Commission and key policy-makers within the apparatuses of the riparian states are employing an outmoded and simplistic notion of ecosystem processes and structures. This older equilibrium-based model ignores critical ecological concepts such as complexity, stress and resilience that must be accounted for in order to generate adaptive management capacities. Despite the admirable 'sustainability' aims presented in Vietnam's

National Plan for Environment and Sustainable Development, very little attention is directed to how ecosystems actually behave and how an understanding of this behaviour might influence development and management goals in, for instance, the Mekong Delta. Increasingly, advocates of ecosystem-based strategies of environmental management are demonstrating the need for management institutions to use flexibility and adaptive management strategies in confronting surprises, perturbations and the essential unpredictability of ecosystems (Lee 1993; Gunderson *et al.* 1995; Holling *et al.* 1998).

The interrelationships between ecological complexity and institutional flexibility also raise crucial questions about the appropriate scale at which river basin institutions operate. On one hand, the inherent complexity of large-scale biophysical systems and need for high degrees of transnational coordination of hydrologic alterations would seem to demand a centralised, politically influential decision-making body along the lines outlined in discussions of international environmental regimes (Young 1989). Ideally, this body would be capable of processing large amounts of information and effectively coordinating the activities of multiple environmental managers, from state agencies to small-scale resource users. On the other hand, there are powerful justifications for a renewed emphasis on supporting or reconstructing 'micro-level' resource management regimes due to their substantial knowledge of local ecosystems and flexibility in adapting to changes (Gadgil *et al.* 1993; Lipschutz 1996: 40–5). Researchers across a spectrum of disciplines have confirmed a strong tendency toward ecological sustainability and participatory decision-making among communities managing resources under some form of common property regime (McCay and Acheson 1990; Ostrom 1990; Baland and Platteau 1996; Berkes and Folke 1998; see also Chapter 5).

Within the context of managing complex river systems such as the Mekong, the challenge is how to coordinate geographically dispersed sets of effective local regimes with large-scale, state-driven economic development activities that are the realm of international regimes. International environmental agreements such as the 1995 Mekong accord in effect serve as umbrellas for many thousands of smaller-scale resource regimes. This can make responsive action to environmental disturbances at an international level extraordinarily difficult. Mechanisms for inducing sustainability at this larger, transnational scale are likely to have little impact on the 'fundamental rules, roles, and relationships that constitute these micro-level regimes' of resource management (Lipschutz 1996: 34). While researchers have initiated discussions of the conceptual linkages between local and transnational resource regimes (Keohane and Ostrom 1994), the dilemmas brought forth by such coordination in practice in the Mekong and elsewhere have not yet been broached. One of the most intractable dilemmas centres on the issue of participation in environmental decision-making.

## *Towards participatory management in the Lower Mekong?*

There is a paradox facing those individuals and groups committed to creating environmental management institutions that result in movement towards ecologically sustainable resource use and sustainable livelihoods for the basin's residents. Are there mechanisms to ensure the participation in environmental decision-making of the people whose lives are sustained by the basin? This question is most acute given the transnational character of the basin delimits management and development decisions almost exclusively to states. Analysts of transnational river disputes and cooperation too often forget that the interests of different social groups that reside in a given basin do not necessarily coincide with those of the states under whose jurisdiction that portion of the basin falls. The history of past water resource development in the Lower Mekong and elsewhere demonstrates that, in far too many cases, the benefits of dam and irrigation projects are disproportionately captured by governments and economic elites. At the same time, the social and environmental costs are disproportionately borne by poor, rural and the indigenous groups (Goldsmith and Hildyard 1984; McCully 1996). To avoid similar mistakes, official institutions at all levels in the Lower Mekong must work to ensure some means of participation for those basin residents whose livelihoods are closely linked to the hydrology and ecology of the basin.

There exists a finely-honed discourse concerning 'participation' apparent in the documents and reports of the MRC and its donors. Despite this rhetoric, genuine engagement in development decision-making forums by the basin's largely rural populace is effectively cut off by political-economic processes at work in individual states. Basin governments argue they are already defending the interests of their citizens by promoting economic development and ensuring that no single riparian state disrupts the integrity of the Mekong River. Advocates of greater participation and ecologically sustainable development patterns in the basin point to past water-related environmental abuses in the region as counter-evidence.[16] NGOs have concentrated on a 'jumping scales' strategy, lobbying for greater accountability from bilateral and multilateral donor institutions, and the MRC itself, which feature 'participatory approaches' prominently in their lending and programme guidelines. Donors and the MRC tend to respond by arguing they can only play an encouraging role in terms of participation within the riparian states (see, for example, an interview with DANIDA in Towards Ecological Recovery and Regional Alliance 1996b); to do otherwise would be an infringement on sovereignty. They further point out that the need for inter-state cooperation to assure holistic river basin development must take priority at the present moment.[17]

Within Vietnam, the opportunities for participation in environmental decisions by those most readily affected by development projects have been negligible. For example, the gigantic Hoa Binh Dam has generated a series of environmental and social problems, especially those related to resettlement of people displaced by the reservoir. The resettlement process has been beset by miscommunications and lack of coordinated planning to ensure evacuees from the reservoir area were able to resume adequate livelihoods (Hirsch 1992). While the focus of this paper has been on the Mekong Delta vis-à-vis upstream development, in some instances Vietnam is itself in an upstream relative to its riparian neighbours. For example, serious questions remain concerning the efficacy of the Yali Falls dam project located on the upper watershed of the Sesan River, which begins in Vietnam's central highlands before descending to Cambodia and conjoining with the Mekong. Over three decades of Mekong Committee-sponsored research has been carried out by consulting firms on the feasibility of an impoundment at Yali Falls, and construction of the project began in the mid-1990s.[18] There has been virtually no consultation with the over 7000 ethnic minority inhabitants who will be resettled due to the reservoir, nor with the downstream communities or authorities in Cambodia. Ironically, there have been no studies on how flow regulation of the Sesan might impact downstream hydrology and fisheries in the Mekong Delta (Lang 1996: 43).

Despite these shortcomings, more optimistic examples of participatory resource development projects that also incorporate elements of ecological sustainability do exist. Beilfuss and Barzen (1994) describe ongoing work in the Tram Chim Reserve of the *Dong Thap Muoi* (Plain of Reeds), a tidal freshwater marsh that suffered severe degradation during the war. An emerging environmental management regime stresses the need to mimic the natural annual water cycle of the area and to draw on the expertise of long-term residents in reconstructing past hydrologic regimes. Other research projects have demonstrated the ways in which restoration of mangrove and cajeput ecosystems can rehabilitate acid sulphate soils more efficiently than the more common and expensive practice of flooding and leaching contaminated areas (Khoa and Roth-Nelson 1994 and Chapter 8).

## Conclusion

Throughout this chapter we have emphasised the contradictions that might arise as a result of balancing ecological and water development concerns. The dynamics of socio-ecological concerns in the Mekong Delta are apparent in saltwater intrusion, flooding, fisheries, and must be balanced with the dynamics of large-scale water resource development within the Mekong Basin as a whole. The preceding discussion of ecological complexity and participatory decision-making highlight some of these contradictions. There is a strong case to be made that current institutions for managing the Mekong at transnational scales are deficient in acknowledging the value of participatory approaches in empowering stakeholders and promoting adaptive capacity. Related to this, present institutions are also deficient in recognising and dealing with the complexity of ecosystems. These deficiencies derive in large part from an uncritical prioritisation of economic development through large-scale projects and the complex hydropolitics of the Mekong. We argue that there is tremendous risk to the ecological systems and livelihoods of the delta if large-scale river development, much of which is driven by economic and political processes at national and transnational scales, is allowed to out-pace the evolution of environmental management institutions which are necessary at a range of local-to-transnational scales.

## Notes

1 We deliberately use the phrase 'environmental managers' to refer to what are often called 'stakeholders' or 'actors' in the environmental management and policy literature. Environmental managers are not exclusively state agents, but include small-scale agriculturists, environmental NGOs, state agencies, international financial institutions, and transnational corporations. This broadening of the traditional definition of 'environmental manager' reflects the great diversity of agents whose decisions both directly and indirectly result in some degree of ecological alteration at different scales (Wilson and Bryant 1997: 6–13).

2 The organizational structure of the Mekong Committee and its administrative body, the Mekong Secretariat, and its rules of operation are described in Binson (1981). The riparian states agreed to substantial organizational restructuring following the 1995 agreement which stipulated creation of the Mekong Committee's successor, the Mekong River Commission (MRC). The MRC consists of three permanent bodies: the MRC Council (representatives of each riparian state at the ministerial level), the MRC Joint Committee (riparian representatives at the department head level), and the MRC Secretariat, the technical and administrative body. The semi-permanent Donor's Consultative Group consists of representatives from donor countries and cooperating agencies (Chooduangngern 1996).

3 Some flavour of this ideology in the context of Mekong development is provided in the document *To Tame a River* which argues the purpose of the Pa Mong dam, a massive hydroelectric and irrigation development scheme slated for the mainstream river upstream of Vientiane but never built, is 'to awaken a sleeping giant . . . harness its might . . . develop its potential for progress . . . provide the energy and water resource foundation for industry and rural development programmes' (US Agency for International Development 1968).

4 This diagram is far from comprehensive and intended only as a rough guide to subsequent discussion in lieu of more detailed analysis of the 'archeology of development' (Watts 1993) in the basin. Several works describe the institutional history of the Mekong Committee in greater detail (Mekong Secretariat 1989; Hori 1993; Jacobs 1996).

5 Throughout the 1950s and 1960s, representatives of the United States – primarily agents of the Bureau of Reclamation, the Army Corps of Engineers and the Agency for International Development – played a crucial role in formulating the dominant development model promoted by the Mekong Committee. All US aid to the Mekong Project was halted in 1975.

6 One salient example involves a series of research visits and seminars jointly sponsored by the Mekong Committee, the US Agency for International Development and the Asia Foundation that culminated in a report titled *Some Environmental Considerations for the Mekong Project*. The primary goal of the report was to develop a research agenda whereby necessary information could be furnished 'to the engineers and planners so that the ecological risks in this development project [the Mekong] could be appropriately weighed' (Challinor 1973: iii). In addition, the Mekong Secretariat established an Environmental Unit in 1976 with United Nations Development Programme support (Mekong Secretariat 1989: 68).

7 The forums include the Asia Development Banks's Greater Mekong Subregion Initiative and Forum for the comprehensive Development of Indochina, the ASEAN-Mekong Basin Development Cooperation, and the Mekong River Commission itself, while the funds include the Asian Development Fund Greater Mekong Capital Fund, Asian Infrastructure Development Company, ASEAN-Mekong Development Fund, MRC Donor Consultative Group, and Mekong Development Bank. How the economic crisis in the late 1990s in Southeast Asia has affected the activities of fora and funds remains unclear.

8 The Hanoi government perceived Johnson's offer largely as an effort to 'buy peace' in the region, and it received little support in the US Congress which saw it as an extremely expensive and risk-filled enterprise (Huddle 1972).

9 Browder (1998) provides a detailed account of these negotiations and a keen analysis of the critical interactions among negotiators and riparian state representatives regarding several difficult issues. The 1975 agreement which followed Cambodia's expulsion from the Mekong Committee stipulated that each riparian state had the right to veto any mainstream projects if the project might threaten downstream flows. By the early 1990s the Thai state, prompted by a long-standing desire to implement interbasin transfer projects on the Mekong for hydroelectric generation and irrigation development, made Cambodia's re-entrance into the Mekong Committee contingent on renegotiation of the Mekong's international regime in order to remove the veto powers of individual states. Thus Article 6 of the new agreement stipulates that during the dry season 'intra-basin use shall be subject to prior consultation' and that any 'inter-basin diversion project shall be agreed upon by the Joint Committee [comprised of representatives from each basin state] through a specific agreement for each project prior to any proposed diversion' (Mekong River Commission 1995a: 5). Veto capability has been effectively replaced by mandatory consultation. These issues remain contentious. The 1995 agreement mandated (1) formulation of a non-binding Basin Development Plan acceptable to the signatory governments that would chart out the future development programme for the basin; (2) negotiation of subsidiary agreements concerning minimum flows; and (3) crafting of procedures for reviewing proposed water uses (Browder 1998: 287). Each of these post-agreement mandates demand continuing negotiations among the basin states. One of the subsidiary agreements was signed in April 2000 (Vietnam News List 2000); the World Bank had designated US$11 million in February 2000 (from the Global Environmental Facilty) for a proposed Water Utilization Project to further refine and implement the terms of the 1995 agreement (Australian Mekong Research Network 2000a).

10 Fish deserve special consideration due to their importance as source of protein and supplemental income to basin inhabitants, but they are not the only fauna likely to be negatively affected by large-scale development of the Mekong. The mainland South East Asian peninsula of which the Mekong Basin is a prominent part 'is probably the most diverse zoogeographic area in South East Asia' (Zakaria-Ismail 1994: 44). A great variety of birds, mammals, reptiles and other animal groups face disruption of habitat in the basin from advancing manifestations of rapid economic development, resource exploitation and population growth. In addition, riparian vegetation, forests and wetlands throughout the Mekong Basin are threatened by the same complex of processes (Dudgeon 1992).

11 The mangrove forest areas in the Mekong Delta have declined from 250 000 ha in 1950 to 94 000 ha in 1988, largely due to conversion to agricultural land (Nedeco 1993). Most recently, approximately 1000 ha of mangrove forest during the 1996 to 1997 period was converted to shrimp aquaculture operations (Phan Nguyen Hong, personal communi-cation). See Chapter 9 for a detailed discussion of mangrove conservation.

12 In spite of these concerns, the existing and planned reservoir storage capacity in Lancang Jiang played a key role in solidifying the 1995 agreement. By assuring augmentation of dry season water supply, the reservoirs enabled the Thai contingent to agree to maintenance of a minimum natural dry season flow because they were secure in the knowledge that they would be able to utilize excess dry season flow afforded by the reservoirs (Browder 1998: 207).

13 These observations point out the fundamental links between the annual flooding cycle and fisheries production. Pantulu (1986b: 723–4) describes how 'annual flooding dominates the fisheries production of the Mekong, inundating soil and vegetation and introducing millions of tonnes of suspended and dissolved solids to the system. These

changes stimulate an explosive annual reproductive outburst.' Because the delta is essentially a vast zone of deposition, fisheries in the region are dependent on the annual cycle.

14 This plan was formulated in 1991 with assistance from the United Nations Development Programme, Swedish International Development Authority, the United Nations Environment Programme and the World Conservation Union (IUCN) (Vietnam State Committee for Sciences *et al.* 1991).

15 This ministry and other resource-based ministries (forestry, agriculture) were reorganized into the Ministry of Agriculture and Rural Development in late 1995.

16 Several regional and international NGOs have assumed prominent roles in debates over Mekong development in the past several years. These include: the Bangkok-based Towards Ecological Recovery and Regional Alliance (TERRA) which publishes the quarterly journal *Watershed: People's Forum on Ecology*; and the International Rivers Network based in Berkeley, California. Oxfam America has also initiated a Mekong Advocacy Programme to focus attention on what large-scale development implies for the basin's largely rural populace. The US-based World Resources Institute has also carried out watershed-oriented programmes in the Mekong region in recent years.

17 Actions are being taken within the Mekong administrative structure to respond to calls for greater participation from non-governmental entities in Mekong decision-making structures. The MRC Secretariat recently launched a study on public participation in the formulation of the Basin Development Plan (Chooduangngern 1996). Independently of the MRC, the Mekong Dialogue on Sustainable Development (MDSD), housed at the University of Can Tho in the Mekong Delta, seeks an advisory role for non-state participants in basin development planning. The MDSD recognizes that 'participation in the Mekong River Commission's planning process will not be adequate if it involves consulting people and groups once decisions, outline plans or "draft plans" have already been finalized' (Vo-Tong Xuan 1996: 272).

18 Tragically, this lack of coordination may have resulted in the loss of life in Cambodia and Vietnam. In December 1999 and again in March 2000, unexpected and heavy rains in the upper catchment areas of the Se San basin resulted in large surges of the river downstream of the Yali Falls Dam. It appears likely that releases of water by the dam's operators, with no warning to downstream residents in both the Vietnamese and Cambodian portions of the basin, resulted in several drownings, crop destruction and the loss of fishing gear (Saroen and Stormer, 2000; cited in Australian Mekong Research Network 2000b).

# 14 Prospects for sustainable development

*W. Neil Adger, P. Mick Kelly and
Nguyen Huu Ninh*

## Introduction

Environmental change, set against a backdrop of pronounced societal trends over coming decades, will undoubtedly test the resilience of Vietnamese society. This final chapter considers particular challenges that Vietnam faces during the early twenty-first century. Drawing on the discussion presented in this book, we consider how emerging trends might affect levels of vulnerability and resilience, taking as examples three particular threats and opportunities resulting from national and international processes, and draw out implications for policy. We consider, first, the impact of globalisation on Vietnam and its environment. We then examine the spatial dimension of the development process within Vietnam, focusing on poverty, patterns of economic growth and population distribution. Finally, we discuss the land allocation process and the consequent trends in access to resources which, as has been argued in many places in this volume, shape patterns of vulnerability and resilience. In the concluding section, we identify three key issues that must be given serious consideration if sustainable and equitable development is to be secured.

## The impact of economic globalisation

It is difficult to pin down what is meant by economic globalisation. In the broadest sense of the term, 'globalisation reflects a widespread perception that the world is rapidly being moulded into a shared social space by economic and technological forces and that developments in one region of the world can have profound consequences for the life chances of individuals or communities on the other side of the globe' (Held *et al.* 1999: 1). Economic globalisation, during the contemporary period, can be characterised by: near-universal participation in the international financial and monetary order; unprecedented gross flows in capital and diversity in these flows; a high transaction velocity, with 24-hour trading; the potential for high, externally-driven impacts within national economies; extensive international surveillance and regulation and the development of private financial infrastructures, all facilitated by modern technology; further growth of multinational banking and the role of the International

Monetary Fund in poorer economies; increasing decentralisation, collaboration and competition; and a shift in the power balance between private finance and states (Held *et al.* 1999: Grid 4.1).

Particular economic aspects of the process discussed in previous chapters have included the trend towards deregulation of capital markets and the freer flow of capital across borders. Foreign direct investment has increased in importance in many Asian developing countries over the past decade and financed many industrialisation strategies (Kim 1997). But, in addition to the movement of capital, there is much international pressure for once-protectionist countries to liberalise trade in good and services. Vietnam has stated its aim, following full accession to the regional economy through ASEAN (Association of South East Asian Nations), to become a full member of the World Trade Organization. In these pronouncements, Vietnam demonstrates a commitment to the rigours of liberalised trade but, wary of the potential social and environmental impacts, is ambivalent in implementing free trade policies (Tongzon 1999).

A further aspect of globalisation is the opening of financial markets. In this respect, as economic renovation continues step by step, Vietnam will increasingly face the challenge of volatile financial and currency markets. The volatility of these markets constituted the major cause of the Asian economic crisis of 1997 (Wade 1998). It is argued that the flows of capital and the volatility of markets lead to other manifestations of globalisation, such as increasing discrepancies between rich and poor. Income is becoming more unequally distributed in many regions, with particular rises in inequality in the former centrally planned economies (e.g. Krugman 1994; Krugman and Venables 1995).

In general, Vietnam has been cushioned from the worst impacts of the Asian economic crisis that began in 1997. But integration with the regional economies and the development of export-led industrialisation can bring unforeseen environmental consequences, as Chapter 11 has shown in the case of Vietnam's neighbours (see also Bezanson 1998). The environmental Kuznets curve hypothesis is not proven – simply dashing for industrial-based economic growth and *hoping* this will lead to increased demand for environmental quality is not a viable option. In fact, Vietnam has been successful in implementing the first stages of environmental protection policies in the 1990s, 'perhaps enabling the economy to undergo industrialisation with comparatively less pollution than the first wave of industrialising countries' (Forsyth 1997: 260). Even this tentative conclusion must be tempered, though, with the evident consequences of rapid growth, unevenly distributed towards major urban areas (Chapter 12). The sheer speed of growth, scale and spatial concentration of economic activity is an important macro-determinant of sustainability.

Even with a late-coming, heavily interventionist industrial policy, many economists believe that Vietnam has the opportunity to benefit from growth in labour intensive sectors such as clothing and textiles, and can move from these to more capital intensive industries. This strategy is depicted as the classic 'flying geese' strategy of East Asian development. Both proponents and critics of these strategies – see, for example, Radelet and Sachs (1997) and Hart-Landsberg and

*Table 14.1* Measures of regional poverty incidence and economic performance

| Region | Percentage of total population | Percentage of regional population below poverty line* | Regional distribution poor (% of total) | Annual economic growth (%) 1991–1993 |
|---|---|---|---|---|
| Northern Uplands | 17 | 59 | 23 | 10.9 |
| Red River Delta | 20 | 49 | 15 | 15.0 |
| North Central | 14 | 71 | 18 | 7.1 |
| South Central | 11 | 49 | 10 | 1.7 |
| Central Highlands | 4 | 50 | 7 | 11.2 |
| South East | 12 | 33 | 7 | 17.5 |
| Mekong Delta | 22 | 48 | 20 | 7.1 |

Source: United Nations Development Programme (1995), World Bank (1995b).

Note
*Poverty line defined as culturally specific level of imputed-income, calibrated by World Bank for each region.

Burkett (1998) – agree that governance is a key issue. Radelet and Sachs argue that:

> Asia has achieved rapid growth despite severe limitations in its governance. Strong central governments control powerful and political bureaucracies that can override local interests, the judiciary and even private property rights . . . leaving local governments weak and unable to address urgent infrastructural and regulatory challenges.
>
> (Radelet and Sachs 1997: 56)

Weak regulation and enforcement can result in significant environmental costs. The Asian Development Bank (1997) has reviewed evidence of environmental degradation in Asia's cities costing billions of dollars as a result of weak and corrupt governance in the major economies. Clearly, there are important lessons to be learnt from Vietnam's Asian neighbours.

## The spatial dimension

Poverty at the national level in Vietnam has a marked spatial dimension (Table 14.1). It is also concentrated within particular sectors in the rural economy such as agriculture and forestry (United Nations Development Programme 1995). The upland regions of the north and central Vietnam, as well as the central coastal regions, have the lowest incomes per capita and the greatest reliance on agriculture. At the national level, poverty is concentrated within the ethnic minorities of Vietnam (United Nations Development Programme 1995).

There are many suggested strategies for the alleviation of poverty in Vietnam, particularly with regard to its high incidence in certain regions. Most of these strategies centre on sustained economic growth (Vietnam Ministry of Science, Technology and Environment 1994; United Nations Development Programme 1995; World Bank 1995b). The regions with the highest incidence of poverty (Table 14.1) have, however, been least successful in stimulating economic growth during the past decade, particularly the central coastal regions. Poverty alleviation strategies advanced by government and aid agencies recommend reduction in the relative isolation of these regions, for example, through road building, alongside investment in targeted social services in education, health and agricultural research and the provision of a social safety net.

Vietnam has a long-standing policy of attempting to achieve balanced regional development through, *inter alia*, redistribution of its population from high-density areas into previously less productive agricultural areas and by restricting movement to cities. The national government policy of encouraging out-migration from the Red River Delta, for example, begun under the 1961–5 Five Year Plan, was motivated by the desire to reduce high population densities and involved direct subsidy, often to levels such as VND 4 000 000 per household, significantly more than the annual income for most families (Xenos *et al.* 1993). People were encouraged to move from the delta regions to more sparsely populated and previously non-agricultural upland areas such as Bac Thai Son La and Lai Chau (Thrift and Forbes 1986: 93). In the period from 1961 to 1975 alone, one million people moved out of the Red River Delta as a result. Although empirical research shows that policy interventions of subsidies and regulations have been successful in encouraging migration, the 'sending' provinces and cities were still experiencing net immigration in the 1980s (Anh Dang *et al.* 1997; and Chapter 12).

In the present-day, the vast majority of land that could be turned over to agriculture has already been converted, and the limited prospects for further expansion of irrigation technology place a major constraint on increasing agricultural production (Pingali *et al.* 1997). Thus, much resettlement in coastal areas involves conversion of coastal wetlands, mangroves and estuarine mudflats to agricultural land or for use in aquaculture; as demonstrated in Chapter 5, households are being encouraged to move without consideration of local institutional and environmental contexts and this has ambiguous impacts on social vulnerability. In upland areas, resettlement often requires forest clearance in fragile ecosystems (Chapters 6 and 7). Sustainability can be difficult to guarantee under both these circumstances. Increasingly, migration in Vietnam is occurring through temporary movements with migrants retaining family ties, and even land, in their home areas, and often sending remittances back to those rural areas, even from large distances, and returning home on a seasonal basis or after an extended period away. Many migrants headed for Hanoi during the late 1980s and early 1990s. They came mainly from the Red River Delta, but also from southern and central provinces (Vu Lu Tap and Taillard 1993). This is classic 'pull' migration, with the attraction of increasing real wages and opportunities in Hanoi and other cities. The easing of restrictions on population movement, alongside market reforms, will undoubtedly increase population mobility and forms of spontaneous migration. A study by Anh Dang *et al.* (1997) suggests that temporary and circular migration will increase as a proportion of all migration as a result of the relaxation of labour laws.

Considering implications for social vulnerability and resilience, one of the most positive effects of migration – indeed, often its prime rationale from the migrant's point of view – is the creation of opportunities for livelihood enhancement and risk sharing, through, for example, remittance income (Locke *et al.* 2000). Reduction of pressure on the natural resource base in the sending areas may be a secondary benefit but must be considered alongside environmental consequences in the migrant receiving areas.

The tension between planned migration policies and the spontaneous demand for increased population mobility is a major challenge both for Vietnam and for many Southeast Asian countries. Migration policies and incentives need jointly to consider the interaction of market opportunities, labour mobility, the nature of available resources, cultural issues and spatial physical planning to minimise risks and enhance benefits. The underlying cause of 'pull' migration in the Vietnamese context is the contrast between the opportunities available in the agrarian rural economy and the greater opportunities for poverty alleviation in the urban economy and its associated growth sectors. The creation of alternative economic growth poles is a feature of government policy and must be a key consideration both in alleviating pressure on Vietnam's burgeoning cities and reducing inequity across the nation. Poverty and access to resources are the keys to understanding vulnerability at the individual level (Chapter 2), and policies for poverty alleviation are, therefore, critically important in this regard.

## Land allocation and access to resources

Perhaps the most important institutional changes in the agricultural areas of Vietnam over the final decade of the twentieth century were those associated with land allocation. The land allocation process began in 1988 and was confirmed with the Land Law in 1993. The process varies according to the type of land. Rice land, for example, is generally allocated on the basis of household membership, while hill land is allocated based on the ability of households to pay cooperatives for the rights to 'improvements' such as planted trees (Rambo 1995a; and Chapters 6 and 7).

Access to land fundamentally determines access to resources in the agrarian economy and, hence, levels of vulnerability and resilience. At this time, though, the net effect of the land allocation process in Vietnam is difficult to assess. For example, some authors argue that, in the areas of provision of public services, such as agricultural research, health and education, the private control of land and consequent increases in private income have been taken as signals to reduce state expenditure on these public goods with ambiguous impacts on the social structure of rural areas acting in different directions (Rambo 1995a; Tipping and Truong Viet Dung 1997). By contrast, Kolko (1997) sees land privatisation, and the parallel emergence of both private credit systems and land rental markets, as having had an overwhelming negative, indeed, cataclysmic impact:

> The basic structure of what had once been the large majority of the nation's relatively stable, organised universe was now placed under severe stress and profoundly traumatised, especially in the northern Red River Delta and along the central coast. It was entirely predictable that the reallocation of land would necessarily affect rural social peace fundamentally for a very long time, creating incalculable risks.
>
> (Kolko 1997: 91).

The negative consequences of land allocation for vulnerability are clearly related to social disruption and stress and to restricted access and entitlement to resources.

Regional disparities in land allocation also have impacts on resilience. In the intensive agricultural lowlands, allocation has been relatively equitable and transparent, and, with leases allocated to households rather than individuals, has avoided gender bias in inheritance patterns (Rambo 1995a). The negative consequences of these reforms on social resilience in these regions have, therefore, been concentrated on the loss of forms of collective action. Although food production has dramatically increased with individual land holding, for example, the reduction in the role of agricultural cooperatives has had negative consequences. According to Rambo:

> Maintenance of communal irrigation networks has become problematic in some areas as there is now no effective institutional arrangements for mobilising the labour to keep the system in repair
>
> (Rambo 1995a: 8)

In the upland areas, full account of the differences in resource use structure, particularly of forest use, has not been incorporated into the land allocation process (Chapter 6). Moreover, as argued in Chapter 7, sustainable development for ethnic minorities requires recognition of the resilience of traditional management practices and 'space' in which to coexist within wider Vietnamese society. Both these factors mitigate against effective and equitable land distribution. In addition, political influence and lack of transparency have conspired to result in more inequitable land transformation in highland areas. In lowland agricultural regions, such practices are somewhat avoided by making the rules clear and unambiguous, and, according to Le Trong Cuc and Rambo (1993), by village level solidarity and autonomy built up over the past three decades. Inequitable allocation and rent-seeking bureaucratic allocation of resources is, however, also observed in the lowlands with respect to common property resources. For resources such as mangroves, existing use by poorer households are overturned by individual allocation of land for reclamation or aquaculture (Chapter 5).

# Conclusions

Sustainable development – ensuring equitable economic growth while protecting the environment on which all depend – necessitates limiting the vulnerability of human populations and natural ecosystems and enhancing their resilience. Human communities under pressure, whether environmental or socio-economic in nature, will likely overexploit the natural resource base and their prospects for economic growth will be ever more limited. It follows that policies and measures directed towards reducing vulnerability to environmental change, strengthening resilience and facilitating the ability to cope with stress and, where appropriate, to adapt must form a key part of any effective strategy for sustainable economic growth.

Checks and balances are required so that trends within the country and at the international scale do not erode the capacity of the Vietnamese people to manage environmental stress. The process of land allocation, for example, can affect social resilience directly by skewing access to resources and indirectly through the weakening of forms of collective action and a reduction in formal social expenditure. Positive tendencies should be strengthened, with compensation promptly deployed to offset adverse effects. A number of authors in this volume have stressed the importance of considering the biophysical and social context, so that measures are not imposed without regard for circumstance and that the benefits of diversity in response are realised. As far as international trends are concerned, Vietnam needs to resist those aspects of economic globalisation that are considered socially undesirable, such as inward investment in polluting industries and opening of certain commodity markets on which poorer sections of the population are dependent, and take advantage of the benefits that greater interaction with the international community can bring. Given the speed of environmental and societal change and our inability to provide accurate forecasts of relevant trends, flexibility, supported by continual monitoring and evaluation, must be a necessary attribute of any sustainable development strategy. This resonates with many aspects of Vietnamese government policy, particularly with regard to step-by-step development planning.

We identify three priority areas that we consider will play a fundamental role in determining prospects for sustainable development in Vietnam. First, there is a strikingly clear need to manage carefully continuing engagement with the process of economic globalisation, reconciling advantages with the loss of self-sufficiency and independence that must follow. Second, we underline the significance of Vietnam's traditional commitment to equity, in terms of both power and access to resources. This commitment provides a very effective basis for limiting social exclusion and promoting sustainable resource utilisation. Third, we emphasise the importance of empowerment at the local level, another historic characteristic of the nation given the traditional basis of the Vietnamese economy in the rural village, and, through this, the realisation of diverse opportunities for development trajectories that are sensitive to local and regional environmental and socio-cultural circumstances.

If there is one lesson for policymakers that emerges above all others from the reviews presented in this book, it is that due consideration must be given to the resilience of individuals and local communities when processes such as land allocation and institutional reform are encouraged on the basis of economic efficiency gains at the national level. If that message is taken seriously, and there are hopeful signs that this is occurring, then the new road of *doi moi* may lead to the reality of sustainable development.

# References

Adger, W.N. (1996) *Approaches to Vulnerability to Climate Change*, Global Environmental Change Working Paper 96-5, Norwich and London: Centre for Social and Economic Research on the Global Environment, University of East Anglia, and University College London.

Adger, W.N. (1999a) 'Exploring income inequality in rural, coastal Vietnam', *Journal of Development Studies* 35, 5: 96–119.

Adger, W.N. (1999b) 'Social vulnerability to climate change and extremes in Vietnam', *World Development* 27: 249–69.

Adger, W.N. (1999c) 'Evolution of economy and environment: an application to land use in lowland Vietnam', *Ecological Economics* 31: 365–79.

Adger, W.N. (2000a) 'Institutional adaptation to environmental risk under the transition in Vietnam', *Annals of the Association of American Geographers* 90: 738–58.

Adger, W.N. (2000b) 'Social and ecological resilience: are they related?', *Progress in Human Geography* 24: 267–84.

Adger, W.N. and Kelly, P.M. (1999) 'Social vulnerability to climate change and the architecture of entitlements', *Mitigation and Adaptation Strategies for Global Change* 4: 253–6.

Adger, W.N. and Luttrell, C. (2000) 'Property rights and the utilisation of wetlands', *Ecological Economics* 35: 75–89.

Adger, W.N. and O'Riordan, T. (2000) 'Population, adaptation and resilience', in T. O'Riordan (ed.) *Environmental Science for Environmental Management, Second Edition*, London: Longman.

Adger, W.N., Kelly, P.M., Nguyen Huu Ninh and Ngo Cam Thanh (1997) *Property Rights and the Social Incidence of Mangrove Conversion in Vietnam*, Global Environmental Change Working Paper 97-21, Norwich and London: Centre for Social and Economic Research on the Global Environment, University of East Anglia, and University College London.

Afsah, S. and Vincent, J. (1997) 'Putting pressure on polluters: Indonesia's PROPER programme', paper presented at the 1997 Asia Environmental Economics Policy Seminar, Harvard Institute for International Development, Cambridge, Mass.

Ahlback, A.J. (1995) *On Forestry in Vietnam, the New Reforestation Strategy and Assistance*, UNDP/FAO Project VIE/92/022, Hanoi: Food and Agriculture Organization.

Aksornkoae, S. (1993) *Ecology and Management of Mangroves*, Bangkok: World Conservation Union.

Alavi, R. (1996) *Industrialisation in Malaysia: Import Substitution and Infant Industry Performance*, London: Routledge.

Amsden, A. (1989) *Asia's Next Giant: South Korea and Late Industrialization*, Oxford: Oxford University Press.

Anh Dang, Goldstein, S. and McNally, J. (1997) 'Internal migration and development in Vietnam', *International Migration Review* 31: 312–37.

Asian Development Bank (1994) *Climate Change in Asia: Vietnam – Regional Study on Global Environmental Issues*, Manila: Asian Development Bank.

Asian Development Bank (1997) *Emerging Asia: Changes and Challenges*, Manila: Asian Development Bank.

Australian Mekong Research Network (2000a) 'World Bank approves GEF "start-up" grant for MRC', AMRN Item 7/2000, 28 February 2000.

Australian Mekong Research Network (2000b) 'Further information on the recent flooding attributed to release of water from Yali Falls Dam', AMRN Item 14/2000, 31 March 2000.

Bach Tan Sinh (1998) 'Environmental policy and conflicting interests: coal mining, tourism and livelihoods in Quang Ninh Province', in P. Hirsch and C. Warren (eds) *The Politics of Environment in South East Asia*, London: Routledge.

Bailey, C. and Pomeroy, C. (1996) 'Resource dependency and development options in coastal Southeast Asia', *Society and Natural Resources* 9: 191–19.

Bakker, K. (1999) 'The politics of hydropower: developing the Mekong', *Political Geography* 18: 209–32.

Baland, J.M. and Platteau, J.P. (1996) *Halting Degradation of Natural Resources: is There a Role for Rural Communities?*, Oxford: Clarendon.

Baland, J.M. and Platteau, J.P. (1997) 'Wealth inequality and efficiency in the commons: (I) the unregulated case', *Oxford Economic Papers* 49: 451–82.

Baland, J.M. and Platteau, J.P. (1998) 'Wealth inequality and efficiency in the commons: (II) the regulated case', *Oxford Economic Papers* 50: 1–22.

Bangkok Post (1997a) 'Six nations look for private investment', *Bangkok Post*, 11 April 1997, Business p. 3.

Bangkok Post (1997b), 'IEAT taken to task over pollution', *Bangkok Post*, 29 November 1997, p. 2.

Barbier, E. (1993) 'Sustainable use of wetlands valuing tropical wetland benefits: economic methodologies and applications', *Geographical Journal* 159: 22–32.

Barbier, E.B. (1994) 'Valuing environmental functions: tropical wetlands', *Land Economics* 70: 155–73.

Barbier, E.B. (1997) 'Introduction to the environmental Kuznets curve special issue' *Environment and Development Economics* 2: 369–81

Barbier, E.B. and Strand, I. (1998) 'Valuing mangrove fishery linkages: a case study of Campeche, Mexico', *Environmental and Resource Economics* 12: 151–66.

Bardhan, P. (1993) 'Analytics of the institutions of informal co-operation in rural development', *World Development* 21: 633–9.

Barker, R. (ed.) (1994) *Agricultural Policy Analysis for Transition to a Market-oriented Economy in Vietnam: Selected Issues*, FAO Economic and Social Development Paper 123, Rome: Food and Agriculture Organization.

Barnston, A.G., Vandendool, H.M., Zebiak, S.E., Barnett, T.P., Ji, M., Rodenhuis, D.R., Cane, M.A., Leetmaa, A., Graham, N.E., Ropelewski, C.R., Kousky, V.E., Olenic, E.A. and Livezey, R.E. (1994) 'Long-lead seasonal forecasts – where do we stand?', *Bulletin of the American Meteorological Society* 75: 2097–114.

Baulch, B. (1996) 'Neglected trade-offs in poverty measurement', *IDS Bulletin* 27, 1: 36–42.

Beilfuss, R.D. and Barzen, J.A. (1994) 'Hydrological restoration in the Mekong Delta, Vietnam', in W.J. Mitsch (ed.) *Global Wetlands: Old World and New*, New York: Elsevier.

Bello, W. and Rosenfeld, S. (1992) *Dragons in Distress: Asia's Miracle Economies in Crisis*, Harmondsworth: Penguin.

Benson, C. (1997) *The Economic Impact of Natural Disasters in Vietnam*, Working Paper 98, London: Overseas Development Institute.

Beresford, M. (1988) *Vietnam: Politics, Economics and Society*, London: Pinter.

Beresford, M. and Fraser, L. (1992) 'Political economy of the environment in Vietnam', *Journal of Contemporary Asia* 22: 3–19.

Bergeret, P. (1995) 'La politique foncière au Vietnam (Land Policy in Vietnam)', in *Proceedings of the International Symposium on Systems-oriented Research in Agriculture and Rural Development*, Montpellier: University of Montpellier.

Berkes, F. and Folke, C. (1998) 'Linking social and ecological systems for resilience and sustainability', in F. Berkes and C. Folke (eds) *Linking Social and Ecological Systems*, Cambridge: Cambridge University Press.

Bernauer, T. (1997) 'Managing international rivers', in O. Young (ed.) *Global Governance: Drawing Insights from the Environmental Experience*, Cambridge: MIT Press.

Bezanson, K.A. (1998) 'Tiger cubs at the crossroads: some policy issues facing Vietnam', *IDS Bulletin* 29, 3: 43–50.

Binson, B. (1981) 'The Lower Mekong Basin development', in M. Zaman (ed.) *River Basin Development: Proceedings of the National Symposium on River Basin Development, Dacca, Bangladesh*, Dublin: Tycooly International.

Birdsall, N. and Wheeler, D. (1992) 'Trade policy and industrial pollution in Latin America: where are the pollution havens?', in P. Low (ed.) *International Trade and the Environment*, World Bank Discussion Paper 159, Washington DC: World Bank.

Blaikie, P. (1993) 'Population change and environmental management: coping and adaptation at the domestic level', in B. Zaba and J. Clarke (eds) *Environment and Population Change*, Liege: Ordina Editions for the International Union for Scientific Study of Population.

Blaikie, P.M. and Brookfield, H. (1987) *Land Degradation and Society*, London: Methuen.

Blaikie, P., Cannon, T., Davis, I. and Wisner, B. (1994) *At Risk: Natural Hazards, People's Vulnerability and Disasters*, London: Routledge.

Blake, S.T. (1968) 'A revision of *Melaleuca leucadendron* and its allies (Myrtaceae)' *Contributions from the Queensland Herbarium*, no. 1.

Bloch, P.C. and Oesterberg, T. (1989) *Land Tenure and Allocation Situation and Policy in Vietnam with Special Reference to the Forest Development Area (Vinh Phu, Hoang Lien Son and Ha Tuyen Provinces)*, Hanoi: Interforest, Swedish International Development Agency.

Bohle, H.G., Downing, T.E. and Watts, M.J. (1994) 'Climate change and social vulnerability: toward a sociology and geography of food insecurity', *Global Environmental Change* 4: 37–48.

Brady, C. (1993) 'South-East Asia: the Mekong River', in C. Thomas and D. Howlett (eds) *Resource Politics: Freshwater and Regional Relations*, Buckingham: Open University Press.

Brandon, C. and Ramankutty, R. (1993) *Toward an Environmental Strategy for Asia*, World Bank Discussion Paper 224, Washington DC: World Bank.

Bray, F. (1986) *The Rice Economies: Technology and Development in Asian Societies*, Oxford: Blackwell.

Brinkman, W.J. and Vo-Tong Xuan (1991) '*Melaleuca leucadendron*, a useful and versatile tree for acid sulphate soils and some other poor environments', *International Tree Crops Journal* 6: 261–74.

Brocheaux, P. (1995) *The Mekong Delta: Ecology, Economy, and Revolution, 1860–1960*, Monograph 12, Madison: Centre for Southeast Asian Studies, University of Wisconsin.

Bromley, D.W. (1989) 'Property relations and economic development: the other land reform', *World Development* 17: 867–77.

Bromley, D.W. (1991) *Environment and Economy: Property Rights and Public Policy*, Oxford: Blackwell.

Bromley, D. and Cernea, M. (1989) *The Management of Common Property Natural Resources*. World Bank Discussion Paper 57, Washington DC: World Bank.

Brookfield, H. and Byron, Y. (eds) (1993) *Southeast Asia's Environmental Future: the Search for Sustainability*, Tokyo: United Nations University Press.

Brookfield, H. and Padoch, C. (1994) 'Appreciating agrodiversity: a look at the dynamism and diversity of indigenous farming practices', *Environment* 36, 5: 6–11, 37–45.

Browder, G. (1998) 'Negotiating an international regime for water allocation in the Mekong River Basin', unpublished PhD dissertation, Stanford University.

Brown, F.Z. (1993) 'The economic development of Vietnam, Laos, and Cambodia', *Journal of Northeast Asian Studies* 12, 4: 3–21.

Brown, H., Himmelberger, J. and White, A. (1993) 'Development–environment inter-actions in the export of hazardous technologies: a comparative study of three multinational affiliates in developing countries', *Technological Forecasting and Social Change* 43: 125–55.

Brown, K. (1995) 'Medicinal plants, indigenous medicine and conservation of biodiversity in Ghana', in T. Swanson (ed.) *Intellectual Property and Biodiversity Conservation*, Cambridge: Cambridge University Press.

Brown, K. (1997) 'Sustainable utilisation: a grand illusion?', in R. Auty and K. Brown (eds) *Approaches to Sustainable Development*, London: Pinter.

Brown, K. and Muchagata, M. (1999) 'Forests and livelihoods of colonist farmers in Eastern Amazonia', in M. Palo and J. Uusivuori (eds) *World Forests: Society and Environment*, Dordrecht: Kluwer.

Bruzon, E. and Carton, P. (1929) *Le Climate de l'Indochine et les Typhon de la Mer de Chine, Deuxieme Partie, Les Typhons de la Mer de Chine*, Hanoi: Imprimerie d'Extrême-Orient.

Bryant, J. (1998) 'Communism, poverty and demographic change in North Vietnam', *Population and Development Review* 24: 235–69.

Bureau of Flood Control (1950) *Flood Damage and Flood Control Activities in Asia and the Far East*, Flood Control Series No. 1, Bureau of Flood Control, Bangkok: United Nations Economic Committee for Asia and the Far East.

Burgess, R., Carmona, K. and Kolstee, T. (1997) *The Challenge of Sustainable Cities*, London: Zed Books.

Burton, I., Kates, R. W. and White, G. F. (1993) *The Environment as Hazard, Second Edition*, New York and London: Guilford Press.

Callison, S. (1983) *Land to the Tiller in the Mekong Delta*, Lanham: University Press of America.

Cameron, D. (1994) *A FAO Statement on the Forest Situation in Vietnam*, Hanoi: Food and Agriculture Organisation.

Cameron, D. (ed.) (1995) 'Socio-economic and Environmental Monitoring Programme in the Forestry Development Area', Vietnam–Sweden Forestry Cooperation Programme, Hanoi: Forestry Research Centre.

Cantor, R. and Rayner, S. (1994) 'Changing perceptions of vulnerability', in R. Socolow, C. Andrews, F. Berkhout and V. Thomas (eds) *Industrial Ecology and Global Change*, Cambridge: Cambridge University Press.

Cao Van Sung (ed.) (1995) *Environment and Bioresources of Vietnam: Present Situation and Solutions*, Hanoi: Thê Giói.

Castella, J.C., Husson, O., Le Quoc Doanh and Ha Dinh Tuan (1999) 'Mise en oeuvre de l'approche ecoregionale dans les montagnes du bassin du fleuve rouge au Vietnam. Le projet systemes agraires de montagne', *Cahiers-de-la-Recherche-Developpement* No. 45: 114–34, Paris: L'Institut de Recherche pour le Developpement.

Center for Environment Research, Education and Development (1993) *Study on Environmental Problems and Management Challenges in Vietnam*, Report to Organisation for Economic Co-operation and Development, Hanoi: Center for Environment Research, Education and Development.

Challinor, D. (1973) 'Preface', in D. Challinor (ed.) *Some Environmental Considerations for the Mekong Basin Project*, New York: Southeast Asia Development Advisory Group of the Asia Society.

Chambers, R. (1989) 'Vulnerability, coping and policy', *IDS Bulletin* 20, 2: 1–7.

Chan, J.C.L. (1985) 'Tropical cyclone activity in the Northwest Pacific in relation to the El Niño/Southern Oscillation phenomenon', *Monthly Weather Review* 113: 599–606.

Chan, J.C.L. (1995) 'Prediction of annual tropical cyclone activity over the western North Pacific and the South China Sea', *International Journal of Climatology* 15: 1011–19.

Chang Yii Tan and Leong Yueh Kwong (1990) 'Industrial and the natural environment policies on pollution', Vol. 8 of UNIDO, *Policy Assessment of the Malaysian Industrial Policy Studies and the Industrial Master Plan*, Vienna: United Nations Industrial Development Organisation.

Chantalakhana, C. (1995) 'Sustainable agriculture: a choice for human survival', in P. Poungsomlee (ed.) *Strategies for Sustainable Agriculture and Rural Development*, Salaya: Mahidol University.

Chapman, E.C. and Daming, H. (1996) 'Downstream implications of China's dams on the Lancang Jiang (Upper Mekong) and their potential significance for greater regional cooperation, basin-wide', in B. Stensholt (ed.) *Development Dilemmas in the Mekong Subregion* (Workshop Proceedings, 1–2 October 1996), Clayton, Australia: Monash Asia Institute, Monash University.

Chooduangngern, S. (1996) 'The Mekong Basin Development Plan', in B. Stensholt (ed.) *Development Dilemmas in the Mekong Subregion* (Workshop Proceedings, 1–2 October 1996), Clayton, Australia: Monash Asia Institute, Monash University.

Choowaew, S. (1993) 'Inventory and management of wetlands in the Lower Mekong Basin', *Asian Journal of Environmental Management* 1, 2: 1–10.

Coase, R. (1960) 'The problem of social cost', *Journal of Law and Economics* 3: 1–44.

Coit, K. (1998) 'Housing policy and slum upgrading in the Ho Chi Minh City', *Habitat International* 22: 223–80

Cole, M., Rayner, A. and Bates, J. (1997) 'The environmental Kuznets curve: an empirical analysis', *Environment and Development Economics* 2: 401–16.

Common, M. (1995) *Sustainability and Policy: Limits to Economics*, Cambridge: Cambridge University Press.

Cooney, C.M. (1999) 'Study alludes to deformities from dioxin in Vietnam', *Environmental Science and Technology* 31: 12.

Copeland, B. (1994) 'International trade and the environment: policy reform in a polluted small open economy', *Journal of Environmental Economics and Management* 26: 44–65.

Copeland, B. and Taylor, S. (1994) 'North–south trade and the environment', *Quarterly Journal of Economics* 109: 755–87.

Crook, C. and Clapp, R.A. (1998) 'Is market-oriented forest conservation a contradiction in terms?', *Environmental Conservation* 25: 131–45.

Cutter, S.L. (1996) 'Vulnerability to environmental hazards', *Progress in Human Geography* 20: 529–39.

Daily, G.C. (1997) *Nature's Services: Societal Dependence on Natural Ecosystems*, Washington DC: Island Press.

Dao Duy Anh (1951) *Viet Nam Van Hoa Su Cuong [A Cultural History of Vietnam]*, Saigon: Xuat Ban Bot Phuong.

Dao The Tuan (1995) 'The peasant household economy and social change', in J.T. Kerkvliet and D.J. Porter (eds) *Vietnam's Rural Transformation*, Boulder: Westview.

Dasgupta, P.S. (1995) 'Population, poverty and the local environment', *Scientific American* February 1995, 26–31.

Davies, S. (1996) *Adaptable Livelihoods: Coping with Food Insecurity in the Malian Sahel*, London: Macmillan.

de Janvry, A., Sadoulet, E. and Thorbecke, E. (1993) 'State, market and civil organisations: new theories, new practices and their implications for rural development: introduction', *World Development* 21: 565–75.

de Melo, M., Denizer, C. and Gelb, A. (1996) 'Patterns of transition from plan to market', *World Bank Economic Review* 10: 397–424.

Dean, J. (1992) 'Trade and the environment: a survey of the literature', in P. Low (ed.) *International Trade and the Environment*, World Bank Discussion Paper 159, Washington DC: World Bank.

Dent, D. (1992) 'Reclamation of acid sulphate soils' *Advances in Soil Science* 17: 79–122.

Department for International Development (1998) *Eliminating World Poverty: A Challenge for the 21st Century*, London: HMSO.

Desloges, C. and Vu Van Me (1996) 'Principaux enjeux de l'allocation des terres foretieres au Vietnam', *Le Flamboyant* 37: 10–13.

Desloges, C. and Vu Van Me (1997) *Methodology for Participatory Land-use Planning and Forest Land Allocation*, Ministry of Agriculture and Rural Development and Department of Forestry, Hanoi: Food and Agriculture Organization.

Diaz, H.F. and Kiladis, G.N. (1992) 'Atmospheric teleconnections associated with the extreme phases of the Southern Oscillation', in H.F. Diaz and V. Markgraf (eds) *El Niño: Historical and Paleoclimatic Aspects of the Southern Oscillation*, Cambridge: Cambridge University Press.

Dierberg, F.E. and Kiattisimkul, W. (1996) 'Issues, impacts and implications of shrimp aquaculture in Thailand', *Environmental Management* 20: 649–66.

Dixon, J.A. and Lal, P.N. (1997) 'The management of coastal wetlands: economic analysis of combined ecologic-economic systems', in P. Dasgupta and K.G. Mäler (ed.) *The Environment and Emerging Development Issues*, Oxford: Clarendon.

Do Dinh Sam (1994) *Shifting Cultivation in Vietnam: its Social, Economic and Environmental Values Relative to Alternative Land Use*, Forestry and Land Use Series Paper 3, London: International Institute for Environment and Development.

Doan Diem (1997) *Participatory land use planning and forest land allocation in Vietnam*, Hanoi: Ministry of Agriculture and Rural Development, Department of Forestry, and Food and Agriculture Organization.

Dollar, D. and Glewwe, P. (1998) 'Poverty and inequality in the early reform period', in D. Dollar, P. Glewwe and J. Litvack (eds) *Household Welfare and Vietnam's Transition*, Washington DC: World Bank.

Drakakis-Smith, D. (1995) 'Third World cities: sustainable urban development I', *Urban Studies* 32: 659–77.

Drakakis-Smith, D. (1996) 'Third World cities: sustainable urban development II – population, labour and poverty', *Urban Studies* 33: 673–70.

Drakakis-Smith, D. (1997) 'Third World cities: sustainable urban development III – basic needs and human rights', *Urban Studies* 34: 797–823.

Drakakis-Smith, D. and Dixon, C. (1997) 'Sustainable urbanisation in Vietnam', *Geoforum* 28: 21–38.

Dubois, O. and Morrison, E. (1998) *Sustainable Livelihoods in Upland Viet Nam: Land Allocation and Beyond*, Forest and Land Use Series, No. 14, London: International Institute for Environment and Development.

Dudgeon, D. (1992) 'Endangered ecosystems: a review of the conservation status of tropical Asian rivers', *Hydrobiologia* 248: 167–91.

Duiker, W.J. (1976) *The Rise of Nationalism in Vietnam, 1900–1941*, Ithica: Cornell University Press.

Dumarest, A. (1935) *La Formation des classes Sociales en Pays Annimities*, Lyon: P. Ferreal.

Duong Lien Chau (2000) 'Lessons from Severe Tropical Storm Linda', in P.M. Kelly, S. Granich and Nguyen Huu Ninh (eds) *Workshop Report, The Impact of El Niño and La Niña on Southeast Asia, 21st–23rd February 2000, Hanoi, Vietnam*, Hanoi: Center for Environment Research Education and Development.

Duong Van Ni (1997) 'Sustainable development in the buffer zones of Tram Chim', in R.J. Safford, Duong Van Ni, E. Maltby and Vo-Tong Xuan (eds) *Towards Sustainable Management of Tram Chim National Reserve, Vietnam*, London: Royal Holloway Institute for Environmental Research.

Duong Van Ni and Vo-Tong Xuan (1998) 'Rural development in a remote area: a case study at Hoa An, Phung Hiep, Can Tho, Vietnam', in *Proceedings of a Conference on Development of Farming Systems in the Mekong Delta of Vietnam, 1997*, Kyoto: Japan International Research Centre for Agriculture Sciences, Cuu Long Rice Research Institute and Farming Systems Research and Development Institute.

Eccleston, D. and Potter, D. (1996) 'Environmental NGOs and different political contexts in Southeast Asia: Malaysia, Indonesia, Vietnam', in M. Parnwell and R. Bryant (eds) *Environmental Change in Southeast Asia*, London: Routledge.

Economic and Social Commission for Asia and the Pacific (1991) *Energy Policy Implications of the Climatic Effects of Fossil Fuel Use in the Asia-Pacific Region*, Bangkok: United Nations.

Economic and Social Commission for Asia and the Pacific and United Nations Centre on Transnational Corporations (1988) *Transnational Corporations and Environmental Management in Asian and Pacific Developing Countries*, Publication Series B 13, Bangkok: Economic and Social Commission for Asia and the Pacific and United Nations Centre on Transnational Corporations.

Economic Intelligence Unit (1996) *Vietnam, Country Profile 1996–7*, London: Economic Intelligence Unit.

Economic Intelligence Unit (1998) *Vietnam, Country Report*, London: Economic Intelligence Unit.

*Economist, The* (1997) 'Rural descent, Vietnam', *The Economist* 344, 8034: 83–4.

Eeuwes, J.G.C.M. (1995) 'Upland use in Vietnam: Institutions and Regulations of Forest Management Systems: Black Thai Co-operatives in Son La Province', Wageningen, Netherlands: Wageningen Agricultural University.

Ekins, P. (1997) 'The Kuznets curve for the environment and economic growth: examining the evidence', *Environment and Planning A* 29: 805–30.

Ellis, F. (1998) 'Household strategies and rural livelihood diversification', *Journal of Development Studies* 35, 1: 1–38.

Emergency Management Australia (1995) *Flood Warning: An Australian Guide*, Canberra: Emergency Management Australia.

England, S.B. and Kammen, D.M. (1993) 'Energy resources and development in Vietnam', *Annual Reviews of Energy and Environment* 18: 137–67.

Ensor, T. and San, P. B. (1996) 'Access and payment for health care: the poor of northern Vietnam', *International Journal of Health Planning and Management* 11: 69–83.

Eriksson, R. (1992) 'Water quality monitoring strategies', in Proceedings of the Water Quality Monitoring Workshop (5–9 October 1992, Phuket, Thailand), Bangkok: Environment Unit, Mekong Secretariat.

Ewel, K.C., Bourgeois, J.A., Cole, T.G. and Zheng, S. (1998a) 'Variation in environmental characteristics and vegetation in high rainfall mangrove forests', *Global Ecology and Biogeography Letters* 7: 49–56.

Ewel, K.C., Twilley, R.R. and Ong, J.E. (1998b) 'Different kinds of mangrove forests provide different goods and services', *Global Ecology and Biogeography Letters* 7: 83–94.

Farnsworth, E.J. and Ellison, A.E. (1997) 'The global conservation status of mangroves', *Ambio* 26: 328–34.

Ferruntino, M. (1995) 'International trade, environmental quality and public policy', Office of Economics Working Paper, Washington DC: US International Trade Commission.

Fforde, A. (1983) 'The historical background to agricultural collectivisation in North Vietnam: the changing role of corporate economic power', Discussion Paper 148, Department of Economics, Birkbeck College, London.

Fforde, A. (1988) 'Specific aspects of the collectivisation of wet-rice cultivation: Vietnamese experience', in J.C. Brada, and K.E. Wädekin (eds) *Socialist Agriculture in Transition*, Boulder: Westview.

Fforde, A. (1990) 'Vietnamese agriculture: changing property rights in a mature collectivised agriculture', in K.E. Wädekin (ed.) *Communist Agriculture: Farming in the Far East and Cuba*, London: Routledge.

Fforde, A. (1993) 'The political economy of "reform" in Vietnam – some reflections', in B. Ljunggren (ed.) *The Challenge of Reform in Indochina*, Cambridge: Harvard University Press.

Fforde, A. (1998) 'Vietnam – culture and economy: died-in-the-wool tigers?', ANU Vietnam Update, Canberra: Australian National University.

Fforde, A. and de Vylder, S. (1996) *From Plan to Market: the Economic Transition in Vietnam*, Boulder: Westview.

Fiedler, P.L., Pickett, S.T.A., Parker, V.T. (1992) 'The paradigm shift in ecology and its implications for conservation', in P. Fielder and S. Jain (eds) *Conservation Biology: the Theory and Practice of Nature Conservation, Preservation, and Management*, New York: Chapman and Hall.

Field, C.B., Osborn, J.G., Hoffman, L.L., Polsenberg, J.F., Ackerley, D.D., Berry, J.A., Bjorkman, O., Held, A., Matson, P.A. and Mooney, H.A. (1998) 'Mangrove biodiversity and ecosystem function', *Global Ecology and Biogeography Letters* 7: 3–14.

Fingleton, J.S. (1990) *Report on Forest Policy and Legislation*, Forest Sector Review, Project FAO/VIE/88/037, Tropical Forest Action Plan, Hanoi: Ministry of Forestry.

Fisher, M. (ed.) (1996) *Proceedings of the Seminar on Approaches to Social Forestry in Vietnam, Hanoi, September 1995*, Hanoi: Ministry of Agriculture and Rural Development.

Fleischhauer, E. and Eger, H. (1998) 'Can sustainable land use be achieved? An introductory view on scientific and political issues', in H.-P. Blum, H. Eger, E. Fleischhauer, A. Hebel, C. Reij and K.G. Steiner (eds) *Towards Sustainable Land Use: Furthering Cooperation between People and Institutions*, Reiskirchen: Catena-Verlag.

Folke, C. and Kautsky, N. (1992) 'Aquaculture with its environment: prospects for sustainability', *Ocean and Coastal Management* 17: 5–24.

Food and Agriculture Organization (1993) *Forest Resources Assessment 1990*, FAO Forestry Paper 112, Rome: Food and Agriculture Organization.

Food and Agriculture Organization (1996) *Workshop on Forestry and Agriculture on Sloping Lands of Northern Vietnam, Phu Ninh, Vinh Phu, Vietnam. 24–28 June 1996*, FAO Field Document No. 8, Hanoi: Food and Agriculture Organization.

Food and Agriculture Organization and International Institute of Rural Reconstruction (1995) *Resource Management for Upland Areas in Southern Asia. An Information Kit*, Bangkok and Sillang: Food and Agriculture Organization Regional Office for Asia and the Pacific and International Institute of Rural Reconstruction.

Forsyth, T. (1997) 'Industrialisation in Vietnam: social change and environment in transitional developing countries', in R.M. Auty and K. Brown (eds) *Approaches to Sustainable Development*, London: Pinter.

Forsyth, T. (1999) *International Investment and Climate Change: Energy Technologies for Developing Countries*, London: Earthscan.

Fu, C.B., Kim, J.-W. and Zhao, Z.C. (1998) 'Preliminary assessment of impacts of global change on Asia', in J. Galloway and J. Melillo (eds) *Asian Change in the Context of Global Change*, Cambridge: Cambridge University Press.

Gadgil, M., Berkes, F. and Folke, C. (1993) 'Indigenous knowledge for biodiversity conservation', *Ambio* 22: 151–6.

Galloway, J. and Melillo, J. (eds) (1998) *Asian Change in the Context of Global Change*, Cambridge: Cambridge University Press.

Ghil, M. and Jiang. N. (1998) 'Recent forecast skill for the El Niño Southern Oscillation', *Geophysical Research Letters* 25: 171–4.

Giddens, A. (1996) 'Affluence, poverty and the idea of a post-scarcity society', *Development and Change* 27: 365–77.

Gill, I. (1996) 'Mekong: dismantling the barriers', *ADB Review*, July–August 1996: 3–8.

Giradet, H. (1996) *The Gaia Atlas of Cities: New Directions for Sustainable Living*, London: Gaia Books.

GKW-Safege (1995) 'Ho Chi Minh City Water Supply Master Plan', GKW-Safege Unpublished Report to Asian Development Bank, Manila.

Gladwin, T. (1987) 'Environment, development and multinational enterprise', in C. Pearson (ed.) *Multinational Corporations, Environment, and the Third World: Business Matters*, Durham: Duke University Press.

Glantz, M.H. (1996) *Currents of Change: El Niño's Impact on Climate and Society*, Cambridge: Cambridge University Press.

Glantz, M.H., Katz, R.W. and Nicholls, N. (eds) (1991) *Teleconnections Linking Worldwide Climate Anomalies*, Cambridge: Cambridge University Press.

Goldman, M. (ed.) (1998) *Privatising Nature: Political Struggles for the Global Commons*, New Brunswick: Rutgers University Press.

Goldsmith, E. and Hildyard, N. (1984) *Social and Environmental Impacts of Large Dams, Volume 1*, San Francisco: Sierra Club Books.

Gourou, P. (1940) *L'Utilisation de Sol en Indochine*, Paris: Centre d'Etudes de Politique Etrangere.

Gourou, P. (1955) *The Peasants of the Tonkin Delta: A Study of Human Geography*, R.R. Miller (translator), New Haven: Human Relations Area Files. First published as *Les Paysans di Deltas Tonkinois: Etude de Geographie Humaine*, Paris: Editions d'Art et d'Histoire, 1936.

Gouvernement Général de l'Indochine (1930) *Dragages de Cochinchine*. Canal Rachgia-Hatien, Saigon: Inspection General de Travaux Publics.

Grabher, G. and Stark, D. (1998) 'Organising diversity: evolutionary theory, network analysis and post-socialism', in J. Pickles and A. Smith (eds) *Theorising Transition: The Political Economy of Post-Communist Transformations*, London: Routledge.

Granich, S.L.V., Kelly, P.M. and Nguyen Huu Ninh (eds) (1993) *Global Warming and Vietnam*, Norwich, London and Hanoi: University of East Anglia, International Institute for Environment and Development and Centre for Environment Research Education and Development. In English and Vietnamese. Also available at http://www.cru.uea.ac.uk/tiempo/floor0/briefing/vietnam/.

Grossman, G. (1994) 'Pollution and growth: What do we know?', in I. Goldin and L.A. Winters (eds) *The Economics of Sustainable Development*, Cambridge: Cambridge University Press.

Grossman, G. and Krueger, A. (1992) *Environmental Impacts of a North American Free Trade Agreement*, Working Paper No. 3914, Cambridge: National Bureau of Economic Research.

Grubb, M., Vrolijk, C. and Brack, D. (1999) *The Kyoto Protocol: A Guide and Assessment*, London: Earthscan.

Gunderson, L., Holling, C.S. and Light, S. (eds) (1995) *Barriers and Bridges to the Renewal of Ecosystems and Institutions*, New York: Columbia University Press.

Gunderson, L.H., Holling, C.S., Pritchard, L. and Peterson, G.D. (1997) *Resilience in Ecosystems, Institutions and Societies*, Discussion Paper No. 95, Stockholm: Beijer International Institute of Ecological Economics.

Gutekunst, C. (1998) *Ökonomische Bewertung von Erosionsschutzmaßnahmen in der Provinz Bac Kan, Nordvietnam (Economic Assessment of Erosion Control Measures in Bac Kan Province, Northern Vietnam)*, unpublished MSc Thesis, Hohenheim University, Germany.

Ha Nghiep (1993) 'The challenge facing Vietnam', in S.L.V. Granich, P.M. Kelly and Nguyen Huu Ninh (eds) *Global Warming and Vietnam*, Norwich, London and Hanoi: University of East Anglia, International Institute for Environment and Development and Center for Environment Research Education and Development. In English and Vietnamese. Also available at http://www.cru.uea.ac.uk/tiempo/floor0/briefing/vietnam/.

Hanhart, K. and Duong Van Ni (1993) 'Water management on rice fields at Hoa An, Mekong Delta, Vietnam', in D.L. Dent and M.E.F. van Mensvoort (eds) *Selected Papers of the Ho Chi Minh City Symposium on Acid Sulphate Soils*, Wageningen: International Institute for Land Reclamation and Improvement.

Hanhart, K., Duong Van Ni, Bakker, N., Bil, F., Postma, I. and van Mensvoort, M.E.F. (1997) 'Surface water management under varying drainage conditions for rice on an acid sulphate soil in the Mekong Delta, Vietnam', *Agricultural Water Management* 33: 99–116.

Hanna, S. and Munasinghe, M. (eds) (1995) *Property Rights and the Environment: Social and Ecological Issues*, Washiangton, DC: Beijer International Institute of Ecological Economics and World Bank.

Hansen, S. (1994) Population: its challenge to economic and social scientists', *International Social Science Journal* 141: 331–42.

Hansen, M. (1999) *Environmental Management in Transnational Corporations in Asia: Does Foreign Ownership Make a Difference? Preliminary Results of a Survey of Environmental Management Practices in 154 TNCs*, UNCTAD/CBS Project, Cross Border Environmental Management in Transnational Corporations, Occasional Paper No. 11, Copenhagen: Copenhagen Business School.

Hardin, G. (1968) 'The tragedy of the commons', *Science* 162: 1243–8.

Hartmann, B. (1998) 'Population, environment and security: a new trinity', *Environment and Urbanization* 10, 2: 113–27.

Hart-Landsberg, M. and Burkett, P. (1998) 'Contradictions of capitalist industrialisation in East Asia: a critique of the "flying geese" theories of development', *Economic Geography* 74: 87–110.

Harvey, D. (1996) *Justice, Nature and the Geography of Difference*, Oxford: Blackwell.

Hatfield Consultants (1998) *Preliminary Assessment of Environmental Impacts Related to Spraying of Agent Orange Herbicide During the Viet Nam War*, report on CD-ROM, available from Hatfield Consultants Ltd, No. 201 1571 Bellevue Avenue, West Vancouver, BC V7V 1A6, Canada.

Hatfield Group (1999) *Studying the Effects of Agent Orange in Vietnam*, report on CD-ROM, CD-ROM, available from Hatfield Consultants Ltd, No. 201 1571 Bellevue Avenue, West Vancouver, BC V7V 1A6, Canada.

Hayami, Y. (1994) 'Strategies for the reform of land policy relations', in R. Barker (ed.) *Agricultural Policy Analysis for Transition to a Market-Oriented Economy in Vietnam: Selected Issues*. FAO Economic and Social Development Paper 123, Rome: Food and Agriculture Organization.

Hayami, Y. and Ruttan, V.W. (1985) *Agricultural Development: an International Perspective, Second Edition*, Baltimore: Johns Hopkins University Press.

Held, D., McGrew, A., Goldblatt, D. and Perraton, J. (1999) *Global Transformations: Politics, Economics and Culture*, Cambridge: Polity Press.

Hettige, H., Huq, M., Pargal, S. and Wheeler, D. (1995) *Determinants of Pollution Abatement in Developing Countries: Evidence from South and Southeast Asia*, Washington DC: World Bank.

Hettige, H., Mani, M. and Wheeler, D. (1998) *Industrial Pollution in Economic Development: Kuznets Revisited*, Development Research Group, Policy Research Working Paper 1876, Washington DC: World Bank.

Hewitt, K. (1997) *Regions of Risk: a Geographical Introduction to Disasters*, Harlow: Longman.

Hiebert, M. (1992) 'New landed gentry', *Far Eastern Economic Review*, 3 December: 61–3.

Hiebert, M. (1994) 'Stuck at the bottom', *Far Eastern Economic Review*, 13 January: 70–71.

Hill, H. (1996) *The Indonesian Economy since 1966*, Cambridge: Cambridge University Press.

Hill, M.T. and Hill, S.A. (1994) 'Fisheries ecology and hydropower in the Mekong River: an evaluation of run-of-the-river projects', Bangkok: Mekong Secretariat.

Hines, D. (1995) *Financial Viability of Smallholder Reforestation in Vietnam*, Hanoi: Ministry of Forestry, Food and Agriculture Organization and World Food Programme.

Hinton, P. (1996) 'Is it possible to "manage" a river? Reflections from the Mekong', in B. Stensholt (ed.) *Development Dilemmas in the Mekong Subregion*, Clayton: Monash Asia Institute, Monash University.

Hirsch, P. (1992) *Social and Environmental Implications of Resource Development in Vietnam: the Case of Hoa Binh Reservoir*, Occasional Paper 17, Research Institute for Asia and the Pacific, University of Sydney, Sydney: Research Institute for Asia and the Pacific.

Hirsch, P. (1997) 'Seeking culprits: ethnicity and resource conflict', *Watershed* 3, 1: 25–8.

Hirsch, P. and Nguyen Viet Thinh (1996) 'Implications of economic reform in Vietnam: agrarian and environmental change in Hien Luong', *Australian Geographer* 27: 165–83.

Hirsch, P. and Warren, C. (eds) (1998) *The Politics of Environment in Southeast Asia*, London: Routledge.

Ho Tai Hue Tam (1985) *Religion in Vietnam: A World of Gods and Spirits in Vietnam: Essay on History, Culture and Society*, New York: Asia Society.

Hoang Cao Trai (1998) Forest land allocation and applied informatics in managing forest and forest allocated land in Thanh Hoa Province, in *Proceedings of the National Workshop on Participatory Land Use Planning and Forest Land Allocation*, Hanoi: Agriculture Publishing House.

Hoang Van Thoi (1999) 'Rational exploitation, maintenance and utilisation of some species in Ca Mau Mangroves', in Phan Nguyen Hong (ed.) *Sustainable and Economically Efficient Utilization of Natural Resources in Mangrove Ecosystem*, Proceedings of National Workshop, Nha Trang City, Vietnam, 1–3 November 1998, Hanoi: Mangrove Ecosystem Research Division.

Hodgkin, T. (1981) *Vietnam. The Revolutionary Path*, London: Macmillan.

Holling, C.S., Schindler, D.W., Walker, B.W. and Roughgarden, J. (1995) 'Biodiversity in the functioning of ecosystems: an ecological synthesis', in C. Perrings, K.G. Mäler, C. Folke, C.S. Holling and B.O. Jansson (eds) *Biodiversity Loss: Economic and Ecological Issues*, Cambridge: Cambridge University Press.

Holling, C.S., Berkes, F. and Folke, C. (1998) 'Science, sustainability and resource management', in F. Berkes and C. Folke (eds) *Linking Social and Ecological Systems: Institutional Learning for Resilience*, Cambridge: Cambridge University Press.

Hopkins, S. (1995) 'Report of a needs assessment: the situation of poor people in Long Vinh and Long Khanh communes, Duyen Hai District, Tra Vinh province', Hanoi: Oxfam.

Hori, H. (1993) 'Development of the Mekong River Basin: its problems and future prospects', *Water International* 18: 110–15.

Houghton, G. and Hunter, C. (1994) *Sustainable Cities*, London: Kingsley.

Houghton, J.T., Jenkins, G.J. and Ephraums, J.J. (eds) (1990) *Climate Change: The IPCC Scientific Assessment*, Cambridge: Cambridge University Press.

Houghton, J.T., Meira Filho, L.G., Callander, B.A., Harris, N., Kattenberg, A. and Maskell, K. (eds) (1996) *Climate Change 1995: the Science of Climate Change*, Cambridge: Cambridge University Press.

Houtart, F. and Lemercinier, G. (1984) *Hai Van: Life in a Vietnamese Commune*, London: Zed Books.

Huddle, F.P. (1972) *The Mekong Project: Opportunities and Problems of Regionalism*, Report prepared for the Subcommittee on National Security Policy and Scientific Developments of the Committee on Foreign Affairs, US House of Representatives, Washington DC: US Government Printing Office.

Hulme, M. and Brown, O. (1998) 'Portraying climate scenario uncertainties in relation to tolerable regional climate change', *Climate Research* 10: 1–14.

Hy Van Luong (1992) *Revolution in the Village: Tradition and Transformation in North Vietnam 1925–1988*, Honolulu: University of Hawaii Press.

Hy Van Luong (1993) 'Economic reform and the intensification of rituals in two north Vietnamese villages 1980–1990', in B. Ljunggren (ed.) *The Challenge of Reform in Indochina*, Cambridge: Harvard University Press.

Hy Van Luong and Unger, J. (1998) 'Wealth, power and poverty in the transition to market economies: the process of socio-economic differentiation in rural China and northern Vietnam', *China Journal* 40: 61–93.

Imamura, F. and Dang Van To (1997) 'Flood and typhoon disasters in Viet Nam in the half century since 1950', *Natural Hazards* 15: 71–87.

Institute of Hydrology, Asian Institute of Technology (1982) *Lower Mekong Basin: Water Balance Study, Phase I Report*, Report prepared by the Overseas Development Administration for the Interim Mekong Committee, Bangkok: Asian Institute of Technology.

Interim Mekong Committee (1992) 'Fisheries in the Lower Mekong Basin (Review of the Fishery Sector in the Lower Mekong Basin): Summary Report', Bangkok: Interim Mekong Committee.

Irvin, G. (1995) 'Vietnam: assessing the achievements of Doi Moi', *Journal of Development Studies* 31: 725–50.

Irvin, G. (1996) 'Emerging issues in Vietnam: privatisation, equality and sustainable growth', *European Journal of Development Research* 8: 178–99.

Irvin, G. (1997) 'Economic transition in Vietnam: will it hurt the poor?', paper presented at the EUROVIET–III Conference, Amsterdam, The Netherlands, 2–4 July 1997.

Jacobs, J.W. (1996) 'Adjusting to climate change in the Lower Mekong', *Global Environmental Change* 6: 7–22.

Jamieson, N. (1993) *Understanding Vietnam*, Berkeley: University of California Press.

Jamieson, N. and Le Trong Cuc and A.T. Rambo (1998) *The Development Crisis in Vietnam's Mountains*, East–West Centre Special Report 6, Honolulu: East–West Centre.

Jenkins, R. (1989) 'Comparing foreign subsidiaries and local firms in LDCs: theoretical issues and empirical evidence', *Journal of Development Studies* 26: 205–28.

Johansson, A., Hoa, H.T., Lap, N.T., Diwan, V. and Eriksson, B. (1996) Population policies and reproductive patterns in Vietnam', *The Lancet* 347: 1529–32.

Jones D.L., Prabowo, A.M. and Kochian, L.V. (1996) 'Aluminium–organic acid interactions in acid soils. II. Influence of solid phase sorption on organic acid–Al complexation and Al rhizotoxicity', *Plant and Soil* 182: 229–37.

Jonsson, J. and Nguyen Hai Nam (1998) 'Land Management Component in Vietnam–Sweden Mountain Rural Development Programme (1996–2000)', in *Proceedings of the National Workshop on Participatory Land Use Planning and Forest Land Allocation*, Hanoi: Agriculture Publishing House.

Kalt, J. (1988) 'The Impact of Domestic Environmental Regulatory Policies on US International Competitiveness', in M. Spence and H. Hazard (eds) *International Competitiveness*, Cambridge: Ballinger.

Kampe, K. (1997) 'Introduction: indigenous peoples of Southeast Asia', in D. McCaskill and K. Kampe (eds) *Development or Domestication? Indigenous Peoples of Southeast Asia*, Chiang Mai: Silkworm Books.

Kasperson, J.X., Kasperson, R.E. and Turner, B.L. (1996) 'Regions at risk: exploring environmental criticality', *Environment* 38, 10: 4–15, 26–9.

Kasperson, R.E. (1992) 'The social amplification of risk: progress in developing an integrative framework', in S. Krimsky and D. Goldring (eds) *Social Theories of Risk*, Westport: Praeger.

Kay, J.J. (1991) 'A nonequilibrium thermodynamic framework for discussing ecosystem integrity', *Environmental Management* 15: 483–95.

Kelly, P.M. (1996) 'Blue Revolution or red herring? Fish farming and development discourse in the Philippines', *Asia Pacific Viewpoint* 37: 39–57.

Kelly, P.M. (2000) 'Towards a sustainable response to climate change', in M. Huxham and D. Sumner (eds) *Science and Environmental Decision-making*, Harlow: Pearson Education.

Kelly, P.M. and Adger, W.N. (2000) 'Theory and practice in assessing vulnerability to climate change and facilitating adaptation', *Climatic Change* 47: 325–52.

Kelly, P.M., Granich, S.L.V. and Secrett, C.M. (1994) 'Global warming: responding to an uncertain future', *Asia Pacific Journal on Environment and Development* 1: 28–45.

Kelly, P.M., Granich, S. and Nguyen Huu Ninh (2000) *Workshop Report, The Impact of El Niño and La Niña on Southeast Asia, 21st–23rd February 2000, Hanoi, Vietnam*, Hanoi: Center for Environment Research Education and Development.

Keohane, R. and Ostrom, E. (1994) 'Local commons and global interdependence – heterogeneity and co-operation in two domains: introduction', *Journal of Theoretical Politics* 6: 403–28.

Kerkvliet, B.J.T. (1995a) 'Rural society and state relations', in B.J.T. Kerkvliet and D.J. Porter (eds) *Vietnam's Rural Transformation*, Boulder: Westview.

Kerkvliet, B.J.T. (1995b) 'Village state relations in Vietnam: the effect of everyday politics on decollectivisation', *Journal of Asian Studies* 54: 396–418.

Kerkvliet, B.J.T. and Porter, D.J. (1995) 'Rural Vietnam in rural Asia', in B.J.T. Kerkvliet and D.J. Porter (eds) *Vietnam's Rural Transformation*, Boulder: Westview.

Kerkvliet, B.J. and Selden, M. (1998) 'Agrarian transformations in China and Vietnam', *China Journal*, 40: 37–58.

Keys, C. (1997) 'The total flood warning system: concept and practice', in J. Handmer (ed.) *Flood Warning: Issues and Practice in Total System Design*, London: Flood Hazard Research Centre, Middlesex University.

Khoa, L.V. and Roth-Nelson, W. (1994) 'Sustainable wetland use in the Mekong River delta of Vietnam', in W.J. Mitsch (ed.) *Global Wetlands: Old World and New*, Amsterdam: Elsevier,

Kilgour, A. (2000) A Study of Low-income Households and their Perceptions of Environmental Problems During Rapid Urbanisation in Hanoi, Vietnam. Unpublished Ph.D thesis, University of Liverpool.

Kim, K.S. (1997) 'Income distribution and poverty: an interregional comparison', *World Development* 25: 1909–24.

Kim, Y.C. (ed.) (1995) *The South East Asian Miracle*, New Brunswick: Transaction Publishers.

King, V.T. (ed.) (1998) *Environmental Challenges in Southeast Asia*, London: Curzon.

Kolko, G. (1995) 'Vietnam since 1975', *Journal of Contemporary Asia* 25: 1–49.

Kolko, G. (1997) *Vietnam: Anatomy of a Peace*, London: Routledge.

Können, G.P., Jones, P.D., Kaltofen, M.H. and Allan, R.J. (1998) 'Extensions of the Southern Oscillation index using early Indonesian and Tahitian meteorological readings', *Journal of Climate* 11: 2325–39.

Koppel, B. (1990) 'Old water, new rice: field notes on agrarian change and Vietnam's *doi moi*', *The Rural Sociologist*, Fall 1990: 2–13.

Krugman, P. (1994) 'The myth of the Asian miracle', *Foreign Affairs* 73, 6: 62–78.

Krugman, P. and Venables, A.J. (1995) 'Globalisation and the inequality of nations', *Quarterly Journal of Economics* 110: 857–80.

Kunstadter, P. and Chapman, E.C. (1978) 'Problems of shifting cultivation and economic development in northern Thailand', in P. Kunstadter, E.C. Chapman and S. Sabhasari (eds) *Farmers in the Forest*, Honolulu: University of Hawaii Press.

Lacoste, Y. (1973) 'An illustration of geographical warfare: bombing the dikes on the Red River, North Vietnam', *Antipode* 5: 1–13.

Lang, C. (1995) 'The legacy of savage development: colonisation of Vietnam's Central Highlands', *Watershed* 1, 2: 36–44.

Lang, C.R. (1996) 'Problems in the making: a critique of Vietnam's tropical forestry action plan', in M.J.G. Parnwell and R.L. Bryant (eds) *Environmental Change in South-East Asia*, London: Routledge.

Le Cong Kiet (1994) 'Native freshwater vegetation communities in the Mekong Delta', *International Journal of Ecology and Environmental Sciences* 20: 55–71.

Le Dien Duc (1989) 'Socialist Republic of Vietnam', in D.A. Scott (ed.) *A Directory of Asian Wetlands*, Gland: World Conservation Union.

Le Dien Duc (1991) *Proceedings of the Workshop on* Melaleuca *rehabilitation and Management, Long Xuyen, An Giang, 14–18 May 1991*, Hanoi: Agricultural Publishing House.

Le Dien Duc (1993) 'Rehabilitation of the *Melaleuca* floodplain forests in the Mekong Delta, Vietnam', in T.J. Davies (ed.) *Towards the Wise Use of Wetlands*, Gland: Ramsar Convention Bureau.

Le Quang Minh, To Phuc Tuong, van Mensvoort, M.E.F. and Bouma, J. (1997) 'Contamination of surface water as affected by land use in acid sulphate soils in the Mekong River Delta, Vietnam', *Agriculture, Ecosystems and Environment* 61: 19–27.

Le Quang Minh, To Phuc Tuong, van Mensvoort, M.E.F. and Bouma, J. (1998) 'Soil and water table management effects on aluminium dynamics in an acid sulphate soil in Vietnam', *Agriculture, Ecosystems and Environment* 68: 255–62.

Le Quang Tri, Nguyen Van Nhan, Huizing, H.G.J. and van Mensvoort, M.E.F (1993) 'Present land uses as basis for land evaluation in two Mekong Delta districts', in D.L. Dent and M.E.F. van Mensvoort (eds) *Selected Papers on the Ho Chi Minh City Symposium on Acid Sulphate Soils*, Publication 53, Wageningen: International Institute for Land Reclamation and Improvement.

Le Thac Can and Vo Quy (1994) 'Vietnam: environmental issues and possible solutions', *Asian Journal of Environmental Management* 2, 2: 69–77.

Le Thac Can, Nguyen Huu Ninh, Hoang Minh Hien and Nguyen Lam Hoe (1994) *Natural Resources and Environment in Vietnam – Issues and Solutions*, Hanoi: Center for Environment Research Education and Development.

Le Thanh Khoi (1955) *Histoire du Vietnam*, Paris: Les editions de Minuit.

Le Trong Cuc (1988) *Agro-forestry Practices in Vietnam*, Environmental and Policy Institute Working Paper No. 9, Program on Environment, East–West Centre, Honolulu: East–West Center.

Le Trong Cuc and Sikor, T. (1996) 'National agricultural development policy and rural organisation', in Le Trong Cuc, A.T. Rambo, K. Fahrney, Tran Duc Vien, J. Romm, and Dang Thi Sy, (eds) *Red Books, Green Hills: The Impact of Economic Reform on Restoration Ecology in the Midlands of Northern Vietnam*, Honolulu: East–West Centre.

Le Trong Cuc, Gilloghy, K. and Rambo, A.T. (1990) *Agroecosystems of the Midlands of Northern Vietnam*, Occasional Paper No. 12, Program on Environment, East–West Center: Honolulu: East–West Center.

Le Trong Cuc and Rambo., A.T. with Gillogly, K. (1993) *Too Many People: Too Little Land. The Human Ecology of a Wet Rice-Growing Village in the Red River Delta of Vietnam*, Occasional Paper No. 15, Program on Environment, East–West Center, Honolulu: East–West Center.

Le Trong Cuc, Sikor, T. and Rucker, M. (1996) 'Village level implementation of economic reform policies in Doan Hung and Thanh Hoa Districts', in Le Trong Cuc, A.T. Rambo, K. Fahrney, Tran Duc Vien, J. Romm and Dang Thi Sy (eds) *Red Books, Green Hills: The Impact of Economic Reform on Restoration Ecology in the Midlands of Northern Vietnam*, Honolulu: East–West Centre.

Lee, H. and Roland-Holst, D. (1993) *International Trade and the Transfer of Environmental Costs and Benefits*, Technical Paper No. 91, OECD Development Centre, Paris: Organisation for Economic Cooperation and Development.

Lee, K. (1993) *Compass and Gyroscope: Integrating Science and Politics for the Environment*, Washington DC: Island Press.

Leonard, J. (1988) *Pollution and the Struggle for the World Product*, Cambridge: Cambridge University Press.

Liljestrom, R., Lindskog, E., Nguyen Van Anh and Vuong Xuan Tinh (1998) *Profit and Poverty in Rural Vietnam: Winners and Losers of a Dismantled Revolution*, London: Curzon.

Lindahl-Kiessling, K. and Landberg, H. (eds) (1994) *Population, Economic Development and the Environment: The Making of Our Common Future*, Oxford: Clarendon.

Lipschutz, R.D. (1996) *Global Civil Society and Global Environmental Governance: the Politics of Nature from Place to Planet*, Albany: SUNY Press.

Lipton, M. (1977) *Why Poor People Stay Poor: A Study of Urban Bias in World Development*, London: Temple Smith.

Ljunggren, B. (1993) 'Market economies under communist regimes', in B. Ljunggren (ed.) *The Challenge of Reform in Indochina*, Cambridge: Cambridge University Press.

Locke, C., Adger, W.N. and Kelly, P.M. (2000) 'Changing places: the impacts of migration on social resilience', *Environment* 42, 7: 24–35.

Lohmann, L. (1991) 'Engineers move in on the Mekong', *New Scientist* 13 July: 44–7.

Low, P. (ed.) (1992) *International Trade and the Environment*, World Bank Discussion Papers, 159, Washington DC: World Bank.

Lucas, R., Wheeler, D. and Hettige, H. (1992) 'Economic development, environmental regulation and the international migration of toxic industrial pollution: 1960–1988', in P. Low (ed.) (1992) *International Trade and the Environment*, World Bank Discussion Paper 159, Washington DC: World Bank.

Lugo, A.E. and Snedaker, S.C. (1974) 'The ecology of mangroves', *Annual Review of Ecology and Systematics* 5: 39–64.

Luong Duc Thiep (1944) *Xa Hoi Viet nam* [*Vietnamese Society*], Hanoi: Han Thuyen.

Mabey, N. and McNally, R. (1999) *Foreign Direct Investment and the Environment: From Pollution Havens to Sustainable Development*, London: World Wide Fund for Nature-UK.

Malarney, S.K. (1997) 'Culture, virtue and political transformation in contemporary northern Vietnam', *Journal of Asian Studies* 56: 899–920.

Malaysia Department of Environment (1995) *Environmental Quality Report, 1994*, Kuala Lumpur: Department of Environment, Ministry of Science, Technology and the Environment.

Malaysia Department of Environment (1996), *Environmental Quality Data, 1992–1995*, Kuala Lumpur: Department of Environment, Ministry of Science, Technology and the Environment.

Malaysia Ministry of International Trade and Industry (1995), *Malaysia International Trade and Industry Report, 1995*: Kuala Lumpur: Ministry of International Trade and Industry.

Mallon, R.L. (1993) 'Vietnam: image and reality', in J. Heath (ed.) *Revitalizing Socialist Enterprise: A Race Against Time*, London: Routledge.

Maltby, E., Burbridge, P. and Fraser, A. (1996) 'Peat and sulphate soils: a case study from Vietnam', in E. Maltby, P. Immirzi and R.J. Safford (eds) *Tropical Lowland Peatlands of Southeast Asia*, Gland, Switzerland: World Conservation Union.

Markandya, A. and Pearce, D.W. (1991) 'Development, the environment and the social rate of discount', *World Bank Research Observer* 6: 137–52.

Marr, D. (1971) *Vietnamese Anticolonialism 1885–1925*, Berkeley: University of California Press.

Marr, D. (1995) *Vietnam, 1945*, Berkeley: University of California Press.

Maurand, P. (1943) *L'Indochine Forestiere*, Paris: Institute Recherché Agronomie de Indochine.

McAdam, L. and Le Nguyen Binh (1996) 'Vietnam's deltas: environmental aspects and investment implications', in B. Stensholt (ed.) *Development Dilemmas in the Mekong Subregion* (Workshop Proceedings, 1–2 October 1996), Clayton: Monash Asia Institute, Monash University.

McAlister, J.T. and Mus, P. (1970) *The Vietnamese and their Revolution*, New York: Harper and Row.

McArthur, H.J., Vu Quyet Thang, Tran Kieu Hanh and Phan Minh Chau (1993) 'Resilience to hazards and natural disasters', in Le Trong Cuc and A.T. Rambo (eds) *Too Many People, Too Little Land: The Human Ecology of a Wet Rice-Growing Village in the Red River Delta of Vietnam*, Honolulu, East–West Center.

McCay, B. and Acheson, J. (1990) 'Human ecology of the commons', in B. McCay and J. Acheson (eds) *The Question of the Commons: the Culture and Ecology of Communal Resources*, Tucson: University of Arizona Press.

McCully, P. (1996) *Silenced Rivers: The Ecology and Politics of Large Dams*, London: Zed Books.

McGee, T.G. (1995a) 'The urban future of Vietnam', *Third World Planning Review* 17: 253–77.

McGee, T.G. (1995b) 'Metrofitting the emerging mega-urban regions of ASEAN: an overview', in T.G. McGee and I. Robinson (eds) *The Mega-urban Regions of Southeast Asia*, Vancouver: UBC Press.

McGee, T.G. and Yeung, Y.-M. (1993) 'Urban futures for Pacific Asia', in T.G. McGee and Y.-M. Yeung (eds) *Pacific Asia in the 21st Century*, Hong Kong: Chinese University Press.

McGregor, G.R. (1994) 'The tropical cyclone hazard over the South China Sea 1970–1989', *Applied Geography* 15: 35–52.

McLean, R.F., Sinha, S.K., Mirza, M.Q. and Lal, M. (1998) 'Tropical Asia', in R.T. Watson, M.C. Zinyowera and R.H. Moss (eds) *The Regional Impacts of Climate Change: An Assessment of Vulnerability*, Cambridge: Cambridge University Press.

Mearns, R. (1996) 'Community, collective action and common grazing: the case of post-socialist Mongolia', *Journal of Development Studies* 32: 297–339.

Mekong River Commission (1995a) *Agreement on the Co-operation for the Sustainable Development of the Mekong River Basin*, Chiang Rai, Thailand, 5 April 1995, Bangkok: Mekong River Commission.

Mekong River Commission (1995b) *Annual Report 1995*, Bangkok: Mekong River Commission Secretariat.

Mekong River Commission (1996) *Annual Report 1996*, Bangkok: Mekong River Commission Secretariat.

Mekong River Commission (1997) *Mekong River Basin Diagnostic Study*, Bangkok: Mekong River Commission Secretariat.

Mekong Secretariat (1988) *Vietnam and the Mekong Committee*, Bangkok: Interim Mekong Committee.

Mekong Secretariat (1989) *The Mekong Committee: a Historical Account (1957–89)*, Bangkok: Mekong Secretariat.

Mekong Secretariat (1994) *Mekong Mainstream Run-of-River Hydropower*, Prepared by Acres International (Canada) and Compagnie du Nationale du Rhone (France), Bangkok: Mekong Secretariat.

Mekong Secretariat (1997) *Mekong News*, May 1997.

Mellac, M. G. (1997) 'Accès aux ressources naturelles et distribution des terres dans un district de montagne du nord Viêt Nam', paper presented at the EUROVIET-III Conference, Amsterdam, The Netherlands, 2–4 July 1997.

Mellor, J. (1978) 'Food price policy and income distribution in low income countries', *Economic Development and Cultural Change* 27: 1–26.

Mellor, J. (ed.) (1995) *Agriculture on the Road to Industrialisation*, Baltimore: Johns Hopkins University Press.

Mellor, J.W. and Ahmed, R. (eds) (1988) *Agricultural Price Policy for Developing Countries*, Baltimore: Johns Hopkins University Press.

Minot, N. and Goletti, F. (1998) 'Export liberalisation and household welfare: the case of rice in Vietnam', *American Journal of Agricultural Economics* 80: 738–49.

Mitchell, M. (1998) 'The political economy of Mekong Basin development', in P. Hirsch and C. Warren (eds) *The Politics of Environment in Southeast Asia: Resources and Resistance*, London: Routledge.

Mitlin, D. and Satterthwaite D. (1996) 'Sustainable development and cities', in C. Pugh (ed.) *Sustainability, the Environment and Urbanisation*, London: Earthscan.

Mitsch, W.J. and Gosselink, J.G. (1993) *Wetlands, Second Edition.*, New York: Van Nostrand Reinhold.

Moise, E.E. (1983) *Land Reform in China and North Vietnam: Consolidating the Revolution and the Village Level*, Chapel Hill: University of North Carolina Press.

Mol, A.P.J. and Frijns, J. (1997) 'Ecological restructuring in industrial Vietnam : the Ho Chi Minh City region', paper presented at the EUROVIET-III Conference, Amsterdam, The Netherlands, 2–4 July 1997.

Mol, A.P.J. and Frijns, J. (1998) 'Environmental reforms in industrial Vietnam', *Asia–Pacific Development Journal* 5, 2: 117–38.

Muldavin, J.S.S. (1996) 'The political ecology of agrarian reform in China: the case of Heilongjiang', in R. Peet and M.J. Watts (eds) *Liberation Ecologies: Environment, Development and Social Movements*, London: Routledge.

Muldavin, J.S.S. (1997) 'Environmental degradation in Heilongjiang: policy reform and agrarian dynamics in China's new hybrid economy', *Annals of the Association of American Geographers* 87: 579–613.

Munasinghe, M. and Cruz, W. (1995) *Economy-wide Policies and the Environment: Lessons from Experience*, World Bank Environment Paper No. 10, Washington DC: World Bank.

Murray, M. (1980) *The Development of Capitalism in Colonial Indochina (1870–1940)*, Berkeley: University of California Press.

Murray, P. and Szelenyi, I. (1984) 'The city in the transition to socialism', *International Journal of Urban Regional Research* 8: 330–50.

Neave, I. and Bui Ngoc Quang (1994) *What Works: Lessons from a Community Forestry Project in Northern Vietnam*, Monograph 5, Hanoi: Care International in Vietnam.

Netherlands Engineering Consultants (Nedeco) (1993) *Master Plan for the Mekong Delta in Vietnam*, Prepared for the Government of Vietnam, State Planning Committee, World Bank, Mekong Secretariat and United Nations Development Programme, Arnhem: Netherlands Engineering Consultants.

Ngo Vinh Long (1973) *Before the Revolution: The Vietnamese Peasant under the French*, Cambridge: MIT Press.

Ngo Vinh Long (1993) 'Reform and rural development: impact on class, sectoral and regional inequalities', in W.S. Turley and M. Selden (eds) *Reinventing Vietnamese Socialism: Doi Moi in Comparative Perspective*, Boulder: Westview.

Nguyen Anh Tuan (1997) 'Energy and environmental issues in Vietnam', *Natural Resources Forum* 21: 201–7.

Nguyen Anh Tuan and Lefevre, T. (1996) 'Analysis of household energy demand in Vietnam', *Energy Policy* 24: 1089–99.

Nguyen Cat Giao and Vu Van Me (1998) 'Guiding principles for forest land allocation as a tool to implement the Government's top priority on land use policy', in *Proceedings of the National Workshop on Participatory Land Use Planning and Forest Land Allocation*, Hanoi: Agriculture Publishing House.

Nguyen Dac Hy (1995) 'Discussions on environmental protection laws', in Cao Van Sung (ed.) *Environment and Bioresources of Vietnam: Present Situation and Solutions*, Hanoi: Thê Giói.

Nguyen Duc Khien (1995) *Current Environmental Conditions in Vietnam*, Hanoi: Ministry for Science and Technology and Environment.

Nguyen Duc Khien (1996a) 'Current environmental conditions in Vietnam', paper presented at the EUROVIET-III Conference, Amsterdam, The Netherlands, 2–4 July 1997.

Nguyen Duc Khien (1996b) 'Modernisation and industrialisation with urban environmental protection in Hanoi', paper presented at the EUROVIET-III Conference, Amsterdam, The Netherlands, 2–4 July 1997.

Nguyen Duc Ngu (2000) *Knowledge on El Niño and La Niña*, Hanoi: Science and Technology Publishing House.

Nguyen Hoang Tri, Nguyen Huu Ninh, Nguyen The Chinh, Tran Viet Lien and Tran Dai Nghia (1997) *Economic Valuation Studies of Mangrove Conservation and Rehabilitation in Nam Ha Province, Red River Delta, Vietnam*, Progress report for SARCS/WOTRO/LOICZ, Hanoi: Mangrove Ecosystem Research Centre and Center for Environment Research Education and Development.

Nguyen Hoang Tri, Adger, W.N. and Kelly, P.M. (1998) 'Natural resource management in mitigating climate impacts: mangrove restoration in Vietnam', *Global Environmental Change – Human And Policy Dimensions* 8: 49–61.

Nguyen Huu Chiem (1993) 'Geo-pedological study of the Mekong Delta', *Southeast Asian Studies* 31: 158–87.

Nguyen Huu Chiem (1994) 'Past and present cropping patterns in the Mekong Delta', *Southeast Asian Studies* 31: 345–85.

Nguyen Huu Dung (1993) 'Kitchen improvement – the combination of traditional and modern', in *Kitchens, Living Environment and Household Energy in Vietnam*, Seminar Report, Urban Building and Energy Project, Lund: University of Lund.

Nguyen Huu Ninh, Tran Viet Lien, Hoang Minh Hien and Luong Quang Huy (2000) *Reducing the Impact of Environmental Emergencies: The Case of El Niño–Southern Oscillation, Vietnam Case Study*, Hanoi: Center for Environment Research Education Research and Development.

Nguyen Huy (1980) 'Ve cac hinh thuc khoan trong hop tac xa trong hop tac xa trong lua' [On types of contract systems in rice growing cooperatives]', *Nghien cuu hinh te* [*Economic Research*] 118: 9–23.

Nguyen Khac Vien (1972) *Traditional Vietnam: Some Historical Stages*, Vietnamese Studies No. 21, Hanoi: Foreign Languages Publishing House.

Nguyen Manh Kiem (1996) 'Strategic orientation for construction and development of Hanoi, Vietnam', *Ambio* 25: 108–9.

Nguyen Ngoc Binh, Raffel, B. and Merwin, W. S. (1975) *A Thousand Years of Vietnamese Poetry*, New York: Alfred A. Knopf.

Nguyen Nhan Quang (1996) 'The Mekong Basin development: Vietnam's concerns', in B. Stensholt (ed.) *Development Dilemmas in the Mekong Subregion*, Clayton: Monash Asia Institute.

Nguyen Quang Ha (1991) *Land Survey and Land Use Planning for Forestry*, Hanoi: Forestry Inventory and Planning Institute, Ministry of Forestry.

Nguyen Quang Ha (1993) *Renovation of Strategies for Forestry Development until 2000*, Hanoi: Ministry of Forestry.

Nguyen Quang My (1992) 'Erosion of hilly land and soil environment in Vietnam', in *Proceedings of the Seminar on Rational Usage of Soil Resources for Development and Environmental Protection*, Hanoi: Association of Soil Scientists.

Nguyen Quang Vinh and Leaf, M. (1996) 'City life in the village of ghosts: a case study of popular housing in Ho Chi Minh City', *Habitat International* 20, 175–90.

Nguyen Thanh Binh (1998*) On Forest Land Allocation and Green Book Issuing in Tuong Duong District*, Proceedings of the National Workshop on Participatory Land Use Planning and Forest Land Allocation, Hanoi: Agriculture Publishing House.

Nguyen The Anh (1985) 'Le famine de 1945 au Nord-Viet-Nam', *The Vietnam Forum* 5: 81–100.

Nguyen The Anh (1995) 'Historical research in Vietnam: a tentative survey', *Journal of Southeast Asian Studies* 26: 121–32.

Nguyen Thi Dieu (1999) *The Mekong River and the Struggle for Indochina: Water, War and Peace*, Westport, CT: Praeger.

Nguyen Thuong Luu, Vu Van Me and Nguyen Tuong Va (1995) 'Land classification and land allocation of forest land in Vietnam: a meeting of the national and local perspective', *Forest, Trees and People Newsletter* 25.

Nguyen Thu Sa (1990) 'Van de ruong dat o dong bang song Cuu Long' [The land problem in the Mekong Delta], in *Mien Nam trong su nghiep doi moi cua ca nuor* [*The South in the Renovation of the Country*], TP Ho Chi Minh: NXB Khao Hoc Xa Hoi.

Nguyen Van Dang (1997) 'Vietnam's forestry on the threshold of the twenty-first century', in *Proceedings of Eleventh World Forestry Congress'* Antalya, Turkey: International Union of Forest Research Organisations 7: 213–16.

Nguyen Van Nhan (1997) 'Wetland mapping in the Mekong Delta and Tram Chim National Reserve area using Geographical Information Systems', in R.J. Safford, Duong Van Ni, E. Maltby and Vo-Tong Xuan (eds) *Towards Sustainable Management of Tram Chim National Reserve, Vietnam*, London: Royal Holloway Institute for Environmental Research.

Nguyen Van Vinh (1961) *Les reformes Agraires au Viet Nam*, Louvain: Universitaire de Louvain, Faculte des Sciences Economiques et Sociales.

Nguyen Xuan Nguyen and Nachuk, S. (1998) 'Brief overview of rural finance in Vietnam', *Vietnam's Socio-Economic Development: A Social Science Review* 14: 24–44.

Nicholls, N. (1992) 'Historical El Niño/Southern Oscillation variability in the Australasian region', in H. F. Diaz and V. Markgraf (eds) *El Niño: Historical and Paleoclimatic Aspects of the Southern Oscillation*, Cambridge: Cambridge University Press.

Nishimori, M. and Yoshino, M. (1990) 'The relationship between ENSO-events and the generation, development and movement of typhoons', *Geographical Review of Japan* 63, A-8: 530–40. In Japanese.

Nordström, H. and Vaughan, S. (1999) *Trade and Environment*, Special Studies 4, Geneva: World Trade Organization.

Norgaard, R.B. (1994) *Development Betrayed: The End of Progress and a Coevolutionary Revisioning of the Future*, London: Routledge.

O'Connor, D. (1994) *Managing the Environment with Rapid Industrialisation: Lessons from the East Asian Experience*, Paris, Organisation for Economic Cooperation and Development Development Centre.

Öjendahl, J. (1995) 'Mainland Southeast Asia: co-operation or conflict over water?', in L. Ohlsson (ed.) *Hydropolitics: Conflicts Over Water as a Development Constraint*, London: Zed Books.

Olson, M. (1965) *The Logic of Collective Action*, Cambridge: Harvard University Press.

O'Riordan, T., Cooper, C.L., Jordan, A., Rayner, S., Richards, K.R., Runci, P. and Yoffe, S. (1998) 'Institutional frameworks for political action', in S. Rayner and E. Malone (eds) *Human Choice and Climate Change: Volume 1 The Societal Framework*, Washington DC: Battelle Press.

Ostrom, E. (1990) *Governing the Commons: the Evolution of Institutions for Collective Action*, Cambridge: Cambridge University Press.

Ostrom, E., Burger, J., Field, C.B., Norgaard, R.B. and Policansky, D. (1999) 'Revisiting the commons: local lessons, global challenges', *Science* 284: 278–82.

Otsuka K. (1999) 'Land tenure and the management of land and trees in Asia and Africa', *Japanese Journal of Rural Economics* 1: 25–38.

Page, J. (1997) 'The East Asian Miracle and the Latin American Consensus: can the twain ever meet?', in N. Birdsall and F. Jaspersen (eds) *Pathways to Growth: Comparing East Asia and Latin America*, Baltimore: Johns Hopkins University Press.

Page, R. and Thanh, M. (1993) *Income Generation for Minority Women in Vietnam: A Case Study*, Hanoi: CARE International.

Panayotou, T. (1993) *Empirical Tests and policy analysis of environmental degradation at different stages of economic development*, Technology and Employment Programme, Working Paper 28, Geneva: International Labour Office.

Panayotou, T. (1997), 'Demystifying the environmental Kuznets curve: turning a black box into a policy tool', *Environment and Development Economics* 2: 465–84.

Panayotou, T. and Naqvi, N. (1996) 'Case study for Vietnam', in D. Reed (ed.) *Structural Adjustment, the Environment, and Sustainable Development*, London: Earthscan.

Pandolfi, L. (1997) 'The new land development process and its social consequences in Hanoi', paper presented at the EUROVIET-III Conference, Amsterdam, The Netherlands, 2–4 July 1997.

Pantulu, V.R. (1986a) 'The Mekong River system', in B.R. Davies and K.F. Walker (eds) *The Ecology of River Systems*, Dordrecht: W. Junk.

Pantulu, V.R. (1986b) 'Fish of the Lower Mekong Basin', in B.R. Davies and K.F. Walker (eds) *The Ecology of River Systems*, Dordrecht: W. Junk.

Parasnis, M. (1999) 'Environmental management in Thailand: Achievements, Barriers and Future Trends', in W. Wehrmeyer and Y. Mulugetta (eds) *Growing Pains: Environmental Management in Developing Countries*, Sheffield: Greenleaf Publishers.

Pargal, S. and Wheeler, D. (1995) *Informal Regulation of Industrial Pollution in Developing Countries: Evidence from Indonesia*, Policy Research Working Paper 1416, Washington DC: World Bank.

Parker, D.J. (1998) *The Condition of the Tropical Cyclone Warning Dissemination System in Mauritius*, London: Flood Hazard Research Centre, Middlesex University.

Parker, D.J., Fordham, M. and Torterotot, J.-P. (1994) 'Real-time hazard management: flood forecasting, warning and response', in E.C. Penning-Rowasell and M.F. Fordham (eds) *Floods across Europe*, London: Middlesex University Press.

Parnwell, M.J.G. and Bryant, R.L. (eds) (1996) *Environmental Change in Southeast Asia: People, Politics and Sustainable Development*, London, Routledge.

Parry, M., Magalhães, A.R. and Nguyen Huu Ninh (eds) (1991) *The Potential Socio-Economic Effects of Climate Change: a Summary of Three Regional Assessments*, Nairobi: United Nations Environment Programme.

Pearce, D.W. and Warford, J. (1993) *World Without End: Economics, Environment and Sustainable Development*, Oxford: Oxford University Press.

Peletz, M.G. (1983) 'Moral and political economies in rural south-east Asia', *Journal of Comparative Studies in Society and History* 25, 4: 731–9.

Pelling, M. (1998) 'Participation, social capital and vulnerability to urban flooding in Guyana', *Journal of International Development* 10: 469–86.

Persson, T. and Tabellini, G. (1994) 'Is inequality harmful for growth?', *American Economic Review* 84: 600–21.

Petr, T. (1994) 'The present status of, and constraints to inland fisheries development in Southeast Asia', in T. Petr and M. Morris (eds) *Regional Symposium on Sustainable Development of Inland Fisheries under Environmental Constraints*, Bangkok, 19–21 October 1994, Rome: Food and Agriculture Organization.

Pham Minh Tuan and van der Poel, P. (1998) 'Land allocation in the Song Da watershed', in *Proceedings of the National Workshop on Participatory Land Use Planning and Forest Land Allocation*, Hanoi: Agriculture Publishing House.

Pham Ngoc Dang and Tran Hieu Nhue (1995) 'Environmental pollution in Vietnam', in Cao Van Sung (ed.) *Environment and Bioresources of Vietnam: Present Situation and Solutions*, Hanoi: Thê Giói.

Pham Quynh Huong (1997) 'Private housing sector and changes in the organisation of urban space', *Vietnam's Socio-Economic Development: A Social Science Review* 12: 64–9.

Pham Tam (1997) *Shelter And Environmental Improvement For The Urban Poor*, Hanoi: International Development Research Centre and Hanoi Architectural University.

Phan Huy Chun (1821) *Lich-troeu Hien-chuong Loai-chi* [A Reference book on the Institutions of Successive (Vietnamese) Dynasties], Presented to Minh-mang in 1821, translation by Luong then Cao Nai Quang (1957), Saigon: Faculty of Law, Saigon University.

Phan Khanh (1984) 'An age-old undertaking', in *Water Control in Vietnam*, Hanoi: Foreign Languages Publishing House.

Phan Nguyen Hong (1994) 'Causes and effects of the deterioration in the mangrove resources and environment in Vietnam', in *Reforestation and Afforestation of Mangroves in Vietnam: Proceedings of the National Workshop, Ho Chi Minh City*, Hanoi: Mangrove Ecosystem Research Centre, Hanoi National Pedagogic University and Action for Mangrove Reforestation.

Phan Nguyen Hong and Hoang Thi San (1993) *Mangroves of Vietnam*, Bangkok: World Conservation Union.

Phan Nguyen Hong, Tran Van Ba, Hoang Thi San, Le Thi Tre, Nguyen Hoang Tri, Mai Sy Tuan and Le Xuan Tuan (1997) *The Role of Mangroves in Vietnam, Planting Techniques and Maintenance*, Hanoi: Agricultural Publishing House.

Phongpaichit, P. and Baker, C. (1996) *Thailand's Boom*, Chiang Mai Silkworm Books.

Pidgeon, N. (1999) 'Risk communication and the social amplification of risk: theory, evidence and policy implications', *Risk Decision and Policy* 4(2): 145–59.

Pidgeon, N., Hood, C., Jones, D., Turner, B. and Gibson, R. (1993) 'Risk perception', in *Risk: Analysis, Perception and Management*, Report of a Royal Society Study Group, London: The Royal Society.

Pingali, P.L. and Vo-Tong Xuan (1992) 'Vietnam: decollectivisation and rice productivity growth', *Economic Development and Cultural Change* 40: 697–718.

Pingali, P.L., Nguyen Tri Khiem, Gerpacio, R.V. and Vo-Tong Xuan (1997) 'Prospects for sustaining Vietnam's rice exporter status', *Food Policy* 22: 345–58.

Poffenberger, M. (ed.) (1998) *Stewards of Vietnam's Upland Forests*, Berkeley: The Asia Forest Network, Center for Southeast Asia Studies, University of California.

Ponce, N., Gertler, P. and Glewwe, P. (1998) 'Will Vietnam grow out of malnutrition?', in D. Dollar, P. Glewwe and J. Litvack (eds) *Household Welfare and Vietnam's Transition*, Washington DC: World Bank.

Popkin, S.L. (1979) *The Rational Peasant: the Political Economy of Rural Society in Vietnam*, Berkeley: University of California Press.

Porter, G. (1995) 'Managing renewable resources in Southern Asia: the problem of deforestation', in Y.C. Kim (ed.) *Southeast Asia's Economic Miracle*, New Brunswick: Transaction Publishers.

Pretty, J. (1998) 'Furthering cooperation between people and institutions', in H.-P. Blume, H. Eger, E. Fleischhauer, A. Hebel, C. Reij and K.G. Steiner (eds) *Towards Sustainable Land Use: Furthering Cooperation between People and Institutions*, Reiskirchen: Catena-Verlag.

Price, R.C. (1966) 'Taming the turbulent Mekong', paper presented at the Commonwealth Club, 15 June 1966, San Francisco.

Primavera, H. (1998) 'Mangroves as nurseries: shrimp populations in mangrove and non-mangrove habitats', *Estuarine, Coastal and Shelf Science* 46: 457–64.

Prosterman, R. and Hanstand, T. (1999) 'Legal impediments to effective rural land relations in Eastern Europe and Central Asia: A comparative perspective', World Bank Technical Paper No. 436, Washington DC: World Bank.

Putterman, L. (1983) 'A modified collective agriculture in rural growth with equity: reconsidering the private uni-modal solution', *World Development* 11: 77–100.

Pyo, H.K. (1995) 'The transition in the political economy of South Korean development: issues and perspectives', in Y.C. Kim (ed.) *The Southeast Asian Miracle*, New Brunswick: Transaction Publishers.

Quang Truong (1987) *Agricultural Collectivisation and Rural Development in Vietnam: A North/South Study, 1955–1985*, Amsterdam: Vrije Universiteit te Amsterdam.

Quiggen, J. (1993) 'Common property, equality and development', *World Development* 21: 1123–38.

Radelet, S. and Sachs, J. (1997) 'Asia's re-emergence', *Foreign Affairs* 76, 6: 44–59.

Rahim, K. and Nesadurai, H. (1996) *ASEAN Sub-Programme on Trade and Environment: National Level Studies – Malaysia*, Kuala Lumpur, Institute of Strategic and International Studies.

Rambo, A.T. (1995a) 'Privatization of communal land in northern Vietnam', *Common Property Resource Digest* 33: 7–9.

Rambo, A.T. (1995b) 'Perspectives on defining highland development challenges in Vietnam: new frontier or cul-de-sac?', in A.T. Rambo, Le Trong Cuc, R.R. Reed and M.R. DiGregorio (eds) *The Challenges of Highland Development in Vietnam*, Honolulu: East–West Centre.

Rambo, A.T. and Le Trong Cuc (1996a) 'Rural development issues in the upland agroecosystems of Vinh Phu Province', in Le Trong Cuc, A.T. Rambo, K. Fahrney, Tran Duc Vien, J. Romm, and Dang Thi Sy (eds) *Red Books, Green Hills: The Impact of Economic Reform on Restoration Ecology in the Midlands of Northern Vietnam*, Honolulu: East–West Centre.

Rambo, A.T. and Le Trong Cuc (eds) (1996b) *Development Trends in Vietnam's Northern Mountain Region*, Honolulu and Hanoi: East–West Centre and Centre for Natural Resource and Environmenal Studies, Vietnam National University.

Rambo, A.T., Reed, R.R., Le Trong Cuc and DiGregorio, M.R. (eds) (1995) *The Challenges of Highland Development in Vietnam*, Honolulu: East–West Centre.

Rasiah, R. (1999) *Transnational Corporations and the Environment: The Case of Malaysia*, UNCTAD/CBS Project, Cross Border Environmental Management in Transnational Corporations, Occasional Paper No. 4, Copenhagen: Copenhagen Business School.

Rasmusson, E.M. and Wallace, J.M. (1983) 'Meteorological aspects of the El Niño Southern Oscillation', *Science* 222: 1195–202.

Rauscher, M. (1997) *International Trade, Factor Movements, and the Environment*, Oxford: Clarendon.

Reardon, T. and Taylor, J.E. (1996) 'Agroclimatic shocks, income inequality and poverty: evidence from Burkino Faso', *World Development* 24: 901–14.

Reed, D. (1996) *Structural Adjustment, the Environment and Sustainable Development*, London: Earthscan.

Rerkasem, B. and Rerkasem, K. (1998) 'Influence of demographic, socio-economic and cultural factors on sustainable land use', in H.-P Blume, H. Eger, E. Fleischhauer, A. Hebel, C. Reij and K.G. Steiner (eds) *Towards Sustainable Land Use: Furthering Cooperation between People and Institutions*, Reiskirchen: Catena-Verlag.

Ribot, J.C., Magalhães, A.R. and Panagides, S.S. (eds) (1996) *Climate Variability, Climate Change and Social Vulnerability in the Semi-arid Tropics*, Cambridge: Cambridge University Press.

Riebsame, W.E., Strzepak, K.M., Wescoat Jr., J.L., Perritt, R., Gaile, G.L., Jacobs, J., Leichenko, R., Magadza, C., Phien, H., Urbiztondo, B.J., Restrepo, P., Rose, W.R., Saleh, M., Ti, L.H., Tucci, C. and Yates, D. (1995) 'Complex river systems', in K. Strzepak and J. Smith (eds) *As Climate Changes: International Impacts and Implications*, Cambridge: Cambridge University Press.

Riedel, J. (1993) 'Vietnam: On the Trail of the Tigers', *World Economy* 16: 401–22.

Rimmer, P. (1995) 'Moving goals, people and information', in T.G. McGee and I. Robinson (eds) *The Mega-Urban Regions of Southeast Asia*, Vancouver: UBC Press.

Robequain, C. (1939) *L'evolution Economique de l'Indochine Française*, Paris: P. Hartmann.

Roberts, T. (1995) 'Mekong mainstream hydropower dams: run-of-the-river or ruin-of-the-river?', *Natural History Bulletin of the Siam Society* 43: 9–19.

Robinson, D. (1988), 'Industrial pollution abatement: the impact on the balance of trade', *Canadian Journal of Economics* 21,1: 187–99.

Rock, M. (1996), 'Pollution intensity of GDP and trade policy: can the World Bank be wrong?', *World Development* 24: 471–80.

Rollett, B. (1963) *Note sur le Vegetation du Vietnam au Sud du 17e Parallele Nord*, Saigon: Archives du Forete Recherché Institute.

Romm, J. and Dang Thi Sy (1996) 'The impact of economic liberalisation on Lap Thach', in Le Trong Cuc, A.T. Rambo, K. Fahrney, Tran Duc Vien, J. Romm and Dang Thi Sy, (eds) *Red Books, Green Hills: The Impact of Economic Reform on Restoration Ecology in the Midlands of Northern Vietnam*, Honolulu: East–West Centre.

Ropelewski, C.F. and Halpert, M.S. (1987) 'Global and regional scale precipitation patterns associated with the El Niño Southern Oscillation', *Monthly Weather Review* 115: 1606–26.

Rothman, D. and de Bruyn, S. (1998) 'Probing into the environmental Kuznets curve hypothesis', *Ecological Economics*, 25: 177–94.

Ruddle, K. (1998) 'Traditional community-based coastal marine fisheries management in Vietnam', *Ocean and Coastal Management* 40: 1–22.

Ruitenbeek, H.J. (1994) 'Modelling economy–ecology linkages in mangroves: economic evidence for promoting conservation in Bintuni Bay, Indonesia', *Ecological Economics* 10: 233–47.

Sachasinh, R., Phantumvanit, D. and Tridech, S. (1992) 'Thailand: Challenges and Responses in Environmental Management', paper presented to the Workshop on Environmental Management in East Asia: Challenge and Response, Organisation for Economic Co-operation and Development Development Centre, Paris, 6–7 August 1992.

Safford, R.J. and Maltby E. (1997) 'The overlooked values of *Melaleuca* wetlands', in R.J. Safford, Duong Van Ni, E. Maltby and Vo-Tong Xuan (eds) *Towards Sustainable Management of Tram Chim National Reserve, Vietnam*, London: Royal Holloway Institute for Environmental Research.

Safford, R.J., Tran Triet, Duong Van Ni and Maltby, E. (1998) 'Status, biodiversity and management of the U Minh wetlands, Vietnam', *Tropical Biodiversity* 5: 217–24.

Sahn, D. (1988) 'The effect of price and income changes on food-energy intake in Sri Lanka', *Economic Development and Cultural Change* 36: 314–40.

Sandler, T. (1997) *Global Challenges: An Approach to Environmental, Political and Economic Problems*, Cambridge: Cambridge University Press.

Sansom, R.L. (1970) *The Economics of Insurgency*, Cambridge: MIT Press.

Saroen, B. and Stormer, C. (2000) 'Viet dam full of lethal surprises', *Phnom Penh Post* 9, 6 (March): 17–30.

Sassen, S. (1994) *Cities in a World Economy*, London: Pine Forge Press.

Satterthwaite, D. (1997) 'Sustainable cities or cities that contribute to sustainable development', *Urban Studies* 34: 1667–91.

Saunders, M.A., Chandler, R.E., Merchent, C.J. and Roberts, F.P. (2000) 'Atlantic hurricanes and NW Pacific typhoons: ENSO spatial impacts on occurrence and landfall', *Geophysical Research Letters* 27: 1147–50.

Schecter, A. (1991) *Dioxins and Related Chemicals in Humans and the Environment*, Banbury Report 35, New York: Cold Spring Harbor Laboratory Press.

Schenk, R. (1998) *Faktoren des Kreditzugangs von Kleinbauern in Nordvietnam (Factors Influencing Smallholders' Access to Credit in Northern Vietnam)*, Masters Thesis, Hohenheim University, Germany.

Schenk, R., Neef, A. and Heidhues, F. (1999) 'Factors influencing access to credit of smallholders in Northern Vietnam', *Vietnam's Socio-Economic Development: A Social Science Review* 18: 56–65.

Scott, J.C. (1976) *The Moral Economy of the Peasant: Rebellion and Subsistence in Southeast Asia*, New Haven: Yale University Press.

Scott, J. C. (1986) 'Everyday forms of peasant resistance', in J.C. Scott and B.J.T. Kerkvleit (eds) *Everyday Forms of Peasant Resistance in Southeast Asia*, London: Frank Cass.

Scott, J. (1998) *Seeing Like a State: How Certain Schemes to Help the Human Condition Have Failed*, New Haven: Yale University Press.

Selden, T. and Song, D. (1994) 'Environmental quality and development: is there a Kuznets curve for air pollution emissions?', *Journal of Environmental Economics and Management* 27: 147–62

Sen, A.K. (1981) *Poverty and Famines: an Essay on Entitlement and Deprivation*, Oxford: Clarendon.

Sen, A.K. (1990) 'Food economics and entitlements', in J. Drèze and A.K. Sen (eds) *The Political Economy of Hunger Vol. 1*, Oxford: Clarendon.

Shafik, N. (1994) 'Economic development and environmental quality: an econometric analysis', *Oxford Economic Papers* 46: 757–73.

Shanmugaratnam, N. (1996) 'Nationalisation, privatisation and the dilemmas of common property management in western Rajasthan', *Journal of Development Studies* 33: 163–87.

Sharp, T. (1982) 'The Mekong dream', *Ceres*, September–October: 31–4.

Shiva, V. (1991) *The Violence of the Green Revolution*, London: Zed Books.

Sikor, T. (1995) 'Decree 327 and the restoration of barren land in the Vietnamese Highlands', in A.T. Rambo, R.R. Reed, Le Trong Cuc and M.R. DiGregorio, (eds) *The Challenges of Highland Development in Vietnam*, Honolulu: East–West Centre.

Sikor, T. (1998) 'Forest policy reform: from state to household forestry', in M. Poffenberger, *Stewards of Vietnam's Upland Forests*, Berkeley: The Asia Forest Network, Center for Southeast Asia Studies, University of California.

Sikor, T.O. and O'Rourke, D. (1996) 'Economic and environmental dynamics of reform in Vietnam', *Asian Survey* 26: 601–17.

Sime, J. (1997) 'Informative flood warnings: occupant response to risk, threat and loss of place', in J. Handmer (ed.) *Flood Warning: Issues and Practice in Total System Design*, London: Flood Hazard Research Centre, Middlesex University.

Sluiter, L. (1992) *The Mekong Currency*, Bangkok: Project for Ecological Recovery/Towards Ecological Recovery and Regional Alliance.

Smart, A. (1998) 'Economic transformation in China: property regimes and social relations', in J. Pickles and A. Smith (eds) *Theorising Transition: The Political Economy of Post-Communist Transformations*, London: Routledge.

Smith, W. (1998) 'Land and the poor: a survey of land use rights in Ha Tinh and Son La Provinces', in *Proceedings of the National Workshop on Participatory Land Use Planning and Forest Land Allocation*, Hanoi: Agriculture Publishing House.

Sneath, D. (1998) 'State policy and pasture degradation in inner Asia', *Science* 281: 1147–8.

Social Forestry Development Project (1994) *Agro-Economic Farm Household Survey in the Song Da Watershed (North-West of Vietnam)*, SFDP Baseline Study No. 3, Hanoi: Social Forestry Development Project.

Stensholt, B. (1996) 'The many faces of Mekong cooperation', in B. Stensholt (ed.) *Development Dilemmas in the Mekong Subregion* (Workshop proceedings, 1–2 October 1996), Clayton, Australia: Monash Asia Institute, Monash University.

Stern, D., Common, M. and Barbier, E. (1996) 'Economic growth and environmental degradation: the Environmental Kuznets Curve and sustainable development', *World Development*, 24: 1151–60.

Stern, N. (1995) 'Growth theories, old and new and the role of agriculture in economic development', Economic and Social Development Paper 136, Rome: Food and Agriculture Organization.

Stren, R., White, R. and Whitney, J. (eds) (1992) *Sustainable Cities: Urbanisation and the Environment in International Perspective*, Oxford: Westview.

Strutt, A. and Anderson, K. 1999, 'Will trade liberalization harm the environment?: the case of Indonesia to 2020', in P. Fredriksson (ed.) *Trade, Global Policy and the Environment*, Discussion Paper No. 402, Washington DC: World Bank.

Sund, T. (1998) 'Environmental protection and foreign investment in Vietnam', in H. Knutsen (ed.) *Internationalisation of Capital and the Opportunity to Pollute*, FIL Working Paper 14, Department of Sociology and Human Geography, Oslo: University of Oslo.

Suphachalasai, S. (1995) 'Export-led industrialisation', in M. Krongkaew, *Thailand's Industrialisation and its Consequences*, London: Macmillan.

Surridge, A.K., Timmins, R.J., Hewitt, G.M. and Bell, D.J. (1999) 'Striped rabbits in Southeast Asia', *Nature* 400: 726.

Tanaka, K. (1995) 'Transformation of rice-based cropping patterns in the Mekong Delta: from intensification to diversification', *Southeast Asian Studies* 33, 3: 81–96.

Tangwisutijit, N. (1995) 'Dioxin released from paper mills a Mekong threat', *The Nation* (Bangkok), 28 October.

Thai Van Trung (1970) *The Forest Vegetation of Viet Nam*, Hanoi: Institute of Forest Inventory and Management.

Thailand Development Research Institute (1994) *Assessment of Sustainable Highland Agricultural Systems, Natural Resources and Environment Program*, Bangkok: Thailand Development Research Institute.

Thrift, N. and Forbes, D. (1986) *The Price of War: Urbanisation in Vietnam 1954–85*, London: Allen and Unwin.

Tilman, D. (1997) 'Biodiversity and ecosystem functioning', in G.C. Daily (ed.) *Nature's Services: Societal Dependence on Natural Ecosystems*, Washington DC: Island Press.

Timmer, C.P. (1993) 'Food policy reform and economic reform in Vietnam', in B. Ljunggren (ed.) *Challenge of Reform in Indochina*, Cambridge: Harvard University Press.

Tipping, G. and Truong Viet Dung (1997) 'Rural health services in Vietnam: their contemporary relevance to other Asian transitional economies', *IDS Bulletin* 28, 1: 110–15.

To Dinh Mai (1987) 'Nhung van de can doi moi de xay dung phat trien tai nguyen rung va nghe rung [Reforms in forestry construction and development]', in Be Viet Dang, Chu Thai Son, Ngo Duc Trinh *et al.* (eds) *Mot so van de kinh te-xa hoi cac tinh mien nui phia bac [Socio-economic Problems in the Northern Mountain Region]*, Hanoi: Social Science Committee, Social Science Publishing House.

To Dinh Mai (1991) 'Organising individual households for forestry development and effective forest land use', in C. Sargent (ed.) *Research Priorities for Sustainable Land Use in Vietnam: Proceedings of a National Seminar 9–13 September, 1991 Forest Science Institute, Hoa Binh*, London: International Institute for Environment and Development.

To Phuc Tuong, Chu Thai Hoanh and Nguyen Tri Khiem (1989) 'Agro-hydrological factors as land qualities in land evaluation for rice cropping patterns in the Mekong delta', in *Proceedings of the Symposium on Acid Soils in the Tropics*, Kandy: Institute of Fundamental Studies.

Tobey, J. (1990) 'The effects of domestic environmental policies on patterns of world trade: an empirical test', *Kyklos*, 43: 191–209.

Tongzon, J. L. (1999) 'The challenge of regional economic integration: the Vietnamese perspective', *Developing Economies* 37: 137–62.

Towards Ecological Recovery and Regional Alliance (1995) 'Mekong politics: "new era", same old plans', *Watershed* 1, 1: 24–9.

Towards Ecological Recovery and Regional Alliance (1996a) 'A "shopping list for donors": Mekong River Commission launches 1996 programme', *Watershed* 1, 2: 22–33.

Towards Ecological Recovery and Regional Alliance (1996b) ' "Putting emphasis on people's participation": Danish aid in the Mekong region', *Watershed* 1(3): 44–6.

Traisawasdichai, M. (1995a) 'Mekong Commission: in search of a purpose', *The Nation* (Bangkok), 17 February 1995.

Traisawasdichai, M. (1995b) 'The Mekong river engineers: four perspectives on developing the mainstream', *The Nation* (Bangkok), 29 December 1995.

Tran Duc Vien and Fahrney, K. (1996) 'An overview of the Midland of Vinh Hi Province', in Le Trong Cuc, A.T. Rambo, K. Fahrney, Tran Duc Vien, J. Romm and Dang Thi Sy (eds) *Red Books, Green Hills: The Impact of Economic Reform on Restoration Ecology in the Midlands of Northern Vietnam*, Honolulu: East–West Centre.

Tran Khai (1994) 'The present land use situation and the strategy for land use from now to the year 2000', in C. Howard (ed.) *Current Land Use in Vietnam*, Proceedings of the Second Land Use Seminar, 22–23 September 1994, Bac Thai, Vietnam, Hanoi and London: Land Use Working Group and International Institute for Environment and Development.

Tran Kim Thach (1980) *Sedimental Geology of the Mekong Delta*, Ho Chi Minh City: General University of Ho Chi Minh City.

Tran The Vinh, (1995) *'Tree Planting Measures to Protect Sea Dike Systems in the Central Provinces of Vietnam'*, paper presented at the Workshop on Mangrove Plantation for Sea Dike Protection, 24–5 December 1995, Hatinh, Vietnam.

Tran Thi Que (1998) 'Microfinance market in mountainous areas: a case study', *Vietnam's Socio-Economic Development – A Social Science Review* 14: 45–61.

Tran Thi Thanh Phuong (1996) 'Environmental management and the policy-making in Vietnam', paper presented at the Seminar on Environment and Sustainable Development in Vietnam, 5–7 December 1996, Australian National University, Canberra.

Tran Thi Van Anh and Le Ngoc Hung (1997) *Women and* doi moi *in Vietnam*, Hanoi: Woman Publishing House.

Tran Thi Van Anh and Nguyen Manh Huan (1995) 'Changing Rural Institutions and Social Relations', in B. J. T. Kerkvliet and D. J. Porter (eds) *Vietnam's Rural Transformation*, Boulder: Westview.

Tran Triet, Safford, R.J., Tran Duy Phat, Duong Van Ni and Maltby, E. (in press) 'Wetland biodiversity overlooked and threatened in the Mekong Delta, Viet Nam: grassland ecosystems in the Ha Tien Plain', Submitted to *Tropical Biodiversity*.

Tran Trong Kim (1964) *Viet Nam su luoc [A Short History of Vietnam]*, Saigon: Tan Viet. First published in Hanoi in 1951 by Tan Viet.

Tran Viet Lien and Nguyen Huu Ninh (1999) 'A preliminary study of the impacts of climate change and sea-level rise on the coastal zone of Vietnam', *Current Topics in Wetlands Biogeochemistry* 3: 161–74.

Tranh Khanh (1994) 'Social disparity in Vietnam', *Business Times*, 24–5 September 1994: 4.

Trin Duy Luan (1995) 'Impacts of economic reforms on urban society', in Vu Tuan Anh (ed.) *Economic Reforms and Development in Vietnam*, Hanoi: Social Science.

Trin Duy Luan (1997) 'Hanoi: the booming of irregular popular housing in the 1990s', paper presented at the EUROVIET-III Conference, Amsterdam, The Netherlands, 2–4 July 1997.

Trin Duy Luan and Nguyen Quang Vinh (1997) *Socio-economic Impacts of Renovation on Housing and Urban Development in Vietnam*, Hanoi: National Centre for Social Sciences and Humanities.

Trinh Le (1992) 'Water pollution and strategy of water pollution control in the Mekong Delta and Central Highlands', paper presented at the Formulation Workshop on Mekong Basin-wide Strategy for Water Pollution Control, Bangkok, 1992.

Trinh Van Thu (1991) 'Advances in forecast dissemination and community preparedness tactics in Viet Nam', paper presented at the Second International Workshop on Tropical Cyclones, 1991.

Tuong Lai (1995) 'Problems of social integration following the 1975 reunification', *Social Sciences* 46, 2: 11–24

United Nations (1999) *Common Country Assessment: Vietnam*, Hanoi: United Nations. Also available at http://www.un.org.vn/undocs.

United Nations Centre for Human Settlements (1996) *An Urbanising World: Global Report on Human Settlements 1996*, Oxford: Oxford University Press.

United Nations Centre on Transnational Corporations (1992) *World Investment Report, 1992*, New York: United Nations.

United Nations Conference on Trade and Development and United Nations Development Programme (1994) *The Interlinkages between Trade and Environment: Thailand*, United Nations Conference on Trade and Development: Geneva.

United Nations Development Programme (1991) *Cities, People and Poverty*, UNDP Strategy Paper, New York: United Nations Development Programme.

United Nations Development Programme (1994) *Report of a Provincial Visit to Hanoi, Strengthening National Capacities to Integrate the Environmental into Investment Planning and Public Policy-Making in Vietnam*, Project VIE/93/981, Hanoi: United Nations Development Programme.

United Nations Development Programme (1995) *Poverty Elimination in Vietnam*, Hanoi: United Nations Development Programme, United Nations Population Fund and United Nations Children's Fund.

United Nations Development Programme (1996) *Promoting Environmental Awareness through Mass Media Campaigns in Vietnam*, Project VIE/93/030, Hanoi: United Nations Development Programme.

United Nations Development Programme (1998) *Human Development Report 1998*, New York: United Nations.

United Nations Development Programme (1999) *Development Co-operation 1998*, Hanoi: United Nations Development Programme.

United Nations Development Programme and Food and Agriculture Organization (1996) *Smallholder Reforestation in Central Vietnam: Experiences from the implementation of UNDP/FAO Project VIE/92/022*, Hanoi: United Nations Development Programme and Food and Agriculture Organization.

United Nations Disaster Management Team and Vietnam Disaster Management Unit (1995) *Emergency Relief and Disaster Mitigation in Viet Nam, Second Edition*, United Nations Disaster Management Team Information Note, Hanoi: United Nations Disaster Management Team and Disaster Management Unit.

United States Agency for International Development (1968) 'To tame a river', Informational pamphlet issued by US AID and US Bureau of Reclamation, Washington DC: US Agency for International Development and US Bureau of Reclamation.

United States Bureau of Reclamation (1956) *Reconnaissance Report: Lower Mekong River Basin*, prepared for International Co-operation Administration, United States Government, Washington DC: US Bureau of Reclamation

United States National Oceanic and Atmospheric Administration (1988) *Constructed Worldwide Tropical Cyclones*, TD9636, Washington DC: National Oceanic and Atmospheric Administration.

United States Navy and National Oceanic and Atmospheric Administration (1994) *Global Tropical/Extratropical Cyclone Climatic Atlas (CD-ROM Version 1.0)*, Washington DC: US Navy and Department of Commerce.

Van Arkadie, B. 1993, 'Managing the renewal process: the case of Vietnam', *Public Administration and Development* 13: 435–51.

van Beers, C. and van den Bergh, J. (1997) 'An empirical multi-country analysis of the impact of environmental regulations on foreign trade flows', *Kyklos* 50: 29–46

van Breemen, N. (1980) 'Acidity of wetland soils, including Histosols, as a constraint to food production', in *Priorities for Alleviating Soil Related Constraints to Food Production in the Tropics*, Los Baños: International Rice Research Institute.

van Breemen, N. (1993) 'Environmental aspects of acid sulphate soils', in D.L. Dent and M.E.F. van Mensvoort (eds) *Selected Papers of the Ho Chi Minh City Symposium on Acid Sulphate Soils*, Publication 53, Wageningen: International Institute for Land Reclamation and Improvement.

van Breemen, N. and Pons, L.J. (1978) 'Acid sulphate soils and rice', in *Soil and Rice*, Los Baños: International Rice Research Institute.

van Keer, K., Thirathon, A. and Janssen, W. (1996) 'To burn or not to burn?', in F. Turkelboom, K. Van Look-Rothschild and K. Van Keer (eds) *Highland Farming: Soil and the Future? Proceedings of the Soil Fertility Conservation Project*, Chiang Mai, Thailand: Maejo University.

Vietnam Department of Agriculture, Forestry and Fishery (1996) *Statistical Data of Agriculture, Forestry and Fishery 1985–1995*, Hanoi: Statistical Publishing House.

Vietnam Disaster Management Unit (1997) *Procedures and Methods for Provincial Damage Assessment*, No. 97-15, Hanoi: Disaster Management Unit, Ministry of Agriculture and Rural Development.

Vietnam General Statistical Office (1995) *Statistical Yearbook 1994*, Hanoi: Statistical Publishing House.

Vietnam General Statistical Office (1996) *Statistical Yearbook 1995*, Hanoi: Statistical Publishing House.

Vietnam General Statistical Office (1997) *Social Indicators in Vietnam 1990–1995*, Hanoi: Statistical Publishing House.

Vietnam General Statistical Office (1998) *Statistical Yearbook 1997*, Hanoi: Statistical Publishing House.

Vietnam Government (1992) *1992 Constitution*, Article 18, Hanoi: Government of Vietnam.

Vietnam Government and United Nations Development Programmme (1989), *Socialist Republic of Vietnam, Socio-economic Development and Investment Requirements for the Five Years 1996–2000*, report to the Consultative Group Meeting, Paris, Hanoi: Vietnam Government and United Nations Development Programmme.

Vietnam Ministry of Agriculture and Food Industry (1993) *Bare Lands in Vietnam: the Existing Situation and the Improving and Using Orientation up to 2000*, Hanoi: Ministry of Agriculture and Food Industry.

Vietnam Ministry of Agriculture and Rural Development and Food and Agriculture Organization (MARD–FAO) (1996) *Land Use Planning and Forest Land Allocation and Other Project Activities*, proceedings of the Review Workshop held in Hue, 15–19 October 1996, Hanoi: Ministry of Agriculture and Food Industry.

Vietnam Ministry of Agriculture and Rural Development and Food and Agriculture Organization (MARD–FAO) (1998) *Proceedings of the National Workshop on Participatory Land Use Planning and Forest Land Allocation*, Hanoi: Agriculture Publishing House.

Vietnam Ministry of Construction (1995) *Urban Sector Strategy Report (Draft)*, Hanoi: Ministry of Construction.

Vietnam Ministry of Construction (1996) *Population and Urban Living Environment in Hanoi City*, Hanoi: National Political Publishing House.

Vietnam Ministry of Forestry (1987) 'Decision No. 1171-QD (30-12-1986) on regulations of production, protection and special use forest [Quyet dinh ban hanh cac loai quy che rung san xuat, rung phong ho va rung dac dung]', in *Regulations of Management of Production – Protection – and Special Use Forest [Quy she quan ly Rung san xuat, Rung Phong ho, Rung dac dung)*, Ministry of Forestry. Hanoi: Agricultural Publishing House.

Vietnam Ministry of Forestry (1990) 'Report on fixed cultivation and settlement (1968–1990) [Bao cao tongket song tac dinh canh dinh cu]', Hanoi: Ministry of Forestry.

Vietnam Ministry of Forestry (1991) *Vietnam: Forestry Sector Review Main Report*, Hanoi: Ministry of Forestry.

Vietnam Ministry of Science, Technology and Environment (1994) *Red River Delta Master Plan Report 8: Soils and Land Use*, Hanoi: United Nations Development Programme and Ministry of Science Technology and Environment.

Vietnam Ministry of Water Resources, United Nations Development Programme and United Nations Department of Humanitarian Affairs (1994) *Strategy and Action Plan for Mitigating Water Disasters in Viet Nam*, Hanoi: Ministry of Water Resources.

Vietnam Ministry of Water Resources, United Nations Development Programme and United Nations Department of Humanitarian Affairs (1995) *First Update of the Strategy and Action Plan for Mitigating Water Disasters in Viet Nam*, Hanoi: Ministry of Water Resources.

Vietnam National Environment Agency and Asian Development Bank (1998) *Viet Nam National Strategy for Hazardous Waste Management*, Hanoi: Vietnam National Environment Agency and Asian Development Bank, Hanoi.

Vietnam National Resources and Environment Research Programme (1986) *Draft National Conservation Strategy*, Hanoi: National Resources and Environment Research Programme.

Vietnam News Agency (1999) 'Vietnamese people's living conditions improved', 29 August 1999, Hanoi: Vietnam News Agency.

Vietnam News Agency (2000) 'Mekong Delta sees thriving rice production after war', 20 May 2000, Hanoi: Vietnam News Agency.

Vietnam News List (2000) 'Mekong four join on river management', 27 April 2000 [cited in vnnews-l, 29 April 2000; http://coombs.anu.edu.au/~vern/vnnews-list.html].

Vietnam State Commission for Sciences (1991) *Maps of Natural Factors of the Cuu Long Delta*, Programme 60-B, Ho Chi Minh City: State Commission for Sciences.

Vietnam State Committee for Sciences, United Nations Development Programme, Swedish International Development Authority, United Nations Environment Programme and International Union of Conservation Unions (1991) *National Plan for Environment and Sustainable Development, 1991–2000*, Hanoi: State Committee for Sciences.

Vietnam State Planning Committee and General Statistical Office (1994) *Vietnam Living Standards Survey 1992–1993*, Hanoi: State Planning Committee and General Statistical Office.

Vietnamese Communist Party (1991) *Documentation of the Seventh National Congress*, Hanoi: Su That Publishing House.

Vincent, J. (1997) 'Testing for environmental Kuznets curves within a developing country', *Environment and Development Economics* 2: 417–31.

Vo Nhan Tri and Booth, A. 1992 'Recent economic developments in Vietnam', *Asian-Pacific Economic Literature* 6, 1: 16–40.

Vo Quy (1997) 'Vietnam: environmental challenges and solutions', paper presented at the EUROVIET-III Conference, Amsterdam, The Netherlands, 2–4 July 1997.

Vo-Tong Xuan (1993) 'Recent advances in integrated land uses on acid sulphate soils', in D.L. Dent, and M.E.F. van Mensvoort (eds) *Selected Papers of the Ho Chi Minh City Symposium on Acid Sulphate Soils*, Publication 53, Wageningen: International Institute for Land Reclamation and Improvement.

Vo-Tong Xuan (1996) 'Mekong dialogue for sustainable development', in B. Stensholt (ed.) *Development Dilemmas in the Mekong Subregion* (Workshop proceedings, 1–2 October 1996), Clayton, Australia: Monash Asia Institute, Monash University.

Vu Lu Tap and Taillard, C. (1993) *An Atlas of Vietnam*, Paris: RECLUS, La Documentation Francais.

Vu Van Me (1997) 'Proposal for the establishment of a working group for forest land allocation in aid projects', in *Workshop on Technical Assistance for Effective Use of External Support to Integrated Agricultural and Rural Development*, Hanoi: Ministry of Agriculture and Rural Development.

Vu Van Me and C. Desloges (1997) 'Methodology for participatory land use planning and forest land allocation', paper presented at the National Workshop on Participatory Land Use Planning and Forest Land Allocation, December 1997, Hanoi.

Wachter, D. (1992) 'Land titling for land conservation in developing countries?', Divisional Working Paper No. 1992-28, Washington DC: Environment Department, World Bank.

Wade, R. (1987) 'The management of common property resources: collective action as an alternative to privatisation or state regulation', *Cambridge Journal of Economics* 11: 95–106.

Wade, R. (1990) *Governing the Market: Economic Theory and the Role of Government in East Asian Industrialisation*, Princeton: Princeton University Press.

Wade, R. (1996) 'Japan, the World Bank and the art of paradigm maintenance: the East Asian Miracle in perspective', *New Left Review* 217: 3–36.

Wade, R. (1998) 'The Asian debt and development crisis of 1997–?: causes and consequences', *World Development* 26: 1535–53.

Walsh, K. and Pittock, A.B. (1998) 'Potential changes in tropical storms, hurricanes, and extreme rainfall events as a result of climate change', *Climatic Change* 39: 199–213.

Wandel, J. (1997) 'Development opportunities and threats to ethnic minority groups in Viet Nam', in D. McCaskill and K. Kampe (eds) *Development or Domestication? Indigenous Peoples of Southeast Asia*, Chiang Mai, Thailand: Silkworm Books.

Watts, M. (1991) 'Entitlements or empowerment? Famine and starvation in Africa', *Review of African Political Economy* 51: 9–26.

Watts, M. (1993) 'Development I: power, knowledge, discursive practice', *Progress in Human Geography* 17: 257–72.

Watts, M.J. (1998) 'Recombinant capitalism: state, de-collectivisation and the agrarian question in Vietnam', in J. Pickles and A. Smith (eds) *Theorising Transition: The Political Economy of Post-Communist Transformations*, London: Routledge.

Watts, M.J. and Bohle, H.G. (1993) 'The space of vulnerability: the causal structure of hunger and famine', *Progress in Human Geography* 17: 43–67.

Wheeler, D. and Martin, P. (1992) 'Prices, policies and the international diffusion of clean technology: the case of wood pulp production', in P. Low (ed.) (1992) *International Trade and the Environment*, World Bank Discussion Paper 159, Washington DC: World Bank.

Wheeler, D. and Martin, P. (1993) 'National economic policy and industrial pollution: the case of Indonesia: 1975–89', paper presented to Workshop on Economy-wide Policies and the Environment, World Bank, Washington DC, 14–15 December 1993.

Wickramanayake, E. (1994) 'Flood mitigation problems in Vietnam', *Disasters* 18: 81–6.

Wiegersma, N. (1988) *Vietnam: Peasant Land, Peasant Revolution*, London: Macmillan.

Wiens, T.B. (1998) 'Agriculture and rural poverty in Vietnam', in D. Dollar, P. Glewwe and J. Litvack (eds) *Household Welfare and Vietnam's Transition*, Washington DC: World Bank.

Williams, P. (1996) ' "A simplistic plumbing problem": a review of the Mekong Mainstream Run-of-River Hydropower Report', *Watershed* 1, 3: 14–15.

Wilson, G. and Bryant, R.L. (1997) *Environmental Management: New Directions for the Twenty-First Century*, London: UCL Press.

Wongsuphasawat, L. (1997) 'The extended metropolitan region and uneven development in Thailand', in C. Dixon and D. Drakakis Smith, *Uneven Development in Southeast Asia*, Aldershot: Avebury.

Woodside, A.B. (1976) *Community and Revolution in Modern Vietnam*, Boston: Houghton Mifflin.

World Bank (1991) *Urban Policy and Economic Development: An Agenda for the 1990s*, Washington DC: World Bank.

World Bank (1992) *World Development Report, 1992: Development and the Environment*, Oxford: Oxford University Press.

World Bank (1993a) *Malaysia, Managing the Costs of Urban Pollution*, Washington DC: World Bank.

World Bank (1993b) *The East Asian Miracle: Economic Growth and Public Policy*, Oxford: Oxford University Press.

World Bank (1993c) *Vietnam: Transition to the Market*, Washington DC: World Bank, East Asia and Pacific Region.

World Bank (1994) *Indonesia: Environment and Development*, Washington DC: World Bank.

World Bank (1995a) *Vietnam: Environmental Program and Policy Priorities for a Socialist Economy in Transition*, Washington DC: World Bank.

World Bank (1995b) *Vietnam: Poverty Assessment Strategy*, Report No. 13442-VN, Washington DC: World Bank.

World Bank (1996a) *Vietnam Water Resources Sector Review*, East Asia and Pacific Region, Report No. 15041-VN, Hanoi: World Bank.

World Bank (1996b) *World Development Report 1996: From Plan to Market*, Oxford: Oxford University Press.

World Bank (1997a) *Can the Environment Wait? Priorities for East Asia*, Washington DC: World Bank.

World Bank (1997b) *Vietnam: Deepening Reform for Growth. An Economic Report.* Hanoi: World Bank.

World Commission on Environment and Development (1987) *Our Common Future*, New York: Oxford University Press.

World Conservation Monitoring Centre (1994) *Biodiversity Profile of the Socialist Republic of Vietnam*, Cambridge: World Conservation Monitoring Centre. Also available at http://www.wcmc.org.uk/infoserv/countryp/vietnam/.

World Meteorological Organization (1999) *The 1997–1998 El Niño Event: A Scientific and Technical Perspective*, WMO–No. 905, Geneva: World Meteorological Organization.

Worster, D. (1994) *Nature's Economy: a History of Ecological Ideas, Second Edition*, Cambridge: Cambridge University Press.

Wyatt-Smith, J. (1963) 'Manual of Malayan silviculture for inland forests', *Malayan Forest Record* 23.

Xenos, P., Tran Thu Phuong, Luu Thi Thao and Vu Xuan Truong (1993) 'Demography and labour', in Le Trong Cuc and A.T. Rambo (eds) *Too Many People, Too Little Land: The Human Ecology of a Wet Rice Growing Village in the Red River Delta of Vietnam*, Occasional Paper No. 15, Environment Program, Honolulu: East–West Centre.

Yates, D. (1999) 'IMF Glum on Vietnam', *Vietnam News*, 26 January 1999.

Yeung, Y.-M. and Lo, F.-C. (1998) *Globalization and the World of Large Cities*, Tokyo: United Nations University Press.

You, J.-I. (1998) 'Income distribution and growth in East Asia', *Journal of Development Studies* 34: 37–65.

Young, O. (1989) *International Cooperation: Building Regimes for Natural Resources and the Environment*, Ithaca: Cornell University Press.

Zakaria-Ismail, M. (1994) 'Zoogeography and diversity of the freshwater fishes of Southeast Asia', *Hydrobiologia* 285: 41–8.

# Index